10 09

© 2005 by The Weekly Standard

## *About the Editor*

WILLIAM KRISTOL is editor of the influential Washington-based political magazine *The Weekly Standard*. Widely recognized as one of the nation's leading political analysts and commentators, he regularly appears on the Fox News Channel. Before starting *The Weekly Standard*, Kristol served as chief of staff to Vice President Dan Quayle and to Secretary of Education William Bennett under President Reagan. Before coming to Washington in 1985, Kristol taught at the University of Pennsylvania and Harvard's Kennedy School of Government.

# THE
# WEEKLY STANDARD

## A READER: 1995–2005

## WILLIAM KRISTOL

### EDITOR

HARPER PERENNIAL

NEW YORK • LONDON • TORONTO • SYDNEY

HARPER ● PERENNIAL

A hardcover edition of this book was published in 2005 by HarperCollins Publishers.

THE WEEKLY STANDARD: A READER: 1995–2005. Copyright © 2005 by The Weekly Standard. All rights reserved. Printed in the United States of America. No part of this book may be used or reproduced in any manner whatsoever without written permission except in the case of brief quotations embodied in critical articles and reviews. For information address HarperCollins Publishers, 10 East 53rd Street, New York, NY 10022.

HarperCollins books may be purchased for educational, business, or sales promotional use. For information please write: Special Markets Department, HarperCollins Publishers, 10 East 53rd Street, New York, NY 10022.

FIRST HARPER PERENNIAL EDITION PUBLISHED 2006.

*Designed by Christine Weathersbee*

Library of Congress Cataloging-in-Publication Data is available upon request.

ISBN-10: 0-06-088285-9 (pbk.)
ISBN-13: 978-0-06-088285-3 (pbk.)

06 07 08 09 10 ❖/RRD 10 9 8 7 6 5 4 3 2 1

*The Weekly Standard* made its debut on September 18, 1995, and is published forty-eight times a year by News America Incorporated.

# Contents

## III. Peace and War at Home and Abroad

## IV.  Books and Arts

## V.  Miniatures

# FOREWORD

## By William Kristol

---

WHEN WE LAUNCHED *The Weekly Standard* ten years ago, I didn't know what I was doing. I'd never actually worked on a magazine before, but I'd grown up watching my father edit a couple of them. And I'd *read* lots of magazines. And I had a great many friends in the business. What's the problem, I figured? How hard can this be?

Today, almost five hundred *Weekly Standard* issues later, it's long since become clear to me how comically naive this must then have seemed to any number of interested observers. And I'll always be grateful to them for not letting me in on the joke at the time—otherwise I might never have had the nerve to carry through on the project. One interested observer in particular deserves thanks here: Rupert Murdoch, chairman and chief executive of News Corporation, the magazine's original and continuing parent company. It was Rupert who committed to publishing the magazine from the start, and he has generously and unfailingly supported it ever since, even while scrupulously fulfilling his pledge to allow us complete editorial independence. Rupert, of all people, surely *did* have an idea exactly what we were getting into with *The Weekly Standard*—and he must just as surely have recognized that I had no idea at all. But he was too polite to say so.

No doubt Rupert was reassured that present with me at the creation of the magazine would be a group of men and women who *did* have many decades of accumulated writing and editing experience: Fred Barnes and John Podhoretz, David Tell and Richard Starr and

Claudia Winkler, David Brooks and Andrew Ferguson and Christopher Caldwell, among others. Rupert's confidence in my colleagues was well placed, of course. Perhaps he had a measure of additional confidence that I could catch up with them as time went by. (Or perhaps he was just in a very good mood.)

In any event, it seems to me that all of us together have learned a thing or two about opinion journalism over the past decade. And it's been *The Weekly Standard*'s singular good fortune to have kept itself quite literally "together" through the whole of its life to date. Among the nine senior-most editorial staffers whose names appeared on the masthead of our maiden issue in September 1995, seven of us are still right here, still working side by side on the same floor of the same office building in downtown Washington, D.C. (John Podhoretz and David Brooks have gone on to become columnists at the *New York Post* and *New York Times* respectively.) And four other notable staffers— Matt Labash, Jonathan V. Last, Victorino Matus, and David Skinner— have also been with us all or almost all the way. To a remarkable extent, in other words, it's the same group of people, whether as editors or writers, who deserve much of the credit—or blame, depending on your perspective—for the roughly ten million words the magazine has already seen into print.

My own perspective is concededly a biased one: I like and admire my *Weekly Standard* colleagues, and I've always been proud of their work. So I will admit that putting together the book you now hold in your hands has proved a gratifying experience. Reviewing a decade's worth of our back issues has reminded me, to be sure, that magazine journalism is a perishable medium, and that political and social commentary is the most perishable magazine journalism of all. The best of it does survive the test of time, however. And it's my sense that in its first decade *The Weekly Standard* has managed to produce a considerable body of writing that will remain worth reading well into the future—much more of it, in fact, than could fairly be represented even in a fat volume like this one.

Choices had to be made, however. More than anything, they've been made with quality in mind. But the selections also reflect an effort to exemplify various aspects of the magazine's development, tempera-

ment, range of interests, and point of view—all in a book of essays that
can be read and enjoyed on its own terms. Certain compromises were of
course required. For example, ours is above all a magazine about politics, and each week's issue thus tends naturally to be absorbed in the
most interesting and important political developments and controver-
sies of the moment. More often than not, however, what's interesting
and important in the news proves somewhat evanescent; that's why they
call it "news." So on the assumption that old news is not what most people will have purchased this book for, its contents deliberately deem-
phasize the shorter, time-sensitive reporting pieces and inside-baseball
items that typically dominate *The Weekly Standard*'s first several pages
and "Scrapbook" section. In their stead are some of our longer, more
reflective political essays. It's a reasonable trade, I think.

Also missing from the present volume is a separate chapter devoted
to our back-page satirical feature, the magazine's always-popular
"Parody." Many of our best Parodies depend for their wit on visual elements that would poorly survive the transition to smaller-format, all-
text book pages. So, with one exception, I've reluctantly left them out.
That exception is an especially celebrated one, though: "*I Disonesti*," a
madcap reimagining of the Monica Lewinsky scandal as an Italian
opera buffa. In any case, the magazine's sense of humor has never been
restricted to its Parody page, and I hope no one will have much trouble
finding it sprinkled throughout this book—among the "Miniatures,"
for instance, a sampling of the short, casual essays we've published
once a week from the very start.

In sum, this is a book, not a magazine, and it's organized accordingly. At the same time, this being a book *about* a magazine, I've also
tried to ensure that it communicates, as best it can, the magazine's
essential history and spirit. Each of the volume's chapters is arranged
chronologically—in the order its selected essays originally appeared in
*The Weekly Standard*. In all but one of those chapters, moreover, the
chronology is pretty evenly distributed across the years. Only in the
section on "Peace and War at Home and Abroad" does this scheme
break down a bit. There the selections are disproportionately drawn
from issues of the magazine published after September 11, 2001—for
reasons I surely needn't waste anyone's time explaining.

Which brings me back to the matter of my own initial innocence about what editing *The Weekly Standard* would involve. As I say, it's been a learning experience. And among the most important lessons I've learned, blindingly obvious though it might at first appear, is one I think applies with equal force not just to Washington journalists but also to the people we write about—and to our readers. It concerns a central, chronic misunderstanding of modern political life. Let's call it "the fallacy of hidden design."

Men and women in public life are nowadays constantly confronted—much to their exasperation, as I recall from my own past life in government—by reporters who have trouble believing in the possibility of a news story whose deepest meaning isn't in some sense a secret. It cannot be, so these reporters suppose, that the president has made his most recent pronouncement or decision simply because he thought it, on balance, the right and timely thing to do. At least it cannot *mostly* be that. There must also be some strategy afoot—probably a cynical or selfish one related to some interest group, polling demographic, or whatnot.

But the president's aides would tell you—correctly, in my experience—that top-level, behind-the-scenes Washington doesn't actually work like this. It can't: People are too busy, there are too many competing agendas to juggle, there's not enough time, in-boxes and calendars are too full, and five big things still have to get done by six o'clock whether you've perfected them or not. Under the circumstances, then, the best and much the safest thing a politician can usually hope to do is play it straight, with minimal calculation.

Mind you, this is not by any means to suggest that I think people in Washington are a tribe of saints. I don't. They aren't. Plenty that goes on here is ill-motivated—or just plain dumb. But at the same time, much of it is also relatively uncomplicated, ingenuous, and transparent. So generally speaking, the clearest and most reliable expressions of a public figure's intentions are his own words and deeds. Put another way: The most accurate and intelligent interpretation of the news tends to be the one that best concentrates its attention not on some imagined, backstage Wizard of Oz, but on what's happening in *front* of the curtain, for all to see and hear.

And as I've discovered these past ten years, it turns out that much the same is true where interpretations of the news *media* are concerned. Back in September 1995, I had some high-falutin' hopes. I thought we'd carefully plan each issue of the magazine, ensuring an ideal balance of subject matter—between the topical and the longer-range, between politics and arts, between foreign and domestic policy. And don't get me wrong, we *have* tried to do that. On balance, over time, I rather think we've succeeded, in fact. But the day-to-day reality of opinion journalism, up close, doesn't look nearly so serene or neat. And a magazine editor gets mugged by that reality on a regular basis. Certain articles you've commissioned never materialize. Others show up but prove to be unusable. Halfway through the week, some unexpected occurrence in the world will inspire one of your writers to dash off a dazzling, must-publish piece, and all of a sudden, the "ideal balance of subject matter" becomes an unobtainable mirage. You're left, instead, with five different articles—each of which is a good, strong, just-right-for-*The Weekly Standard* piece, but all of which happen also to be about, say, developments in the Middle East.

There's no avoiding it sometimes. And, while I'd hardly make a claim for it as a landmark insight into human nature—life is messy; problems and opportunities pop up out of the blue; all you can do is try your hardest; often there is *less* on view than meets the eye—the fact remains, I think, that surprisingly few people seem able to remember these truths when the conversation turns either to politics or to media commentary. It was more common in the magazine's first few years, I suppose, but it still happens all the time: Some not insignificant number of people always assume that *The Weekly Standard* isn't really published in English, but in code—that its contents are designed to advance a surreptitious political agenda. *The Weekly Standard* is a conservative magazine, of course. We make no bones about it. And ours tends toward a particular kind of conservatism; our pages are its home, we like to think. But that's the point: The distinctive point of view in question has been worked out—and is still being worked out—on paper, in public, over the long haul. And it's also the case that in those very same pages we have consistently and more or less routinely run authors who manifestly don't agree with one

another. I often get asked: "Why are you printing this particular argument about that particular subject at this particular time?" And just as
often the honest answer is that someone's recently offered us the article
in question and we've decided we like it. Simple as that. Sometimes a
magazine is really just a magazine.

And sometimes a magazine about the news is *driven* by the news.
Much that's happened at *The Weekly Standard* we never—and probably
couldn't have—expected. On Labor Day, 1995, when we were just getting under way, Washington's "Gingrich Revolution" was still in full
swing. Already in our first issue we were spotting weaknesses in the
speaker and his footsoldiers. But overall, we were fairly enthusiastic
about them, just the same, and the cover of that premiere issue carried
a picture of Newt Gingrich as a martial, confident Tarzan—under the
soon-to-be-laughable headline "Permanent Offense." Two months
later, history will record, Gingrich and the congressional Republicans
were on permanent *defense*, having alienated the country with an ill-
conceived government shutdown. And a few weeks after that,
President Clinton decided (belatedly in our view) to launch a military
intervention in Bosnia. One of our contributing editors, Charles
Krauthammer, thought and wrote that Clinton was wrong to do so.
But the magazine editorially supported the president, in that same
issue and subsequently—for which sin a not-insignificant chunk of our
original subscribers immediately cancelled out on us.

So it has gone over the subsequent weeks and months and years.
Once upon a time, I see from an early cover article, we apparently
believed it possible to "Crush the Internet!" Oops. One of us, who
shall remain nameless, argued in late 1995 that Colin Powell would
make an excellent president. Some of us thought the Republicans
should nominate John McCain in 2000; others of us thought the
McCain idea was nuts. And in each of these and dozens of other
instances, lots of readers got irked with us. Early in the magazine's history, I remember mentioning to a friend that I seemed to have made
more enemies in one year at *The Weekly Standard* than I had during my
previous ten years in government and politics combined.

Then there was 9/11, the ultimate in unexpected developments.
Not everyone was taken entirely by surprise, it's important to note—

and here I would refer you to "A Cowering Superpower," the eerily prescient Reuel Marc Gerecht essay on Osama bin Laden from our July 30, 2001, issue. But I think it fair to say, at very least, that before the fact of 9/11, my colleagues and I would never have anticipated that *The Weekly Standard*'s pages were soon to be so thoroughly and persistently dominated by coverage of a global-scale war on Middle Eastern terrorism and despotism. And since 9/11, I don't suppose any of us would think we had much serious choice in the matter.

Which ultimately speaks—if a former political philosophy professor may be forgiven a pseudo-Platonic moment—to the *weekliness* of weekly journalism. A weekly comes out every week. A weekly *has* to come out every week. Early on, back in the "Permanent Offense" days, I remember lamenting aloud at an editorial meeting that there was one short article in the otherwise terrific issue we'd shortly be printing that seemed to be of less than highest quality. My friend and colleague Fred Barnes immediately set me straight about this. Perhaps it wasn't the best imaginable piece of writing, Fred said of the essay in question. But the best imaginable piece of writing, in this case, did not exist, he pointed out. And the piece we had in hand, by contrast, possessed what Fred considered among the most important journalistic qualities: "the quality of doneness."

Wiser pseudo-Platonic words were never spoken.

Eager as I am to achieve the quality of doneness with this introduction—so that you might pass your attention to the essays on offer in the pages that follow—I could not forgive myself if I did not first pay appropriate tribute to certain other people not yet mentioned, who've made major contributions to this book in one form or another.

Though he'd not yet come aboard at the magazine's inauguration, and though he's recently left us to take over leadership of the journal *First Things*, Joseph Bottum was responsible, as our longtime "Books & Arts" editor, for supervising almost all the fine cultural and literary coverage collected in this volume—and for much else *The Weekly Standard* has published over the years, besides.

Many colleagues at the magazine have shared their advice and insights about how to put together this volume. Our publisher, Terry

Eastland, and our publicity director, Catherine Titus Lowe, have done special double duty in making the book a reality. David Tell has led the detailed editorial work on the book project. Matthew Continetti did a great deal of the back-issue research required to select its contents. And our patient and careful intern, Jordan Fabian, has attended to all the necessary keyboard and computer business.

At HarperCollins, Jane Friedman, David Hirshey, and Nick Trautwein cannot be praised highly enough. They've been a joy to work with.

And finally, however inadequately, I'd like to thank the many thousands of other people—colleagues, contributors, subscribers, online browsers, friends, and critics alike—who've helped make *The Weekly Standard* such an enjoyable and engrossing adventure throughout the past ten years.

# I

---

# APPRECIATIONS

# FREDERIC BARNES SR.

## By Fred Barnes

*(Originally published, as "My Father's Day,"
in the July 1, 1996, issue)*

A CONCEIT OF THE MODERN AGE is that we're free and independent-thinking people who decide, wholly on our own, how to live our lives. Stereotype plays on this theme. One is the rebellion of children of conservative parents who transformed themselves into counterculture radicals in the 1960s. Another is the leftie kids donning suits and dresses and becoming Reaganites and Newtoids in the 1980s and 1990s.

I'm dubious, particularly because my case and that of practically everyone I know is quite the opposite. I'm almost completely the creation of my father, Frederic Barnes (I'm Jr.). Or to put it another way, I'm a lagging indicator: When his interests and ideas and obsessions changed, mine changed, only a few years later.

I didn't realize this until a year or two ago. A friend of mine, John Yates, mentioned that his older brothers had gone into business because in their formative years, their father's life was concentrated on his career. But when John hit those years, his father was focused more on having a deeper Christian spiritual life for himself and his family. So John became an Episcopal priest.

My dad's shifts in focus were played out on only one person: me. Of course, I thought each time I was acting of my own free will. Maybe I operated under that illusion because my father never told me what to do. He simply made clear what he thought was most important in life, and I eventually came around.

My first interest was the military. My father had gone to West Point, class of 1934. He served in the cavalry and intelligence, then transferred from the Army to the Air Force after World War II. West Point was a big part of his life. As a kid in St. Louis, he had scrambled to line up an appointment. He barely made it under the age limit. By the time classes started, he was 22, probably the oldest plebe that year.

Going to West Point became my goal. When I got an appointment at age 18, however, my dad's interests had changed. He'd become a tough-minded anti-Communist, a fan of the House Un-American Activities Committee, and, naturally, a conservative Republican. He was a charter subscriber to *National Review* in 1955. He loved *Human Events*. He read the *Wall Street Journal* every day, especially enjoying the political reporting of a young writer named Robert Novak. After a bit, I was reading newspapers and magazines regularly. And when he began writing letters to the editor, I soon followed suit.

I turned down the chance to be a cadet. My dad was livid, I'm told, at least for a spell. But he never criticized my decision to my face. After all, he had to know I wasn't rejecting him; I was only following his lead. His chief interest was now politics, not the military. When I got out of college, the only jobs I sought were in journalism, and a few years later, I began covering politics.

After he retired from the Air Force and spent 10 years as a stock-broker, he and my mother moved to Vero Beach, Fla. He had always been a devout Christian, though not a demonstrative one. But in Florida he experienced a powerful religious awakening. An evangelical, charismatic faith pervaded his life. He talked a lot about Jesus Christ. He prayed in public. He and my mom began counseling people in trouble. When my wife, kids, and I visited, there was often some stranger there. Folks dropped by at all hours for advice and prayer and warm attention and fellowship.

I was amazed and, initially, appalled. My 70-year-old father had a new personality. He took over as head of Christian education at his church. He went on retreats. He proselytized. My family and I were prime targets. Visit after visit, he and my mother told us about the joys and rewards of their heightened Christian faith. They prayed for us and with us.

My sister and her husband were first to accept my parents' renewed faith, but I resisted for six or seven years. I refused to go to church with my parents. I balked at reading the Christian pamphlets and testimonies pushed my way. I failed to bond with young Christians my dad sent my way in hopes their example would prompt my conversion. But one example was overpowering to me—my father. I saw how his life was changed for the better. And since his faith had become the most important thing in his life, I was sure to follow. In 1980, my wife and I knelt in our living room and became Christians.

My father died on June 14 at age 87. I don't know if he was aware of how much he'd shaped my life. He never claimed to have swayed anybody's life (if credit was to be given, God got it). But he changed mine more than once, and I'll never be able to thank him enough for it.

# DOROTHY LAMOUR

## By David Gelernter

*(Originally published, as "Dorothy Lamour: An Elegy,"
in the November 18, 1996, issue)*

A PERFECT FACT TO REMEMBER us by, we "post-moderns" or whatever we are: When we say that entertainment is "adult," we mean it is infantile. The pictorial spelling-out of exactly what happens when a couple goes to bed is in the "Billy learns how to tie his shoes" spirit of edifying books for toddlers. Vulgar words and obscene pictures are the quintessence of teenage-boyness. Compared with the typical modern flick, Dorothy Lamour's best pictures have a paradoxical superiority: They are better for children, and vastly more adult.

She died September 23 at 81. It is a painful era for lovers of American culture—we are losing the remaining heroes and heroines of the golden age of the 1930s, '40s, and '50s. A few months back, Gene Kelly, a wonderful song-and-dance man when he reined in his pompous, arty urges. Ginger Rogers last year; when you add up the crunchy wisecrack comedy (*Fifth Avenue Girl*, *The Major and the Minor*), the solid acting (*Stage Door*, *Kitty Foyle*), and the sublime art of her dances with Astaire, she was the greatest lady of American cinema —a fact the obituaries forgot to mention. As for Miss Lamour, she was no Ginger Rogers. But she had great charm, a certain grace, and the marvelous art of keeping things in perspective. A heroine she was, of the wonderful age when the movies were grown-up solid citizens instead of (as the mood dictates) sullen prima donnas, foul-mouthed children, or raving nut cases.

People remember her seven "Road" films with Bing Crosby and Bob Hope (from *The Road to Singapore* of 1940 through *The Road to Hong Kong* of 1962) as corny. And they are—corny and funny. Funny movies are still produced nowadays, but hers were more sophisticated than today's average comedy insofar as it is harder to be funny when you limit yourself to inoffensive words and inoffensive situations. Vulgarity has always been the shortest route to a cheap laugh. I don't claim there is no such thing as a modern-yet-sophisticated movie, or that Miss Lamour's pictures were sophisticated in absolute terms. No one called them sophisticated at the time. But the trend lines are interesting, and the sophistication trend is sharply down.

There is a wonderful, pure essence-of-Lamour moment in the second worst *Road* movie, *The Road to Bali* of 1952. (*Hong Kong* is dead last.) She's a South Pacific princess, and Bing and Bob are adventurers who get lured to her island by a sneering bad guy who needs stooges to help him with his dirty work. She throws them a party. They sit cross-legged on the floor in kilts (never mind how the kilts got in there) on either side of her, and there follows a triple-decker, three-voiced fugue of a scene. The plot calls for the princess to demonstrate South Pacific sorcery, and so she makes a rope hang in air, a girl wriggle out of a vase. On top of that, the usual irrelevant Hope-and-Crosby clowning: "It's mass hypnosis!" "Where'd you learn a word like that?" "Don't listen to him, Princess, he got kicked out of kindergarten for cheating at finger painting." And superimposed on the whole thing is Miss Lamour's serene, priceless smile. In character she thinks these two adventurers are silly but sort of sweet. As an actress she thinks all three of them on that crazy set—Bing, Bob, and Dorothy—are silly but sort of sweet.

Her smile has the facts down perfectly. Bing is as unprepossessing a romantic lead as the movies ever produced. He *always* gets the girl (is the one-and-only actor ever to star with Astaire and get *Fred*'s girl), but the basis of his romantic appeal is cornball ballads and nice guyness. He mocks his own singing style and big ears. Hope is the only comedian ever to succeed in being funny on the basis of really, really trying. And Miss Lamour, supposed focus of double-barrel Hope-and-Crosby romantic passion, is herself no spring chicken circa 1952; she has (eh-hem) put on a few pounds, and happens to be got up in a head-dress that looks like the large transformers you see outside power plants, except diamond-spangled. The whole scene is just micro-inches shy of ridiculous, but she rescues it with the sheer knowing sweetness of her smile; because of that smile it is not ridiculous at all but funny and even a little bit touching.

She was the sarong girl, her unvarying title at the height of her career in the first half of the '40s. Her early pictures (*Jungle Princess* of 1936, *Her Jungle Love* of 1938, among others) established the sarong theme in an age when a movie goddess was supposed to speak softly but carry a big shtick, Homeric-epithet style. (Brunette-headed Dorothy, sarong-girl to the strong-greaved Achaeans.) With the first of the *Road* movies she hit her stride. These pictures, it is easy to forget, had sub-stantial sexual content; it was no small responsibility being *the* sarong girl. They were deceptive cognac-filled bonbons compared with the shot-of-Thunderbird approach we favor today, and I would rate some of them higher in sexual-proof than the hauntingly subtle, mysterious allure of *Sharon Stone—Naked!!* or whatever they call movies nowadays. ("So don't miss *Watch Demi Moore Undress*, coming soon to a theater near you!")

We might talk about "the sexual subtext of the *Road* shows," except that there is nothing sub about it. The sex is high-octane, more X than R. The first time you see Miss Lamour in *The Road to Singapore* she is wearing a filmy dress plus tasseled bra, and she is part of a night-club act that also stars her nasty keeper and a horsewhip. As the act gets underway she holds a cigarette between her lips; he whips it in two. More neat tricks follow. She makes up to Bing and Bob, and when she heads off in their direction the bad guy slaps the whip round her belly and reels her in. She rolls her big eyes tragically and hustles out with the visiting Americans, follows them home, and later on lets her hair down and gets into her sleeping sarong for a nighttime-and-moonlight number. In case you missed the point, she launches her next *Road* picture (*Zanzibar*, 1941) by getting sold as a slave. This time she has the sarong on from the start (what else would you wear to a slave sale?), a tattered little off-the-shoulder number; her hands are manacled and she glances round the room in tiny peeps like an anxious kitten. The sale turns out to be a scam—she and a girlfriend induce gullible romantics like Bing and Bob to buy her in a phony auction and set her free, and the ladies split the take with the auctioneer. But who cares? The audience got what it paid for. "She seems so unconscious of her deshabille," said the *New York Times* review of *Singapore*, "you just know her director and camera man were not for a minute."

Yet it was all harmless; if pornography happened, it was only in the privacy of your own mind—a place that, so far as today's Hollywood is concerned, doesn't even exist. Miss Lamour provoked not lust so much as wistful desire. During the war she was "the girl," E. B. White reported, "above all others desired by the men in Army camps." She was busy during the war; she had more on her plate than being desirable. She toured the country selling war bonds. She was so good at it, the government put a private railcar at her disposal. She sold 300-odd million dollars' worth and got a commendation from the Treasury Department twenty years late, in 1965. Flogging bonds wasn't fun-and-games—"a typical day for Miss Lamour," said a news story, "embraces about ten hours of war work." But when she showed up at the Martin Airplane factory in Baltimore, the management refused admittance and she had to make her pitch out front. No insult intended, the company explained; just that "when any good-looking woman walks through the plant it costs us 1,000 man-hours of labor. Dorothy Lamour might cost us half a bomber." Understood. "So she stepped aside," the story reports, "like a good patriot."

E. B. White, I would judge, had a thing for her. He analyzes the universal truths of maleness in terms of her status as Number One Dreamboat—tongue in cheek, but with acute interest. "If you know what a soldier wants, you know what Man wants." Which is? "A beautiful, but comprehensible, creature who does not destroy a perfect situation by forming a complete sentence." (She speaks with "studied native-girlishness," noted the *Times* reviewer, quoting Miss Lamour's remark about a rival for Bing's affection: "She is ver-ree prit-tee, no?")

"Man's most persistent dream," White continues, getting serious, "is of a forest pool and a girl coming out of it unashamed, walking toward him with a wavy motion, childlike in her wonder, a girl exquisitely untroubled, as quiet and accommodating and beautiful as a young green tree." We are not supposed to write like that anymore.

Now we have revoltingly vicious, wildly obvious rap songs, patronized by suburban youth and defended by yammering morons in the name of art. Is that a good swap? Satisfactory? Swinish obscenity is okay, disgusto-puko pornography is okay, but polite society is scandalized by the word "girl." Hence movie titles like *Pretty Woman*, which hit home with the overwrought phoniness of Victorian euphemism run amok. The oblivious headline of a recent news story about female reporters: "The Boys on the Bus Are Women."

A constant and besetting phoniness—our whole society rings with it like churchbells. In abolishing the idea of flirtatious or sensitive or shy or graceful or delicate or romantic or girlish femininity, replacing it with the lie of the manly woman, we have made our culture—look around you!—hideous. Joyless. Graceless. Ugly as sin.

Miss Lamour's best movies were no masterpieces; they were merely lovely and still are. "Casual and refreshing spontaneity" said a reviewer of her work in *Zanzibar*, which is exactly right. In *Singapore*, she races into a village square, Bob chases after and grabs her.

Bob: "Why, she's *got* it!"

Dorothy (tragic, sobby): "I have not!"

Bing, wrestling Bob to free her: "*What* has she got?"

Bob: "She's got . . ." And all three break into "An Apple for the Teacher," a song-and-dance number. Turns out idea was to draw a crowd and gyp some money out of them. Dorothy hams it up between the two hams-in-chief with her angelic-suave-arch smile that always looks as if she is about to burst out giggling. In *The Road to Utopia* of 1945, the best of the series, she sings a little number that became a hit—*Don't tell me I'm smart, tell me how you like my personal-it-tee.* . . . She is supposedly a vampy femme fatale, which she realizes is ridiculous, given her always-considerable difficulty not giggling in people's faces; and once again her smile takes it all in, so knowing and yet so sweet. An interviewer asked her if she had ever studied acting or music. "No," she said, "can't you tell?" In her graciousness she achieved lasting dignity, and we will remember her as a woman who did her country good. Sold a bunch of bonds, made fine, funny movies, cheered us up, heartened us, brightened the dreams of a lot of lonely GIs.

Bob Hope survives, though evidently his health is not so hot. I wish I could meet Hope and tell him he meant a lot to us; that my young boys love his *Road* pictures and we watch them together all the time. I won't ever, but it's gladdening to think that other people have told him. And her friends, I am sure, told Dorothy Lamour the same thing.

A survey discovered that, nowadays, "movie star" is the bottom-ranked career American parents want for their children. Today's stars are richer and a lot more self-important than Dorothy Lamour, but the sense of having buoyed the country just a little is a thing all their money will never buy them.

# BEN HOGAN

## By Jay Nordlinger

*(Originally published, as "Hogan Hero,"
in the August 11, 1997, issue)*

WHEN BEN HOGAN DIED on July 25, the golf world seemed slightly stunned. He was 84 and had been sick for several years, but he was always a hovering presence around the game, a necessary part of its self-image. Not that he ever talked to anyone. He kept to himself at Shady Oaks Country Club in Fort Worth, smoking his cigarettes, staring out the window of the men's grill. Rule No. 1 at Shady Oaks was "Don't bother Mr. Hogan." But everyone revered him, and the staff of the club, during his long, final absence, kept a sign on his table that said "Reserved for Mr. Hogan."

He had been the greatest golfer in the world, the sport's most mysterious hero. The first American star was Bobby Jones—scion of Atlanta society, Harvard educated, the epitome of the gentleman golfer. Then came Hogan, who could not have been more different: hardscrabble, maniacal, obsessive about everything he touched. Next there was Arnold Palmer, golf's first television idol, who melted the screen with his charisma and approachability. And after him came Jack Nicklaus, the finest player ever, as even Hogan partisans will admit.

But it was Hogan who did most to develop the modern game. Before him, golf had been a "feel" sport, all art and no science, dominated by grizzled Brits and talented good-time Charlies like Walter Hagen. Hogan determined to make golf systematic and knowable. He was the first pro to make a religion out of practice. Hour after hour he stood on the shag range, experimenting with his swing, "digging it out of the dirt," as he said. He was a man utterly controlled by golf, and eventually he learned to control it. He could place his shots wherever he wanted, producing a "fade," a gentle left-to-right motion conducive to accuracy. He won 63 tournaments, including nine "majors," the tournaments that really count. In his banner year of 1953, he won three of the four majors—the Masters, the U.S. Open, and the British Open (all but the PGA)—an achievement still unequaled. At the time, he was the most famous athlete in the country, along with Joe DiMaggio and Ted Williams. All in all, Hogan defined a new standard and invited his opponents and imitators to meet it.

Every student of golf is familiar with the details of Hogan's life. He was born in 1912 in Stephenville, Texas, the son of a blacksmith. Ben was 9 when his father took a gun and killed himself, with Ben in the room. Soon, Ben found a job as a caddy at a local club called Glen Garden. There, he threw himself into the game with a desperate abandon. He "practiced until his hands bled" (as innumerable fathers have told their sons). He had no friends to speak of, only an imaginary companion named "Hennie Bogan," who sat on his shoulder and admonished him to do better. When night came, Ben slept in the course's sand bunkers. He announced to his mother that he would make himself a champion golfer or die. His boyhood was almost completely devoid of comfort or joy, but he later said, "I feel sorry for rich kids, I really do. They're never going to have the opportunity I had."

At 17, he dropped out of school and turned pro. Yet he was far from a brilliant golfer. He was adequate, and burned with a desire, never quenched, to get better. He failed, repeatedly, for some 15 years. He was so poor that he stole fruit from orchards and vegetables from gardens. In one well-known instance, he was robbed of the tires on his car in Oakland, California. He pounded on a brick wall and sobbed to another golfer, "This is the end. I can't move another inch." But he made his way to the course, wrapped himself in a mental cocoon, and shot 67, earning him a $285 check, the largest he had ever seen. He could go on.

Sometime in 1946, according to lore, Hogan had a revelation. In one version of the story, it came in a dream; in another, it came during one of his incessant practice sessions. He had always been plagued with a "hook"—a right-to-left running shot that leaves a golfer feeling helpless—but now he figured out how to hit a soft, manageable fade. This was Hogan's "secret," a much-debated insight about which Hogan himself was endlessly coy. (Sam Snead once remarked, disgustedly, "Anybody can say he's got a secret when he won't tell what it is.") Whatever he glimpsed, Hogan began to win, and win consistently.

There emerged a mighty triumvirate of Snead, Hogan, and Hogan's boyhood acquaintance Byron Nelson. The three men were markedly dissimilar: Nelson a near-saint; Hogan a bitter perfectionist; Snead a crude, extravagantly gifted country bumpkin. It seemed that one of them would win every tournament on tour. But in time, by some unfathomable force of will, Hogan pulled ahead. He was unstoppable. The rest of the field would look at him and, demoralized, simply know that he would not falter. At 5 feet, 8 inches, 140 pounds, "Bantam Ben" was the most feared competitor in golf. In January 1949, *Time* magazine put him on its cover, with the legend, "If you can't outplay them, outwork them."

One month later came "the Crash," as it is known in golf history. Hogan and his wife Valerie were returning home to Fort Worth from a tournament in Phoenix. The fog around El Paso was thick. A Greyhound bus, not seeing the Hogans' car, tried to pass a truck and barreled straight toward them. A second before impact, Hogan hurled himself across his wife in an effort to protect her. His action probably spared his own life, as the car's steering column was propelled through the driver's seat. Valerie was relatively unharmed, but Hogan was close to death. For two months, the nation's attention was riveted on the hospital. Word was that, even if he survived, he would be an invalid. On April 1, he was taken from his bed on a stretcher and placed on a train back home. There, slowly, in extraordinary pain, Hogan began to sit up and later to walk. Cards, letters, and telegrams poured in to him from every state. No longer was he viewed as a cold, distant golfing machine, but as a valiant, lion-hearted battler. Everyone—for a change—was rooting for him. Hogan had never succumbed to anything, and he would not, in fact, succumb to the Crash.

He first swung a club again in the autumn. In December, he played 18 holes, with the help of a motor scooter. Two weeks later, he entered the Los Angeles Open. Amazingly, despite his aching and fatigued body, he played Snead to a tie. He lost in the playoff, but, as Grantland Rice famously wrote, he really "didn't lose—his legs simply weren't strong enough to carry his heart around." In June the next year, 16 months after the car crash, Hogan won the U.S. Open at Merion near Philadelphia, an event Dan Jenkins called "the most incredible comeback in the history of sports." On the 72nd hole—the final hole of the tournament—Hogan laced a 1-iron to the green to cinch the championship. The photograph taken of Hogan's follow-through on that shot—with Hogan ideally posed, wearing his trademark "Hogan cap"—is a totem of the game, displayed on nearly every golf-shop wall.

In 1951, Hollywood made a movie about Hogan: *Follow the Sun*, starring Glenn Ford. In 1953, he sailed to Carnoustie in Scotland to participate in the British Open, the only time he did so. The Scots, astonished at the precision and concentration of the peculiar Texan who captured their tournament with ease, dubbed him "The Wee Ice Man." When his ship docked in New York, the city gave him a ticker-tape parade down Broadway, the first since General MacArthur's. The next year, Hogan founded a club-manufacturing company, which bore his name and which he was to oversee until 1993. In 1957, he contributed a series of instructional articles to *Sports Illustrated*, which became the bestselling *Five Lessons: The Modern Fundamentals of Golf*—a book that, though effective, confused many with its barely comprehensible talk of "pronation" and "supination." Hogan played creditable golf into the late '60s, but his putting—the bane of any golfer's advancing years—gave out on him, rendering his always-superb ball-striking moot.

While an admirable man, Hogan was not a pleasant one. In fact, many would say—even in a time of eulogy—that he was intolerable. Once, when he was sitting alone at his table for eight at Shady Oaks, someone cracked, "There's Hogan, with all his friends." Gary Player supposedly called him up from South America one day, suffering from a slump and seeking help. "What clubs are you playing?" asked Hogan. "Dunlop," answered Player. "Then call Mr. Dunlop," Hogan replied, hanging up. Nick Faldo once asked him what it took to win the U.S.

Open. Hogan answered, with impeccable logic, "Shoot the lowest score." Similarly, when someone complained, "I'm having trouble with my long putts," Hogan came back with, "Why don't you hit them closer to the hole?" He once teased a golfer who yearned to know how to play a particular shot by saying, ludicrously, "I try to hit it on the second groove." And then there was the time, when President Eisenhower phoned, that Hogan barked to his secretary, "I'm not going to play with that g—d— hack." Hogan acknowledged no power above the ability to hit a golf ball soundly and to prevail in important tournaments.

"The Hawk" (this was another of Hogan's nicknames) was not the kind of hero that we have come to expect: the hero of the *Donahue* age, telling interviewers of his joys and sorrows, his triumphs and defeats, wearing his emotions on his sleeve. He once said—explaining his refusal to make public appearances, even to inaugurate the minor-league circuit christened the Hogan Tour—"Not everyone wants publicity, you know." Shrewd man that he was, he probably recognized the dangers of overexposure and the benefits of silence. He gave only one significant interview in the last decades of his life, in 1987 to a golf magazine: "The Hawk Talks!" the cover blared. His biographer, Curt Sampson, writes, "Insular types such as Bennie Hogan have always been drawn to golf, a sport requiring an ability to concentrate for long periods of time but with no mandate for cooperation or closeness with a teammate. He also enjoyed the utter fairness of the game, the way it compelled him to accept all the credit or all the blame. He loved its solitude, the way it absorbed him."

Hogan was unwilling to play the Senior Tour—on which Snead, Palmer, and others love to entertain and soak up the applause— because he could not stand for the public to see him at less than his best. But he still hit balls, never stopped practicing, never allowed his hands to grow uncallused.

Hogan—almost unique among professionals—did not play for glory (though he achieved it) or for money (though he earned it). He played in order to conquer the game, to solve its riddles, to bring it, at long last, to its knees. Upon hearing of Hogan's death, Ben Crenshaw said, "He defined the inner will that lives within us." No, it manifestly does not live within all of us—even dormant—but it lived unappeasably within Hogan, and because of it he was a great player and a great man.

# ELIA KAZAN

## By Stephen Schwartz

⟿

*(Originally published, as "The Rehabilitation of Elia Kazan,"
in the February 8, 1999 issue)*

ON MARCH 21, a long-standing and bitter injustice will be rectified:
That evening, the Academy of Motion Picture Arts and Sciences is
scheduled to award a special Oscar to the 89-year-old director Elia
Kazan. How the glittering audience at Oscar Night will greet this controversial presentation is hard to predict. The award is a direct rebuke
to the American Film Institute and other movieland institutions that
have snubbed Kazan repeatedly since the 1970s, although he was once
among their brightest lights. Now, what amounts to Kazan's rehabilitation after decades of blackballing and smears marks a notable breach
of the Iron Curtain that has long surrounded Hollywood's collective
memory.

No figure in American popular culture this century is more
deserving of honor for a lifetime of achievement than Kazan. The son
of immigrants from the Ottoman Empire, he was successful as an
actor, stage director, and novelist; and in the movies, he created masterpieces like *A Streetcar Named Desire* (1951), *Viva Zapata!* (1952), and
*On the Waterfront* (1954). Yet Kazan saw his reputation savaged in a
witch hunt—not the infamous hunt for Communists in Hollywood,
but the later and far more destructive unofficial inquisition loosed
against anti-Communists.

To understand Kazan's emblematic fate is an exercise in cultural
archaeology. It requires sifting through the ruins of the intellectual left

for clues to the bizarre anxieties attached to the figure of the anti-Communist "informer." For that is the term, drawn from the lingo of the gangster, that leftists and "liberals" attached to Elia Kazan. More than any other personal journey, his life shows how, in the aftermath of America's confrontation with Stalinism, history demonstrated its capacity for producing contradictory outcomes and claiming human sacrifices.

Elia Kazanjoglous was born in 1909 to a Greek family in what was then Constantinople. Four years later, his father moved the family to New York and opened what became a prosperous carpet business. The young Kazan graduated from Williams College and studied drama at Yale. Along the way, he picked up the nickname "Gadget" or "Gadg."

In 1932, with the political and economic storms of the Depression raging, Gadg Kazan joined the Group Theater, in New York. The encounter would influence a generation of American performers. Those were days when, even with the New Deal in full swing, the fear was widespread and real that the country could succumb to a red revolution. The Group Theater had been founded by the playwright Clifford Odets and other young leftists, along with such nonpolitical figures as Lee Strasberg. In line with the excitements of the time, most left-wing theater consisted of agit-prop skits on the sidewalks of New York, in furtherance of Communist propaganda.

All that changed one night in 1935, when Odets's new play *Waiting for Lefty* opened. Unlike other left-wing dramatists, Odets was a born playwright, and his talent was fortified by his collaboration with Strasberg and Kazan. *Waiting for Lefty* was Art; not the greatest achievement in the history of the stage, but Art, nonetheless.

On the stage sat a group of men, the leaders of a taxi drivers' union. The action developed around the progress of a meeting called to consider a strike. In front of the stage, between the stage and the audience, actors conjured up the past, the inner lives and secret strivings of the drivers. All present waited for "Lefty," the charismatic rank-and-file leader without whom the strike could not begin.

At the play's unexpected conclusion, the young Elia Kazan, planted in the audience, burst to the front of the theater and shouted that Lefty's body had been found at the taxi barn with a bullet in his head. Other actors seated among the spectators leapt to their feet, shouting as one, "Strike! Strike! Strike!" In a crescendo of protest, filled with sympathetic fury at the death of the proletarian hero, the audience was swept into the chorus.

It was unforgettable. It was a revolution. The American theater had been changed forever.

Outside the theater, revolution failed to materialize in America, and Clifford Odets never fully realized his abilities. But in the late 1930s, performances of *Waiting for Lefty* were packed, and many young people who saw it started reading the Communist weekly *New Masses*, and some of them eventually joined the Communist ranks.

The ultimate failure of Odets's career was part and parcel of the withered hopes of the radical intellectuals of his time. Notwithstanding the stirring slogans of solidarity purveyed in performance and leaflet and song, as the grim decade wore on, Soviet communism perverted and betrayed the enthusiasm of its adherents. The young Elia Kazan, who had joined the Communist party in 1935, left it disillusioned within about a year and a half.

The horrors of Stalin's forced collectivization and the ensuing famines were covered up (by, among others, Walter Duranty of the *New York Times*). But in 1936, the Great Purges of old Bolsheviks began in Moscow, very publicly, with the trial of Grigory Zinoviev and Lev Kamenev. As a preliminary to their execution, these sometime companions of Lenin were forced to abase themselves with false confessions of counterrevolutionary activity.

The Spanish Civil War broke out the same year, and the international left ardently embraced the cause of the Spanish Republic. But Soviet intervention on the side of the Republic led to the murder of revolutionaries guilty of the fatal error of opposing Stalin. It led, too, to the left's defeat. As veteran Spanish radical Joaquim Maurín put it, once Spaniards came to see the war as a struggle between Stalin and General Francisco Franco, the brutal incipient dictator, the Republic was doomed, for Franco at least was a Spaniard. The Republic collapsed in 1939.

Within six months, Stalin hatched an alliance with Hitler, and the two mass murderers began carving up Poland. These undeniable horrors—the purges, the betrayal of Spain, and the Hitler-Stalin pact—soured most of the young people who had been so stirred by Odets's play.

Elia Kazan, meanwhile, had become a journeyman actor and a rising director on Broadway. He soon started acting in movies and directing short films. After World War II, his movie-directing career took off in earnest, with *A Tree Grows In Brooklyn* (1945), and just two years later, he won his first Oscar—best director, for *Gentlemen's Agreement*.

A film starring Gregory Peck that attacked anti-Semitism, *Gentlemen's Agreement* was a landmark of early political correctness. It caused an uproar. Talky and dated though it seems now, it established Kazan as one of Hollywood's left-wing talents. Also in 1947, Kazan joined Lee Strasberg to found the Actors Studio, first in New York, then in Los Angeles. Actors Studio taught "Method" acting, developed by the Soviet stage director Konstantin Stanislavsky. Among the school's products were Marlon Brando and Marilyn Monroe.

In 1949, Kazan's production of Arthur Miller's *Death of a Salesman* won him plaudits as Broadway's finest director. But he and Brando were poised for much greater attainments—reached in 1951, with the film of *A Streetcar Named Desire*. Lyrical, corrosive, and heartbreaking, Tennessee Williams's creation explored the shadow side of American romantic illusions with a profundity that Miller never rivaled. *Streetcar*'s frank sexuality—especially Brando's rendition of an incoherent yet charismatic masculinity—brusquely ended the era of prim Hollywood censorship.

It was then that Kazan, at the height of his fame, was drawn into the controversy over reds in Hollywood.

The U.S. government's investigations of Communist influence in Hollywood had begun in 1947, at a time when politically attentive Americans were caught up in the emerging Cold War. For patriotic citizens, it was a frightening period. Stalin increasingly reminded them of Hitler. Since the war, Soviet armies had stayed on in Eastern and Central Europe, keeping an eye on the puppet regimes Moscow had installed. And around the world, Communists manipulated a fraudulent "peace" movement.

This last was central to the Communists' strategy toward the United States. Back when Hitler and Stalin had been allied, from

1939 to 1941, American Communists, in tandem with Nazi agents, had deployed an array of pseudo-pacifist slogans—"The Yanks Are Not Coming!" "No Imperialist War!"—exploiting traditional American isolationism. After 1945, the Soviet dictatorship went beyond borrowing arguments and tactics from the Nazis and actually adopted the role and methods of the fascists in its confrontation with the democracies.

Young American "fellow-travelers," hypnotized by the Communist peace offensive, seemed to know nothing of even this recent past. Kazan, by contrast, vividly recalled the Stalinist betrayals of the '30s and the phony pacifism of the Hitler-Stalin pact. Regardless of his popularity among "liberals" and his own continuing leftist sympathies, he saw communism as the enemy of everything he valued.

In April 1952, Kazan took a public stand. The previous January, he had been subpoenaed to testify before a closed executive session of the House Un-American Activities Committee in Washington, holding hearings on Communists in Hollywood. He had appeared but had refused to identify his former comrades—that is, he had refused to break the silence imposed on Communists by the party's conspiratorial discipline and on ex-Communists by the manipulation of guilt.

But in the ensuing months, Kazan changed his mind. He came to believe that the secrecy imposed by the party was inappropriate in America and that the Communists' demand for protection had been indulged too far. No previous radicals in this country had ever claimed the protection of the law for their clandestine activities; no other society in history had offered its citizens rights behind which to shield their political subversion. How could a revolutionary movement merit constitutional protection when its very purpose represented a repudiation of the U.S. Constitution?

On April 10, 1952, Kazan appeared before the committee a second time, at his own request, and "named names" in open session. Interestingly enough, while he knew the entire Hollywood Communist milieu in great detail, he concentrated on the Group Theater—the Communists he had known during the revolutionary period in the mid-'30s when he himself had been a party member.

He identified nine members of the cell to which he had belonged:

Odets; the late actor J. Edward Bromberg; the actor Morris Carnovsky, who had appeared before the committee and pled the Fifth Amendment; actress Phoebe Brand, whom Kazan had helped recruit; Paula Strasberg, wife of the anti-Communist Lee; actor Tony Kraber; party functionary Ted Wellman (alias Sid Benson), who with Kraber had recruited Kazan; Lewis Leverett, co-leader of the cell; and an actor named Art Smith.

Kazan recounted how party activities in the theater world had been directed by cultural commissar V. J. Jerome and Andrew Overgaard, a paid official of the Communist International. His prepared statement also mentioned three photographers, Paul Strand, Leo Hurwitz, and Ralph Steiner, as well as a playwright, Arnaud d'Usseau, the deceased actor Robert Caille, and four members of a Communist front, the League of Workers Theaters.

The Group Theater had been saved from Stalinist control, Kazan testified, by the firm stance of three anti-Communists: Lee Strasberg, critic Harold Clurman, and acting teacher Cheryl Crawford. Kazan had quit the party in 1936 because he had "had enough of regimentation, enough of being told what to think, say and do, enough of their habitual violation of the daily practices of democracy to which I was accustomed."

A month later, Odets made a similar voluntary appearance before the committee and named Kazan, along with five of those Kazan had mentioned; the two had discussed their testimony before appearing. None of the names they mentioned offered any surprise; all but the trio of photographers had been prominent and unapologetic in their defense of Stalinism during the '30s, although Mrs. Strasberg, like Kazan, had subsequently become an anti-Communist.

Elia Kazan had decided where his loyalties lay, and he would never draw back. Interestingly, he suffered no immediate rejection by the Hollywood left. In the broader scheme of things, the party and its supporters were clearly on the ropes. Stalin still ruled in Moscow, and war was raging in Korea, with Soviet pilots in action against U.S. and Allied forces.

In 1954, Kazan cast Brando in *On the Waterfront*, which took a bouquet of Oscars including best director. From the beginning, Kazan made clear that the film—about a union member who defies peer pressure and chooses to testify against labor racketeers—was inspired by his own decision to speak out. "A story about man's duty to society" was the description he offered the press. The screenplay was written by another ex- and anti-Communist, Budd Schulberg. In some respects, the story paralleled and completed the message presented in *Waiting for Lefty* almost twenty years before.

Yet even after the defiant *On the Waterfront*, Kazan was spared the full force of leftist hatred. He continued to produce great work—*East of Eden* (1955), with James Dean, another of his discoveries; *A Face in the Crowd* (1957), about the rise of a radio entertainer to political power; and *Baby Doll* (1956) and *Splendor in the Grass* (1961), two more demonstrations of his skill at handling complex, intimate subjects. With *America, America* (1963), he began a series of projects overtly concerned with his own life, including his marvelous memoirs (not published until 1988). He also continued to direct for the stage and wrote successful novels like *The Arrangement* (1969).

It was only during the 1970s, in the aftermath of the political convulsions of the '60s, that a revived leftist fundamentalism more virulent even than that of the '30s emerged and found a target in Elia Kazan.

Two decades after the House Un-American Activities Committee probe of Hollywood, a new witch hunt developed in the land. It was led by "liberal" intellectuals holding that "stool-pigeons" are worse than Stalinists. Why this logic did not prompt them to vilify those Americans who had turned in supporters of the Nazis—or, for that matter, "informers" who testified in murder trials—was never explained. From this point on, Kazan was dogged by a drumbeat of insults and carping gossip.

The worst damage to his reputation was done in the late 1970s by a man dedicated to defending Communist spies, Victor Navasky, publisher of the *Nation*. In full moral-absolutist cry, convinced that the Vietnam tragedy had forever justified Communist pretensions, Navasky set out to write a kind of dual biography of Elia Kazan (bad) and Arthur Miller (good, for keeping silent before the House committee). But Kazan's refusal to apologize for his actions or to assist Navasky with his project stirred Navasky's rage. The result was a book called *Naming Names* that appeared in 1980.

Up until this time, Kazan's creative work had carried more weight in most quarters than the left's contempt for him. But the young aspiring screenwriters who read Navasky in the early '80s had no grasp of Kazan's extraordinary achievements. Their only concern was to punish him for straying from a rigid defense of the global left. The handful of former Communists he had named to the committee— most of whom had left the film industry before he testified—were transformed in his critics' minds into hundreds of victims hounded out of the business.

From this point on, the contrasting trajectories of Kazan's reputation and those of the Stalinists he opposed say a great deal about the meaning of conscience in Hollywood. While Kazan was shunned, denied work, and otherwise humiliated, the Hollywood Ten—the cell of hard-core Stalinists who sought to turn the 1947 House committee hearings into something approximating a congressional riot, and paid for it with prison sentences—were lionized. Not only were their reputations restored, but institutions like the Hollywood talent guilds fawned over the Ten and repudiated their own supposed complicity with the establishment. While the Ten (who became Nine when the courageous Edward Dmytryk broke with the group) were acclaimed by "liberals" for what amounted to Soviet patriotism, Kazan's achievements were routinely dismissed in such venues as the American Film Institute, where his American patriotism was an embarrassment.

The thick varnish of sentimentality coating Hollywood's romance with Stalinism long remained intact, impervious even to extensive revelations about clandestine Soviet activities in the United States from the Russian and American archives. We now know from the Venona decryptions released by the National Security Agency, for example, that Mikhail Kalatozov, a Soviet director and cinema functionary prominent in Hollywood during World War II, was a high-ranking KGB agent. When Kalatozov's name was brought up in the House committee hearings, the Stalinists jeered, claiming that this Soviet operative had only come to the legendary city to buy prints of movies to show back in the motherland.

But we see from the Venona traffic that Kalatozov—who would later direct the famous 1957 Soviet war film *The Cranes Are Flying*—was a spy reporting directly to Grigory Kheifitz and Grigory Kasparov, the two NKVD station chiefs in San Francisco during World War II. (*The Cranes Are Flying* was shown to great fanfare in the Eisenhower White House.) Indeed, Venona evidence establishes beyond doubt that Hollywood was a major target of KGB operations in the United States, fully justifying the congressional inquiry.

The latest landmark in Hollywood's shunning of Kazan came in 1996, when the Los Angeles Film Critics Association dropped him from consideration for its career achievement award. Instead of Kazan, the honor was presented to Roger Corman, producer of, among other films, *Attack of the Crab Monsters*.

Reviewing this dismal history, one marvels that the Motion Picture Academy has broken down at last and decided on the special Oscar to be given in March. Reportedly, the actor Karl Malden, a star of *On The Waterfront*, argued the case for Kazan before the academy's board, to no dissent whatever. Industry sources point out that the crusade to exalt the Hollywood Ten has been mainly an enthusiasm of screenwriters, who tend to be leftists, while directors, producers, and actors always valued Kazan's art. Indeed, among the young generation in these fields, there is a surprising adulation of directors like Kazan, Samuel Fuller, and Robert Aldrich, despite their political incorrectness. Outside Hollywood, too, it may be a sign of the times that Navasky himself, while intransigent on the cases of Alger Hiss and the Rosenbergs (!), told the *New York Times* that Kazan's age, the passage of time, and the excellence of his work have softened Navasky's views. "It's a human thing," he said. "He's not physically well, and he made this great cinematic contribution."

By contrast, Abraham Polonsky, a Hollywood writer who would never have been heard of had he not received a House committee subpoena long ago, met a reporter with a snarl: "Has [Kazan] ever said, 'Gee, I'm sorry. I shouldn't have done that. I was wrong'?"

Well, no. Kazan has refused, over the past decade, to elaborate on these matters beyond the dignified statement in his memoirs: "I did what I did because it was the more tolerable of two alternatives that were, either way, painful, even disastrous, and either way wrong for me. That's what a difficult decision means: Either way you go, you lose." No explanation whatever, of course, comes from Polonsky and others so long devoted to Joseph Stalin. Has Polonsky ever said he regretted enthusiastically supporting the Soviet dictatorship that created Joe McCarthy?

In the late '40s and early '50s, many people, when called upon to choose between the House committee and Stalin, chose the committee. Today, belatedly, others may be starting to see the wisdom of that judgment. It may even be that the thaw begun in the Soviet Union when Khrushchev was premier is finally reaching the sunny precincts of Hollywood.

# JOE DIMAGGIO

## By Donald Kagan

*(Originally published, as "Joe DiMaggio,
Baseball's Aristocrat," in the March 22, 1999, issue)*

ON MARCH 8, 1999, Joe DiMaggio died in his 85th year, a baseball legend, but also an American hero who represented the virtues and ideals of his era. His achievements as a player were extraordinary: a lifetime batting average of .325, with a seasonal high of .381. He had seasonal highs of 167 runs batted in and 46 home runs, with a lifetime total of 361. Twice he led the league in homers, twice in RBIs, and twice again in slugging percentage.

But hitting was only part of the story, for DiMaggio was a complete player, a great fielder and a brilliant, if unobtrusive, base runner. No one played a shallower center field, which permitted him to cut off looping singles and short line drives; no one raced back more swiftly, covered more ground, judged more truly, or threw with greater power and accuracy. He had good but not outstanding speed as a base runner. Stealing bases was no part of Yankee strategy, but his judgment and skill on the bases caught the attention of discerning observers. George Will, that connoisseur of the game, notes with admiration that Joe was never thrown out while going from first to third on a base hit.

Of all his achievements, however, the one that best accounts for his unique status as an American hero is his feat in 1941 of hitting safely in 56 consecutive games, a record that remains on the books after 57 years. A first step to comprehending the magnitude of this feat is to know that of the thousands of batters who have played in baseball's century of major league play, the closest anyone has come is 44, and only a handful have approached that. DiMaggio's record has never been seriously threatened. For that there are reasons. All record-breakers face mounting pressure as their achievement mounts. But a home-run hitter can go for days, even weeks, without success, knowing that there will be time for a hot streak when the slump is over.

By contrast, consider the consecutive-game record-breaker. Each day he must succeed or the race is over. The pitcher he faces may be unhittable that day or so wild as to withhold a good pitch. The fielders may outdo themselves and steal hits with great plays. Never mind—he still must hit safely every day. In a fine book on the streak, Michael Seidel caught its heroic character: "The individual effort required for a personal hitting streak is comparable to what heroic legend calls the *aristeia*, whereby great energies are gathered for a day, dispensed, and then regenerated for yet another day, in an epic wonder of consistency." There is nothing like it in baseball, and DiMaggio's steadiness, cool determination, and brilliant ability caught the nation's sustained attention in 1941 as no athletic event had done before. The whole country asked each day, "Did he hit?" and rooted for the streak to continue.

But baseball is a team game, and individual statistics are important chiefly insofar as they contribute to victory, and here DiMaggio was supreme. In his 13 years in the majors his team won the pennant 10 times and the world championship 9. To be sure, the Yankees of his day were an outstanding team, but his contribution as a player, quiet leader, and exemplar was essential to its greatness. These qualities were never so evident as during the hitting streak of 1941, when DiMaggio's exploits had meaning not for himself alone, but carried and inspired his companions, as the deeds of true heroes do. During the streak, Johnny Sturm, Frank Crosetti, and Phil Rizzuto, none of them normally a great hitter, each enjoyed a lesser streak of his own. At the beginning of Joe's streak, the Yankees were in a terrible slump and five and a half games out of first place. At its end, they had destroyed the will of the opposition, were safely in first place, and on their way to clinching the pennant on September 4, the earliest date in history, 20 games ahead of the next best team. That summer a song swept the nation:

*From coast to coast, that's all you hear*
*Of Joe the One-Man Show,*
*He's glorified the horsehide sphere,*
*Joltin' Joe DiMaggio.*
*Joe . . . Joe . . . DiMaggio . . .*
*We want you on our side.*

*He'll live in baseball's Hall of Fame,*
*He got there blow-by-blow,*
*Our kids will tell their kids his name,*
*Joltin' Joe DiMaggio.*

So did the raging Achilles inspire his fellow Achaeans against the Trojans, and, so, at somewhat greater length, did Homer sing of Achilles' deeds.

But there is more still to true heroism: the qualities of courage, suffering, and sacrifice. These DiMaggio displayed most strikingly in 1949. Before the season, he had a bone spur removed from his heel. The pain was so great as to keep him out of every game until the end of June. The Yankees were going up to Boston for a three-game series against the team they had to beat. DiMaggio blasted four home runs in three games, batting in nine runs as New York swept the series. The importance of that *aristeia* was very clear on the day before the end of the season. The Red Sox came into Yankee Stadium for two last games. Had they won but one of the three snatched from them in June, the championship would have been theirs already. Instead they had to win one of the remaining two. DiMaggio had missed the last couple of weeks, felled by a case of viral pneumonia. Once again, the ailing warrior returned to the field of battle. Weak as he was, he managed two hits and led his mates to victory. The next day, the staggering Joe managed to run out a triple and last until the ninth inning before weakness and leg cramps forced him from the field. The inspired Yankees won the game and the championship. The 1949 season was only the most dramatic instance of the heroic power DiMaggio's example brought to the efforts of his team.

His major league career of 13 years was cut short by 3 years of service as a soldier in the Second World War, by the wear and tear of injuries suffered from time to time through his career, and, perhaps, by a quiet but powerful pride that forbade him to play beneath the level of excellence he had established. When asked why he had quit, he replied that it was because he had standards. The memory of even great baseball players generally fades quickly. Within a few years, only true fans and a few others remember them, but Joe's retirement in 1951 somehow did not end the remarkable connection he had made with the American people. Over time, it became clear that he was more than a great former ballplayer, that he had become a hero whose rare public appearances brought thunderous applause and respectful awe generations after he stopped playing. Why was that?

The answer lies in the way he played the game and the manner in which he conducted himself on and off the field. The words always applied to him are grace and style and class, words that carry the values of aristocracy more than democracy.

Class, after all, derives from the Latin word that means rank or social standing; unmodified, it means "of high rank and standing." Webster's dictionary rightly tells us that in slang or common American use it means "excellence, especially in style." That is what made DiMaggio stand out in his time. On the field, he played the game hard and to win, but with the gentlemanly grace that does not call attention to itself, that makes difficult plays look easy. We do not remember him leaping or diving but gliding easily to reach the ball. After a great play or key hit he never cavorted or capered but simply looked down while the crowd roared. He never argued with umpires or fought with opponents. Off the field he spoke to the press as little as possible and rarely gave them an opinion. He did all his talking on the field with legs, bat, and glove. Off the field, he insisted on his privacy and maintained a quiet dignity that was rare even in its day. On the field, he employed his unique talents not to polish his self-esteem but to bring victory to the team.

And his day was not ours. America was a democracy, but of a different kind. Its people were more respectful of excellence, both of matter and manner, prepared to follow the leadership of those they deemed superior in achievement and "class." People wanted to behave according to a higher and better code because they believed that in doing so they would themselves become better, worthier, "classier." Those who are too young to remember should look at the movies and photographs of games at Yankee Stadium in DiMaggio's day. The men wore white shirts and ties under coats and hats, the proper attire in public, even at a ball game. People were more conscious of the opportunity American society gave them to move into a better way of life than they were of the indignity of not being there already. They were not insulted by the notion that another way of life might be better than their own.

American democracy in DiMaggio's day reached a point in its development where the common man had the power to decide and chose to look up. The people respected and elected their betters in the expectation of reaching the heights themselves. In much of DiMaggio's day, the leader of the democracy was Franklin D. Roosevelt, an American aristocrat if there ever was one, with an accent rare even at Harvard and a cigarette holder characteristic of the classes, not the masses. Ordinary Americans admired these markers of class as they admired the aloof elegance and dignity of the Yankee Clipper. Joe was the son of a poor Sicilian fisherman, not the scion of Dutch patroons from the Hudson Valley, yet his classic grace and style seemed to raise him above the crowd, a model of class and excellence for others to emulate.

In those days such qualities led not to envy and the charge that he had abandoned his roots and heritage. Instead, Italian-Americans all over the country glowed with pride and felt elevated by his success. He himself never referred to his family's origins, much less did he try to use them to any advantage. He was simply an American who quietly went to serve his country when called to war, like other Americans. That is the way they wanted it, no special attention, no privileges or compensations, merely the opportunity to achieve respect, maintain their dignity, and improve themselves and their families.

DiMaggio was a democratic hero when American democracy was closer to the Periclean ideal, when the goal was understood to be the forging of a single people in pursuit of an excellence which all respected and to which all could aspire. It aimed to raise its citizens to a higher level by providing splendid models and the opportunity for all to seek to emulate them. But history seems to show that democracies change and become less respectful of high standards. Then democracy's leaders and exemplars shy from any hint of superiority, seeking to win support by claiming identity with the least of the citizens. They resort to flattery rather than the elevation of the common man, corrupting the culture and the polity by appealing to the masses at the lowest level. If a man of genteel origins is elected, he tries to speak in the inelegant tone and language of the common man and pretends to like eating pork rinds. So has the last half-century changed America, and yet the fame and celebration of DiMaggio have never been greater. His death was a major national story, leading the front pages and the network news, the subject of innumerable encomia.

What is it that explains this continued veneration in such a different world? It appears that all eras need true heroes, superior models of qualities that we admire, whether or not they are fashionable. The shining image of DiMaggio, even in a degenerate age, reminds people of a higher ideal, half-forgotten but impossible to ignore. Half a century after his retirement, people who never saw him play somehow retain an idea of his special character, of what he meant to the Americans of his day, and they are elevated by his example.

# EDWARD C. BANFIELD

## By James Q. Wilson

*(Originally published, as "The Man Who Knew
Too Much," in the October 18, 1999, issue)*

IN THE INCREASINGLY DULL, narrow, methodologically obscure world
of the social sciences, it is hard to find a mind that speaks not only to
its students but to its nation. Most scholars can't write, many can't
think. Ed Banfield could write and think.

When he died a few days ago, his life gave new meaning to the old
saw about being a prophet without honor in your own country. Almost
everything he wrote was criticized at the time it appeared for being
wrong-headed. In 1955 he and Martin Meyerson published an account
of how Chicago built public housing projects in which they explained
how mischievous these projects were likely to be: tall, institutional
buildings filled with tiny apartments built in areas that guaranteed
racial segregation. All this was to be done on the basis of the federal
Housing Act of 1949, which said little about what goals housing was
to achieve or why other ways of financing it—housing vouchers, for
example—should not be available. This was heresy to the authors of
the law and to most right-thinking planners.

Within two decades, high-rise public housing was widely viewed
as a huge mistake and efforts were made to create vouchers so that poor
families could afford to rent housing in the existing market. Local
authorities in St. Louis had dynamited a big housing project there
after describing it as a hopeless failure. It is not likely that Ed and
Martin's book received much credit for having pointed the way.

In 1958, Ed, with the assistance of his wife, Laura, explained why a backward area in southern Italy was poor. The reason was not government neglect or poor education but culture. In this area of Italy, the Banfields said in *The Moral Basis of a Backward Society*, people would not cooperate outside the boundaries of their immediate families. These "amoral familists" were the product of a high death rate, a defective system for owning land, and the absence of any extended families. By contrast, in a town of about the same size located in an equally forbidding part of southern Utah, the residents published a local newspaper and had a remarkable variety of associations, each busily involved in improving the life of the community. In southern Italy, people would not cooperate; in southern Utah, they scarcely did anything else.

Foreign aid programs ignored this finding and went about persuading other nations to accept large grants to build new projects. Few of these projects created sustained economic growth. Where growth did occur, as in Singapore, Hong Kong, and South Korea, there was little foreign aid and what existed made little difference.

Today, David S. Landes, in his magisterial book that explains why some nations become wealthy while others remain poor, offers a one-word explanation: culture. He is right, but the Banfield book written forty years earlier is not mentioned.

In 1970, Ed published his best-known and most controversial work, *The Unheavenly City*. In it he argued that the "urban crisis" was misunderstood. Many aspects of the so-called crisis, such as congestion or the business flight to the suburbs, are not really problems at all; some that are modest problems, such as transportation, could be managed rather well by putting high peak-hour tolls on key roads and staggering working hours; and many of the greatest problems, such as crime, poverty, and racial injustice, are things that we shall find it exceptionally difficult to manage.

Consider racial injustice. Racism is quite real, though much diminished in recent years, and it has a powerful effect. But the central problem for black Americans is not racism but poverty. And poverty is in part the result of where blacks live and what opportunities confront them. When they live in areas with many unskilled workers and few jobs for unskilled people, they will suffer. When they grow up in families that do not own small businesses, they will find it harder to move into jobs available to them or to meet people who can tell them about jobs elsewhere. That whites treat blacks differently than they treat other whites is obviously true, but "much of what appears . . . as race prejudice is really *class* prejudice."

In 1987, William Julius Wilson, a black scholar, published his widely acclaimed book, *The Truly Disadvantaged*. In it he says that, while racism remains a powerful force, it cannot explain the plight of inner-city blacks. The problem is poverty—social class—and that poverty flows from the material conditions of black neighborhoods. Banfield's book is mentioned in Wilson's bibliography, but his argument is mentioned only in passing.

Both Wilson and Banfield explain the core urban problems as ones that flow from social class. To Wilson, an "underclass" has emerged, made up of people who lack skills, experience long-term unemployment, engage in street crime, and are part of families with prolonged welfare dependency. Banfield would have agreed. But to Wilson, the underclass suffers from a shortage of jobs and available fathers, while for Banfield it suffers from a defective culture.

Wilson argued that changing the economic condition of underclass blacks would change their underclass culture; Banfield argued that unless the underclass culture was first changed (and he doubted much could be done in that regard), the economic condition of poor blacks would not improve. The central urban problem of modern America is to discover which theory is correct.

Banfield had some ideas to help address the culture (though he thought no government would adopt them): Keep the unemployment rate low, repeal minimum-wage laws, lower the school-leaving age, provide a negative income tax (that is, a cash benefit) to the "competent poor," supply intensive birth-control guidance to the "incompetent poor," and pay problem families to send their children to decent day-care programs.

*The Unheavenly City* sold well but was bitterly attacked by academics and book reviewers; Wilson's book was widely praised by the same critics. But on the central facts, both books say the same thing, and on the unknown facts—What will work?—neither book can (of necessity) offer much evidence.

Ed Banfield's work would probably have benefited from a quality he was incapable of supplying. If it had been written in the dreary style of modern sociology or, worse, if he had produced articles filled with game-theoretic models and endless regression equations, he might have been taken more seriously. But Ed was a journalist before he was a scholar, and his commitment to clear, forceful writing was unshakable.

He was more than a clear writer with a Ph.D.; everything he wrote was embedded in a powerful theoretical overview of the subject. "Theory," to him, meant clarifying how people can think about a difficulty, and the theories he produced—on social planning, political influence, economic backwardness, and urban problems—are short masterpieces of incisive prose.

His remarkable mind was deeply rooted in Western philosophy as well as social science. To read his books is to be carried along by extraordinary prose in which you learn about David Hume and John Stuart Mill as well as about pressing human issues. To him, the central human problem was cooperation: How can society induce people to work together in informal groups—Edmund Burke's "little platoons"—to manage their common problems? No one has ever thought through this issue more lucidly, and hence no one I can think of has done more to illuminate the human condition of the modern world.

A few months ago, a group of Ed's former students and colleagues met for two days to discuss his work. Our fondness for this amusing and gregarious man was manifest, as were our memories of the tortures through which he put us as he taught us to think and write. Rereading his work as a whole reminded us that we had been privileged to know one of the best minds we had ever encountered, a person whose rigorous intellect and extraordinary knowledge created a standard to which all of us aspired but which none of us attained.

# PAUL COVERDELL

## By Matthew Rees

*(Originally published, as "The Gentleman from Georgia," in the July 31, 2000, issue)*

A SENATOR'S WORK IS A NEVER-ENDING series of committee hearings, caucus meetings, floor votes, flights, fund-raisers, and constituent service. It is, in many ways, a dreadful job that inevitably produces burnout. Yet Paul Coverdell, who died suddenly and much too young last week at the age of 61, may have been the only senator in U.S. history who had more spring in his step after seven years in office than at the beginning of his tenure.

In this way and many others, Coverdell defied the senatorial stereotype. He didn't come from a prominent family, he wasn't particularly handsome, and he had a speaking style that, as the joke went, looked and sounded like someone imitating Dana Carvey imitating George Bush. He was never much of a back-slapping glad-hander. His campaign slogans—"Paul Coverdell Means Business" and "Coverdell Works"—reflected his simple, can-do approach to politics.

Before coming to the Senate in 1993, Coverdell spent more than 20 years building an insurance company and the Georgia Republican party. (The party was so small when he began—he was one of just four Republicans in the state senate in 1971—that he used to joke about its meeting in a phone booth.) He proved a spectacular success at both tasks, and the experiences taught him a skill noticeably lacking among today's senators of both parties: how, against all odds, to get things done.

Coverdell lived by the simple creed of a Boy Scout: Be prepared. His work ethic was the stuff of Senate lore. If he wasn't sleeping, he was working. He once took his briefcase to an Atlanta Braves game and worked from the third inning on. Trent Lott, the Senate Republican leader, was so enamored of Coverdell and his work habits that he put him in charge of countless task forces and working groups. He also dubbed him "Mikey," a reference to the kid in the Life cereal commercials who was always willing to eat anything.

Yet Coverdell was not a grind who worked for the sake of working. His labors flowed from his deeply held belief in the value of freedom. A few years ago he wrote that "ensuring freedom is to me the highest possible goal of a political party. Why freedom? Because human experience has shown that the greatest practical good for the greatest number is achieved by free people through free elections and free markets."

Tagged as a moderate upon his election in 1992—he was mildly pro-choice, and his chief Republican primary opponent had been Bob Barr—Coverdell emerged as one of the more conservative members of the Senate, and certainly one of the most effective conservatives. Asked about this seeming ideological shift, he downplayed it, saying he'd simply become "more concerned about government's intrusion into our lives."

Coverdell's concern jelled in 1993, with the release of the Clinton plan to remake the American health care system. While many Senate Republicans were squeamish about raising objections, Coverdell told them in one now-famous meeting precisely what was at stake: "Think of this as 1939. We have to choose whether to be Chamberlain or Churchill." To that end, he began organizing meetings attended by Senate staffers, activists, and interest-group representatives, and these meetings quickly became the nerve center of the opposition. Coverdell's boundless energy, and willingness to do the organizational scut work his colleagues couldn't be bothered with, yielded this apt characterization of his time on Capitol Hill: "The best staffer in the Senate." In the end, he may have done more to defeat the Clinton health care plan than anyone else in Congress.

Coverdell was not, however, a wild-eyed partisan. He struck close alliances with Democratic senators like Bob Torricelli and Dianne

Feinstein. And though a journalism major in college, Coverdell rarely made for good copy. His loyalty to his colleagues prevented him from revealing much of anything in interviews.

Phil Gramm, who delivered a moving tribute to Coverdell on the Senate floor, once pinpointed a secret behind his close friend's success: "People like to put Paul in leadership positions because he makes other people look good." Indeed, for all that Coverdell had already accomplished, there was a widespread belief among Republicans that his best years were ahead of him. Slated to move into the number three position in the Senate GOP hierarchy next year, replacing the retiring Connie Mack, he was widely expected to continue his rapid ascent of the greasy pole. As a confidant of George W. Bush and a close friend of Bush-*père*, Coverdell would have been even more influential in a Bush presidency. As it is, his success in the Senate stands as proof that Washington does occasionally reward, rather than punish, talent and effort. And decency. Tom Daschle, the Senate minority leader, in a tribute last Wednesday, called Coverdell "a gentleman." So he was.

It is fitting that Coverdell's last floor speech in the Senate, on July 13, was devoted not to advancing an arcane piece of legislation but to honoring another friend of freedom, Ronald Reagan. Fitting because Coverdell had emerged as a latter-day Reaganite who, like Reagan, was more interested in what was accomplished than in who got the credit. In his speech, Coverdell described the former president as someone who "preferred to see himself as a simple citizen who had been called upon to aid the nation he so loved." That's a good description of Ronald Reagan. It's also a good description of the late Paul Coverdell.

# RONALD REAGAN

## By Jeffrey Bell

—◆—

*(Originally published, as "The Candidate and the
Briefing Book," in the February 5, 2001, issue)*

IT WAS 1975, and I found myself in the middle of a struggle of wills
between John Sears and Ronald Reagan. In retrospect, this may sound
interesting, but at the time it was anything but enjoyable. Sears was
the most brilliant political strategist I've ever known. Reagan was the
greatest man I've ever known, though to be honest I had no inkling of
this yet.

Not for the first or last time, Sears and Reagan were furious at each
other, so furious that I didn't know what to do or what to make of it.

The issue was the briefing book Sears had instructed me to write
for Reagan, to prepare him for his upcoming primary challenge to
President Gerald Ford. I was writing it, but Reagan wasn't reading it.
This was not a morale builder for me, but to Sears it was infuriating.
Sears's everyday demeanor was droll and understated, but when he was
angry, most people who knew him found him frightening, even on
occasion Reagan, who normally seemed afraid of no one.

To Sears and to me, the gold standard of presidential politics was
the Nixon campaign of 1968. Objectively speaking, that campaign
and that candidate had made quite a few mistakes. But from the perspective of 1975, it was the only time in almost a half century that the
Republican party had taken over the White House without running a
war hero. And it was our formative experience.

In 1968, Sears at a precocious 27 was a top-level Nixon political operative. At 24, I was a lowly research assistant, just out of the Army, running errands for Pat Buchanan and a policy/issues team that included (a partial list) Alan Greenspan, Richard Allen, William Safire, Martin Anderson, Ray Price, Richard Whalen, plus (junior aides like me) John Lehman, Kevin Phillips, Ken Khachigian, and (following Nixon's defeat of Nelson Rockefeller) George Gilder and a handful of other liberal Republicans.

All of these staffers, and others, contributed to Nixon's briefing book, which was maintained and constantly updated by Buchanan. The briefing book, written in question-and-answer format, was enormous, of biblical proportions, growing and evolving as the campaign progressed. The reason so much effort was put into it is that Nixon wanted it that way. He spent countless hours poring over it, and knew it well, because he never wanted to be surprised by the nastiest question his worst enemy could think of. His writers were kept energized by Nixon's commitment to the briefing book, which meant that at any time, and without warning, they were likely to hear the former vice president using the exact words they had written to fend off somebody's question. Those words had better be accurate and defensible, or the writers knew they might find themselves invited into the crewcut, uncompassionate presence of campaign chief of staff H. R. Haldeman, Nixon's designated bad cop.

Now, eight years later, post-Watergate spending limits had arrived, and what Nixon's awe-inspiring stable of writers and policy advisers had labored to produce had fallen largely to me, as research director of Citizens for Reagan, with help from a corporal's guard of outside volunteers. Aside from a willingness to write in an authoritative tone about subjects I knew little or nothing about—a necessity in these matters—my main virtue in Sears's eyes was that I had watched Buchanan continually update Nixon's briefing book and therefore had some idea of how the process was supposed to work.

But Reagan seldom looked at the book. To Sears, this meant one thing: Reagan was intellectually lazy and would be unprepared for what awaited him in his challenge to Ford.

Sears had earlier been picked to manage the anticipated presidential campaign of Vice President Spiro Agnew, and I was hoping to work on the issues staff, when a scandal erupted that eventually forced Agnew to resign his office in October 1973 to avoid an indictment for bribery. Sears had therefore shot his way into the leadership of Reagan's campaign against Ford quite late, and now his worst fears about the aging Hollywood actor, the very fears that had caused him to prefer Agnew as the next conservative standard-bearer, seemed in danger of being realized. Reagan was too much a lightweight to bother to read his own briefing book. The liberal press and Ford strategist Stuart Spencer would combine to eat Reagan alive.

They didn't quite do that, but Ford did of course beat Reagan, and Sears and I (along with many others) thought Reagan's preparation on issues was a factor. A lot of people, it should be mentioned, blamed Reagan's loss on me, for persuading Reagan to advocate an overly ambitious $90 billion decentralization plan that he found difficult to defend. Part of my private defense was my contention that I and the others who worked on the plan had anticipated most of the attacks by the liberals and the Ford campaign, but that Reagan wouldn't read that part of his briefing book. (The only virtue of this defense is that it was so ineffective I soon gave up on it.)

Key to Ford's victory was his come-from-behind 49–48 percent win in New Hampshire, where he pounded on the $90 billion plan and on Reagan's long-standing advocacy of a voluntary Social Security system. Ford increased his margins in several subsequent primaries, and the Reagan campaign ran out of money. On Tuesday, March 23, Reagan completed his campaigning in North Carolina, and defeat was so universally expected that he and his traveling party took off in the campaign plane before any results were known, planning to concede the nomination to Ford a day or two later.

But two strong-willed and extremely stubborn Reagan backers from North Carolina, a freshman senator named Jesse Helms and his political manager, Tom Ellis, had raised enough money to buy local television time for a 30-minute speech by Reagan denouncing the Ford-Kissinger policy of détente with the Soviet Union. The speech had been taped weeks earlier in the studio of a Florida station that had offered all the presidential candidates a half hour of free time. Although there was widespread agreement that the foreign-policy theme was beginning to click, Reagan's national staff was skeptical that large numbers of voters would listen to 30 minutes of any politician, particularly a tape that had never been intended as professional advertising, and looked it.

Helms and Ellis would not take no for an answer. Eventually, Ellis cut a deal with Sears dedicating all the money the North Carolina Reagan committee managed to raise to more air time for the Florida videotape.

Word came to the Reagan traveling party in mid-air that against every expectation, Reagan had defeated Ford in North Carolina. At first, Reagan himself refused to believe the news. But not only had he won, when he went on national television a week later to resurrect his campaign, so much money came in that the campaign couldn't spend it all.

Everyone could see the flaws in Reagan and his 1976 campaign, including mistakes made by Sears and me. Yet somehow it all added up to more than the sum of its parts. Reagan ended up defeating Ford in 12 primaries, and came within one or two delegations of winning the nomination at the Kansas City convention. When Jimmy Carter defeated Ford, Reagan emerged as the real GOP winner of 1976. He was well positioned to run again in 1980.

After a bloody series of power struggles with Reagan's California staff, Sears emerged as the campaign manager again. Reagan was such an overwhelming front-runner for the nomination going into 1980 that many of his advisers, including Sears, hoped he could avoid the kind of near-death experience he had lived through the night of North Carolina. This hope was not to be realized. Not surprisingly, the crisis of the 1980 nomination fight was triggered by a resumption of the tension between Reagan and Sears over the quality of Reagan's issue preparation: In a sense, the briefing book, again.

Sears's most fateful decision, the one that armed his many enemies in Reagan's orbit, was to keep Reagan out of the *Des Moines Register* debate right before the Iowa caucus in January 1980. At the time, this seemed reasonable. Reagan, who had been well known in Iowa since his days as a young sports announcer in the 1930s, seemed to have a solid lead over a field of challengers far less well known than he. More important, Sears feared Reagan wasn't ready to debate. He concluded that the risk of a Reagan embarrassment in the debate was greater than the risk of losing Iowa. Indeed, Sears thought Reagan might win Iowa and still see his campaign begin to unravel if he looked ill-prepared, out of touch, and therefore too old to be president.

So Reagan skipped the debate. Bush won Iowa, and was instantly transformed from an unknown, single-digit New Hampshire candidate to a solid front-runner. The closeness of Bush's win in Iowa made it appear that if Reagan had attended the debate, he would have at least edged Bush in the Iowa caucus and remained the front-runner in New Hampshire. In the Reagan camp, Sears got full blame for a possibly fatal blunder.

It was clear to Reagan and those close to him that Sears's decision to bypass the Iowa debate was a vote of no confidence in Reagan's issue preparation. Moreover, issue preparation was a subtext of Sears's factional wars against Reagan's California veterans. Sears earlier had forced Martin Anderson, Lyn Nofziger, and (most shockingly) Mike Deaver to leave the campaign staff. And he and his allies increasingly blamed Reagan's only surviving California adviser, Ed Meese, for inadequate preparation of the candidate.

Following Iowa, the 69-year-old Reagan was counted out by many in the national press. The view that Reagan was just too old and too right-wing to become president was back in full force. George Bush, now a nationally known figure, exulted about the "Big Mo" and flew confidently to New Hampshire with a lead in the nation's first primary of 10 points or more. Poor cash-flow management had left the Reagan campaign perilously close to its legal spending limits not just in New Hampshire, but in the nomination fight as a whole. This would make it difficult if not impossible for Reagan to make a stand in a subsequent state should he lose New Hampshire.

Reagan needed a North Carolina-style resurrection, but this time the roles of Jesse Helms and Tom Ellis would have to be played by two men who now were barely on speaking terms—Ronald Reagan and John Sears. Amazingly, they both proved equal to the occasion. In their utterly different ways, they began operating in a kind of political overdrive I've never seen equaled before or since.

This was the last political cycle in which there was a five-week interval between Iowa and New Hampshire. Reagan set himself a dawn-to-late-night schedule, and kept it. He performed well in press interviews and candidate debates, seeming to relish the role of underdog. He gave greater emphasis to New York congressman Jack Kemp's tax cut proposal as a defining issue against Bush, who opposed the tax cut.

Sears, for his part, probed relentlessly for weaknesses in Bush's disciplined, risk-averse team. After a couple of weeks, Sears began an elaborate series of ploys revolving around a candidate debate scheduled for the Friday before primary day and sponsored by the *Nashua Telegraph*. First, the Reagan campaign proposed a debate involving only the two front-runners. The *Telegraph* and the Bush campaign eagerly accepted. Then, when the other four candidates actively campaigning in New Hampshire protested, Sears executed a sudden reversal, positioning Reagan as the candidate of inclusion, while the Bush campaign and the newspaper attempted to stick to the earlier agreement. All of the other candidates, together with much of the press, began attacking Bush as a snob and elitist for—what? For having accepted Sears's original proposal for a two-man debate and sticking to it.

I was not on the 1980 campaign staff, but in late 1979 Sears and his chief deputy, Charlie Black, fired Reagan's Madison Avenue advertising firm. Black called me and asked me to supervise the making of new commercials centering around Reagan's advocacy of an across-the-board 30 percent cut in federal income tax rates, modeled on advertising themes I had used in a Senate run two years earlier in New Jersey. By the time I flew to Los Angeles with Philadelphia ad man Elliot Curson in late January to make the new spots, Reagan had lost Iowa.

Reagan had been attracted to supply-side arguments long before they bore that label, and had praised Jack Kemp's proposed tax cut from the time it was unveiled in 1977. But more than anyone else, it was Sears who pushed the tax cut as a centerpiece of 1980 strategy and had promoted Kemp's increasing prominence among Reagan's advisers, over considerable opposition from Sears's critics in California and elsewhere.

I arrived in North Andover, Massachusetts, the site of the hotel being used by the Reagan traveling party, on the weekend before the primary to take part in the final drilling of Reagan for the Nashua debate. I was unaware of much that had been happening. In quick succession, I learned that the campaign's private polling showed that the tax-cut spots were working; that Reagan had retaken the lead over Bush in the state by about 10 points; and that Sears, Black, and Lake had something else up their sleeve which they couldn't or wouldn't tell me about. This was underlined by their absence from the briefing session with Reagan.

Already encouraged by what I had heard, I was elated by Reagan's performance in the debate drill. He was at the top of his game, confident and well-versed on the issues, foreign and domestic, that had been thrown at him by the 20 or so staff members and outside advisers sitting around a large conference table.

I sought out Sears to tell him how impressive the candidate had been. Sears fixed me with a withering, almost angry smile, and said with unmistakable sarcasm, "Is that right?" He walked away without another word. Only then did it hit me that Sears was so alienated from Reagan that he seemed incapable of accepting good news about him.

At the Nashua debate that night, Sears sprang his final trap on Bush by orchestrating the appearance of the four also-ran candidates. As is well remembered, Reagan uttered the legendary line, "I paid for this microphone, Mr. Green," while Bush froze. Though it wasn't televised and New Hampshire voters saw no more than a few sound bites, Reagan devastated Bush in the debate.

On primary day, the tense, tight Reagan-Bush primary collapsed into a rout, 50 to 23 percent in favor of Reagan. The nomination fight was effectively over. Before the polls closed, Reagan called in Sears and his top lieutenants, Black and Lake, and fired them on the spot. William Casey was named the new campaign manager, and Reagan's California team returned, one by one, to the inner circle.

Sears's stormy partnership with Reagan was at an end, ironically at the absolute peak of its success, yet irreparable. But the fundamental question between them—was Reagan adequately preparing himself to run for president and, ultimately, to be president?—was to continue in one form or another, without Sears, for the rest of that campaign, indeed for the rest of Reagan's career.

That October, most of Reagan's advisers vehemently opposed allowing him to debate one-on-one with President Jimmy Carter, fearing the worst. James Baker, who had been the campaign manager for Reagan's principal opponents in the nomination fights of 1976 and 1980, argued that Reagan should debate Carter. He was right and wound up as Reagan's White House chief of staff. In 1984, White House aide Richard Darman was attacked for "overbriefing" Reagan for his first debate with Walter Mondale, on the unstated assumption that Reagan, by then at 73 the oldest president ever, was not up to absorbing much if any information. And prior to almost every G-7 or superpower summit Reagan attended, State Department and other officials were invariably heard to complain that Reagan would be taken to the cleaners if he didn't pay more attention to *their* briefing books.

At this remove, it is easier to understand why Nixon needed and used his briefing book than why Reagan had so little interest in his. Nixon had a gift for absorbing details but no overarching belief system. To him each question was independent of every other and—given his view of his enemies—a potential land mine. He had a hunger to know and think through, as a discrete matter, every question that he and his advisers thought might arise. He lacked an ideological organizing principle to help him do this, so he needed the briefing book.

By contrast, Reagan held an intense, compelling vision of America and the world that did not seem to depend on detailed knowledge. The puzzle was famously summed up in eight words by one of his national security advisers, Robert McFarlane: "He knows so little and accomplishes so much."

Reagan's detractors have always put their emphasis on the first part of the sentence, his admirers on the second. But each side knows that the full McFarlane sentence has weight, as does the paradox at its heart.

What accounts for the paradox of Reagan? Isaiah Berlin's metaphor of the fox and the hedgehog—based on Archilochus' dictum that the fox knows many things and the hedgehog one big thing—offers one possible solution. Some politicians—Bill Clinton comes to mind—are clearly in the fox category. Reagan seems more like a hedgehog—until you try to figure out what was the one big thing he knew. Was it that tax rates must come down? Or was it that the Soviet Union was far more fragile, far more vulnerable to outside pressure than anyone else realized? Or was it that Americans are still capable of seeing their country as a shining city on a hill, capable of changing the world by force of example and advocacy? It's hard to say.

What does seem to be the case is that Reagan had an extraordinarily high batting average on the judgment calls that came across his desk. He never seemed to know as much as his advisers about any one thing, but this didn't stop him from being right again and again, including on issues where all his advisers thought he was just this side of insane.

Reagan, for a political leader, had a unique way of looking at politics. Most politicians love political gossip. Reagan had no interest in it. He didn't care who the chairman of the Ohio GOP was, or what he thought, or who he was sleeping with. Instead, Reagan would spend endless hours reading and answering his personal mail. When I was on his staff, I thought this was a waste of time. I now believe it was at the heart of his populism. It gave him a vivid window on how voters think. This may explain some of his success. But again and again, Reagan made the right call on subjects he never got mail about.

There are other theories about Reagan that verge on the mystical. The secular version is that he had extraordinary intuition, or luck. Reagan himself appeared to have genuine humility about his success, whether it was due to luck or something deeper. After his shooting in 1981, he seemed to feel that his life had been spared to do the will of God.

I believe Reagan's religious beliefs gave him an extraordinary inner peace, and theology teaches us that God can use human beings to work his will. But even if true, what was it about Reagan that made this so difficult to see while it was happening? The other indisputable world-historical figure of the era, Pope John Paul II, has no less humility and no less willingness to serve God. But I have never met anyone who was in the immediate presence of the pope who doubted that he was in the presence of greatness. And I doubt many people who saw Winston Churchill at close range between 1940 and 1945 were oblivious to his extraordinary political gifts. What gave the seemingly far less gifted, far less sophisticated Reagan his political edge?

I find myself going back to Reagan's political ideology, which was post–World War II American conservatism. Is there a possibility that this belief system gave Reagan an effective tool, a framework that enabled him to make good decisions without a lot of particular knowledge—without a detailed briefing book?

At first glance this seems absurd, especially in view of the widespread suspicion that this ideology has cracked up, has run its course, however well it may have been suited to its time. After all, if ideology was key, shouldn't Reagan have had more in the way of imitators and successors? But none of the major politicians who succeeded, or attempted to succeed, Reagan on the national scene has had his combination of beliefs. Those who shared his economics have almost always played down his social conservatism. Those who shared his social beliefs have tended to lack his optimism about America's role in the world.

Perhaps that is more the fault of his successors than of his ideology. It is striking that the unfinished parts of Reagan's agenda have an odd way of bubbling back to the surface. Consider: Today's major debate in foreign and defense policy is deployment of the Strategic Defense Initiative. And if Reagan's reduction of the top tax rate from 70 to 28 percent was the greatest policy event of the 1980s, the repeal of the federal welfare entitlement will almost certainly be remembered as the biggest (and most surprising) policy event of the 1990s. The second event has as much a Reaganite stamp as the first.

At a national governors' conference in the early 1970s, a motion was offered to have Washington completely take over Aid to Families with Dependent Children. The motion carried, 49 to 1. Reagan, of course, was the no vote. He argued that, instead, the program should be returned completely to the states. When this more or less happened, more than two decades later under President Bill Clinton, the prime legislative strategist for the decentralizers was Robert Carleson—the man who had served as Reagan's commissioner of welfare in California.

Is this all simple happenstance? Or is it possible Reagan operated from an ideological framework that is deeply relevant and persuasive— and that is, or could be, as alive today as it was in the 1980s, when he came to dominate the politics of the nation and the world?

Beginning in the 1950s and continuing through his presidency,

Reagan was a voracious consumer of conservative ideas, often through his subscriptions to *Human Events* and *National Review*. He was a follower of classical economics and supported the gold standard. He scoffed at the mythical "trust fund" often claimed for Social Security and favored a voluntary system. He was always attracted to a simple, low-rate tax system and to decentralization of programs being handled badly by Washington.

Reagan had no interest in the isolationist strain that dominated postwar conservatism in the 1940s and early 1950s. He never lost the Wilsonian commitment to the spreading of American democracy he held in his years as an active Democrat. As an alumnus of the (anti-Communist) Hollywood left, he resonated to the view of the world held by former Communists like Frank Meyer, men and women whose messianic devotion to saving the world through revolution had been transferred to a commitment to America as an idea. Needless to say, this ambitious, optimistic brand of conservatism is a polar opposite to the older strain of pessimistic, quasi-aristocratic European conservatism exemplified by thinkers like Russell Kirk.

Influenced though he was by libertarian thought in economics, Reagan in political office was a strong supporter of state and police power on behalf of the social order. Legalization of narcotics, and the guaranteed annual income as a substitute for welfare—proposals flirted with by many libertarian-leaning conservatives—held no appeal for him.

On social issues, Reagan was firmly on the side of traditional values. He felt he had been sold a bill of goods when he signed what proved to be a permissive abortion law in his first year as governor in 1967, and he became fervently pro-life in the years following *Roe* v. *Wade* in 1973. He caused the first strongly pro-life plank to be inserted in the Republican platform in 1980, and as president even published a pro-life book, *Abortion and the Conscience of the Nation*, in 1984. A few days before he left office, Reagan told the *New York Times* that his greatest regret about his presidency was that he was unable to do more to protect the unborn, and said that America will not be "completely civilized" as long as abortion is legal.

Reagan was quite capable of using the bully pulpit, and often spoke of the need for cultural renewal, but I agree with one of his biographers, Dinesh D'Souza, that "he would not have endorsed the right's effort to achieve this end by abjuring the use of state power. . . . Reagan understood that the way to change the culture is to change law and public policy."

While the view of Reagan as a great communicator is incontestable, I have come to believe that it is profoundly misleading. The picture we are invited to have is that Reagan was such a superb speaker that he could get people to believe virtually anything. Once he left the political scene, this logic goes, his views resumed their status as bizarre or extreme, losing their relevance to serious political debate.

I believe the truth is very different. The striking thing to me, thinking back about what it was like to work with Reagan when he was making political decisions, is not how persuasive he was at the time, but how often he proved to be right in retrospect. His judgment on matters of substance was astoundingly good, including and perhaps especially on matters where his advisers and others around him were completely unpersuaded, in not a few cases completely baffled.

Impressive as Reagan's communications skills were, in other words, his decisions about what to communicate were even better. This most certainly included his leadership of the diverse, inchoate movement of revolt against the left that we have come to know as the postwar conservative movement. Reagan invariably gravitated toward the aspects of American conservatism that were optimistic not cynical, populist not elitist, egalitarian not hierarchical, moral not relativistic—in short, toward what is distinctively American in American conservatism.

At the end of this road was the vision that moved Ronald Reagan most of all: America as a shining city on a hill, exerting magnetic power on the rest of the world. As D'Souza puts it, "his American exceptionalism was inextricably united with American universalism." And as a *Washington Post* editorial once noted in a rare moment of bemused respect, when Reagan ventured abroad he found not just the nation but the world was his oyster.

As we observe Reagan's 90th birthday on February 6, then, we should avoid nostalgia for what it was like to serve under a great leader: Most of us didn't know he was at the time. Or for the unity of

purpose Reagan's leadership supposedly provided: Many of us on Team
Reagan often found ourselves at each other's throats.

Above all, we should put to rest the idea of the Great Communicator:
What Reagan told the American people about who we are, or who we
should be, resonated far more deeply than any inflection of his voice. And
we should therefore stop assuming that his success is unrepeatable. If we,
American conservatives, take his belief system seriously, as a guide to the
challenges of the present as well as the past, the greatest successes of the
Reagan era may still lie ahead of us.

# SETH BENARDETE

## By Harvey Mansfield

*(Originally published in the December 3, 2001, issue)*

SETH BENARDETE WAS A SCHOLAR, a philosopher, and a most extraordinary man. His post in life was to be a classics professor at New York University, but he was not an especially prominent professor. Nor was he much known in the world of public intellectuals, a realm he never tried to enter. He wrote books on Greek poetry and philosophy, and before he died on November 14, 2001, at the age of seventy-one, he was the most learned man alive—and, I venture to assert, the deepest thinker as well.

To me, he was both friend and hero. The hero got in the way of our friendship because he was in every way my superior, and the best I could offer him was my unspoken admiration.

I first met him in 1957 when he arrived at the Society of Fellows at Harvard, a group of very bright or highly praised young persons who are given the run of the university for three years. He had received his B.A. in classics from the University of Chicago in 1949 and his Ph.D. from the Committee on Social Thought in 1955, with a dissertation on "Achilles and Hector: The Homeric Hero." I was introduced to Benardete by his fellow student at Chicago and our common friend Allan Bloom. All three of us were in the company of those who saw something quite remarkable in the teaching of Leo Strauss.

Bloom in his brilliance went on to become a bestselling author and a figure of renown. Benardete did not. Because of his obvious gifts he received high honors when he was young, but then he settled in as a professor at NYU in 1964. When in 1984 his books began to appear in a steady stream, he was largely ignored. Nonetheless, he was held in awe by some Straussians, and he had a select following among students from the courses on Plato that he taught over the years at the New School for Social Research—as well as devotees elsewhere who sensed his greatness.

Not surprisingly, the classics profession never gave him recognition or honor. Classicists are only somewhat more insular and thick-headed than most professors, and their neglect did not bother Benardete. He left the task of punishing lesser scholars to others. His books have no anger in them. They are there for people who want to fly to strange places without buying a ticket and without being frisked by security guards.

Actually, in Benardete's view it's very important that flights to strange places are protected by security guards. Benardete was extremely learned in the details of philology, more so indeed than those who know nothing else and are proud of it. But his specialty was the *whole* of things—the whole that is depicted to us by poetry and explained to us by philosophy. The depiction by poets tells us the extra-large-sized beliefs we need to hold in order to live as we do. Philosophers call these beliefs into question and, to the extent possible, try to replace them with rational explanations.

This might sound like "the old quarrel between philosophy and poetry" featured in Plato's *Republic*. But without denying the existence of such a quarrel, Benardete found philosophy in poetry and poetry in philosophy.

That was the theme of his books on Homer, Plato, and Sophocles. Poetry with its image-making aims at, and depends on, the nature of things that is the object of philosophy. And philosophy with its logic cannot simply reject the conceits and the plotting of the poet. It must "learn from our mistakes"—not so much to avoid them as to see why we make them. This relearning is what Benardete called, following Plato, the "second sailing": It is at the heart of all serious thinking.

My summary does not convey the adventurous sparkle of Benardete's prose as he alternately plunges into the deep and returns to the surface. His books have been published by the University of Chicago Press, a faithful friend to him and his readers. Those who have never read Seth Benardete might begin with a volume of his essays, *The Argument of the Action*, published last year. Soon to come is a book of reminiscence and self-summary called *Experiences in Reflection: Conversations with Seth Benardete.*

He was a family man—husband of Jane, father of Ethan and Alexandra Emma—and a scholar who worked seven days a week. When he died, he left the world, as the best human beings always do, richer for his having lived and poorer for his being gone.

# DANIEL PATRICK MOYNIHAN

## By William Kristol

◆

*(Originally published, as "Daniel Patrick Moynihan, 1927–2003,"
in the April 7, 2003, issue)*

THE WORLD HAS NO NEED for another contribution to the fitting stream of tributes to Daniel Patrick Moynihan's extraordinary life and work. But I hope a brief personal reminiscence will not be amiss.

Everyone knows about Pat Moynihan's political and intellectual accomplishments. What is perhaps less well known is Pat's humanity. The last time I spoke with him was in early November 2002. My father had recently had a major operation, and was home from the hospital. Pat had been in touch throughout with my mother, and had spoken with my father since the operation—but he didn't want to bother them at their apartment by calling when they might be resting or busy. So he called me at home on a Saturday afternoon to ask how my father was doing, and to pass on his and Liz's best wishes and love.

The call was characteristic of Pat in a couple of ways. It was kind and thoughtful in its intention with respect to my parents. It was also unusual in its execution. Pat called, and our 15-year-old son answered the phone. Rather than simply ask for Susan or me, Pat engaged Joe in a discussion of his school and other activities.

Joe was a bit awed to be speaking to the famous Mr. Moynihan; he also had some difficulty understanding Pat, whose speech patterns were, one recalls, a bit unusual. And the phone connection, for some reason, wasn't very good. So by the end of their exchange, Joe was a little rattled, though proud to have had a real conversation with a world-historical figure—a kindness Pat knew he was performing, but performed naturally. (Susan reminded me, as we reminisced about this incident after hearing of Pat's death, that Joe was the child who was in utero when Pat exclaimed to her, "I love pregnant women. They look as if, if you dropped them from a tall building, they would bounce.")

That day, after I got on the phone and reassured Pat about my father, we had a longish conversation about *The Weekly Standard*. Pat had particularly enjoyed one piece in the most recent issue, he said. What was that? I asked, assuming he would praise David Brooks on "Saddam's Brain" or Max Boot on deterrence, or perhaps even Gary Anderson on Norman Podhoretz's book on the Prophets. No. Pat wanted to praise, at some length, Joe Epstein's review of the new biography of Max Beerbohm.

It turned out Pat Moynihan was a great fan of Max Beerbohm. As a young man studying in London in the early 1950s, it seems, Pat had made a pilgrimage to visit the elderly Beerbohm. But, if I recall the story accurately, after taking trains and buses to arrive at Beerbohm's house, Pat had approached the gate, felt suddenly intimidated at the thought of meeting the great man, and returned to London.

Having myself been intimidated (through no fault of his) by Pat Moynihan for the almost forty years I'd known him, I was amused at the idea of Pat's being intimidated by anyone. But I was reminded, as he dilated brilliantly on Beerbohm's works and his meaning to the young readers and writers of Pat's generation, how unusual—how unique— was the range of Moynihan's interests, knowledge, and enthusiasm.

*The Weekly Standard* had the honor to publish Pat Moynihan once. He reviewed—generously and enthusiastically—Norman Podhoretz's memoir, *Ex-Friends*, in February 1999. He offered only what he called "one quiet reservation" about Norman's "thrilling" book. Surely, Pat wrote, "Lionel Trilling and his wife Diana were never truly ex-friends. Indeed in the closing paragraph of the chapter on Trilling, [Podhoretz] records, 'I think about him a lot, always with admiration, gratitude, and indeed love.' That is as it should be." It was characteristic of Pat that even in a book review, he would want to soften a rupture, to heal a break.

I first met Pat when, as a 12-year-old, I did a bit of volunteer work in his campaign for New York City Council president in 1965. I then worked for him in the summer of 1970 in the Nixon White House, and in his 1976 Senate primary race against Bella Abzug. While we subsequently drifted apart politically, I always remained proud to claim some relationship of debt and obligation to him. He was a kind benefactor and a gentle instructor, who put friendship ahead of partisanship, generosity ahead of ideology. I will think about him a lot, always with admiration, gratitude, and indeed love.

# JOHNNY CARSON

## By Larry Miller

———

*(Originally published, as "So Long, Johnny,"
in the February 7, 2005, issue)*

WHEN A PROMINENT AMERICAN in any field passes on, it's front-page news. Some sneer at this and say, "The same thing happens to everyone. Why is it bigger if it happens to a star?" But I think it is bigger. Yes, thousands probably die in the same way at the same time, and each is a sorrow, but the passing of a beloved icon makes us all stop and think and reflect and remember, and gives a country with too little in common a great deal in common, if only briefly. So it is with Johnny Carson. Even in the hard-edged world of politics, for instance, I like to think that, when they heard the news, Howard Dean and Karl Rove and everyone in between stopped strategizing for a minute and thought, "Boy, I really loved that guy."

I've always felt the things written about comedians after they're gone come up short. "The low-key Nebraskan" is a phrase that's been bandied about already, which sounds a little like all the other low-key Nebraskans could've made America laugh for 30 years, too, if they felt like it.

Well, I loved the guy, and I mean, first, as a fan. I feel sorry for the younger folks who never saw him, who too often have to absorb their entertainment today in cynical bites, and think humor means anger and audacity and graphic descriptions of this and that. They will never know what it means when you take talent and hard work and mix it with grace, joy, class, respect, and forbearance.

I'm not any cleverer than the good reporters who've already written so much about him. For my part, I thought you might like to hear a story from my times on the *Tonight Show*, starring Johnny Carson.

There were a bunch of other shows, and those of us doing stand-up jumped at the chance to do any of them, but the *Tonight Show* was the one you wanted, period. The others were important, and good exposure, and big steps forward, but there were only two groups, really, B.C. and A.C.: Before Carson, and After Carson. A lot of good comics never got a shot, but I was one of the lucky ones.

My first time was in 1986, and I guess I was on 15 or 20 times till he left in 1992. As many of you know, there was a special place in Johnny's heart and on his show for young comedians, and there were a bunch of traditions surrounding those appearances. Every comic wanted them all.

You probably know about "The Big Okay." It was after you finished, and not everyone got it. We knew there was no way we were going to be invited over to sit on the couch—that was for another time; you didn't just go from captain to colonel, you have to be a major first. But what you wanted was to bow and say thank you to the audience, and look over to Johnny. If he liked you, you'd get a smile, and if he really liked you, you'd get a smile and a wink, and if he really, really liked you, you'd get a smile and a wink and The Big Okay. Once in a blue moon he liked someone so much—Steven Wright was one, I think—he'd wave you over on your first shot. I got the smile and the wink and The Big Okay, and that was heaven on earth right there.

A few appearances later, I got called over to the couch, and there's a bit of a story to that. A friend of mine had noticed I was wearing the same clothes on dozens of other shows and said, "You're doing the *Tonight Show* now, idiot. You need a better outfit." This was fine with me, and he took me to a fancy joint in Beverly Hills, one of those places that doesn't even have a name on it, you just pull around in back and someone lets you in. So they hooked me up with a black, double-breasted Armani suit, and a sharp shirt and tie, and I'd still be embarrassed, 19 years later, to tell you how much it cost. My pal was on the road on the day of the show, so my agent at the time, Tom

Stern, went with me. We picked up the suit at the store, and drove to NBC.

Now, I'm always early for things, so there was plenty of time to walk out onstage while the studio was empty, and run the material, and check my notes, and have some coffee, and get made up. I said hello to Jim McCawley (the segment producer who hired all the comics; he passed away some years ago), and he said, "I'll see you in a few," and the band struck up, and the show began. I watched Johnny's monologue from backstage, and then strolled back to the dressing room, the picture of calmness, ready to roll. I took my sneakers and casual pants off during the first guest, and put the white shirt and tie on, and the dress socks and the shoes, and watched the intro for the second guest in my underwear. (A good comic never puts the suit on too soon: It wrinkles.) Then, cool and happy, I unzipped the bag from the store, took out the jacket, and stared at the other side of the bag for a few seconds.

There were no pants.

I turned to Tom, and said, "No pants." On the TV in the background, the second guest came out and shook hands with Johnny. Tom ran out to find Jim.

I picked up the phone in the dressing room and called the store, and when the salesman came on I repeated my new mantra, "No pants." He found them in the back and said, "Don't worry, I'll bring them right over."

I hung up and looked in the mirror at my fancy new shirt and tie and boxer shorts, and the high socks and wingtips, and wondered how the Armani jacket was going to look over the pair of beige painter's pants I had worn to the studio. I was grateful the salesman was going to try and bring them over, but, please, Beverly Hills to Burbank on a Thursday at 5:33 p.m.? There was no way. By missile, in the middle of the night, it's still 20 minutes to Burbank. But I'll tell you, I don't remember being scared. In fact I was as calm as a vat of whiskey. Of course, maybe I was just in deep shock.

This outer calm hadn't yet translated to my speech center, though. I kept saying "No pants" every few seconds. Like the good agent and friend he was, Tom didn't want me to see him throw up, or scream, so he pulled himself together and turned back out down the hall for a breath. Unfortunately, the direction he chose dead-ended in a wall nine inches later. He hit his head so hard it made a sound and instantly grew a lump the size of another, smaller head.

Now Jim came running back with the wardrobe guy; they had found a pair of very nice black dress pants which went with the jacket. Jim had a big smile of victory, and it made me feel ungrateful and churlish to point out there was just one tiny problem: The pants had, apparently, last been worn by William Conrad, and were at least 75 inches in the waist. We looked at each other: Jim, the wardrobe guy, me, Tom, and the lump on Tom's forehead (which I'll call Tom Jr.). The wardrobe guy said, "Try them on. I can nip them in the back." I did, and he could, but the nip was two and a half feet long. I don't want to judge, but I think that's too long for a nip.

I took them off and we all glanced up at the television as Johnny went to commercial. "Okay," Jim said, "Johnny's going to do one more segment. Let's get you behind the curtain. If the pants don't get here in time from the store, you'll go out in these." Fine, and off we all went, Tom and Tom Jr. bringing up the rear, their motor skills still noticeably impaired.

I got behind the curtain, and the guy holding it didn't even blink when he saw me in my underwear and Jim holding a pair of enormous clown pants. After all, this was the same guy who had pulled the curtain for Tiny Tim. Tom and his lump said, "I'm going to wait for the guy from the store," and off he limped. I admired his spunk, since I didn't see any chance at all that either the pants or Tom would make it to the front gate.

"I think you better put on the pants," said Jim. The star being interviewed was wrapping up his last story, and Johnny was laughing. I held a finger up and tried desperately to remember what my first line was. They went to commercial, the band kicked up, and Jim said, "Okay, I really think you have to put on the pants." I finally remembered my first line, and Jim said, "Larry, please put on the pants." Well, I had no choice, and I pulled them over my shoes as the band came back from commercial, and my heart sank a little as I buckled the front and felt the tent-sized piece of material in the back. I was going to look like Quasimodo in a jet pack. And then . . .

Suddenly something crashed, someone screamed, and around the corner came poor, dazed Tom shouting, "I GOT THE PANTS. I GOT THE PANTS." The only reason Johnny and the audience couldn't hear him was because the band had just blasted out their last big, long brass lick. We all looked at each other like Easter Island statues. The only one who took it in stride was the guy at the curtain.

A low moan started in my throat, became a shudder, and ended in a shrug. I ripped the fat pants off as fast as I could and grabbed the new ones just as we heard Johnny saying, *"Our next guest is a very funny young man . . ."* I pulled one leg on and started the other, *". . . who's been with us before . . ."* The other leg went in and I stuffed the shirt down. *"He'll be appearing at The Punchline in Atlanta on the 25th . . ."* I clipped the pants, looked at the belt in my hand and threw it away hard, stage right, and heard Jim mutter, "Oh, Jesus. Oh, Jesus." I started buttoning the inner button on the double-breasted jacket, and missed it, and tried it again, and missed it again, and the stage hand started pulling the curtain and put his hand on my shoulder. *"So please welcome . . ."* I growled at the inner button and abandoned it. *"Larry . . ."* The hand began to push, and I closed the jacket and fastened the outer button. *". . . Miller!"*

I guess I love this business and always have, and maybe we're all a little nuts, and maybe you have to be, but all I remember is strolling around that curtain without a care in the world and having a great time and a great set, and when I bowed and waved goodbye, that's when I turned to Johnny, and that's when he waved me over to the couch for the first time.

It was pretty great. I shook his hand, which was even better, and he said, "Good stuff," which was the best of all, and I shook hands with the other folks on the couch, and Ed (who I loved as soon as he smiled at me), and then they went to commercial. And as the band started playing, Johnny leaned over and said, "Funny. Funny stuff." (Can't you just hear that sharp rhythm of his saying those short phrases?) And I said thank you and then he picked up a pencil and started tapping it on the desk to the music—remember how he used to do that?—and then we came back from commercial, and he wrapped it up, and thanked everyone, and we shook hands again, and everyone stood, and the band played, and the credits rolled. And he came around the desk, and I guess I was looking down at my suit, kind of amazed at how close it had all been, and I saw him coming over, and I shrugged and said, "New suit." And he said, "Sharp."

You all know how good Carson was. In an era that seems to grow coarser each day, he radiated manners. Everyone used to say that Walter Cronkite was the most trusted man in America when he was on, but I think it just might have been Johnny Carson. I know who I'd trust more, and I sure know who was funnier.

There's an old Jewish saying that every man's heaven or hell is determined by what people say about him after his death. It's a good thought, and if it's true, Johnny Carson is soaring very, very high.

I guess Ed is supposed to lead us now in shouting, "How high is he?" Johnny would have a good line for that one.

One thing's certain. I'll bet God just waved him over to the couch.

# JOHN PAUL II

## By Joseph Bottum

*(Originally published, as "John Paul the Great, 1920–2005,"
in the April 18, 2005, issue)*

HISTORY LABORS—A WORN MACHINE, sick with torsion, ill-meshed—
and every repair of an old fault ruptures something new. Or so it
seems, much of the time. Our historical choices are limited, constrained by the poverty of what appears possible at any given moment.
To be a good leader is, for most figures who walk the world's stage,
merely to pick the best among the available options—to push back
where one can, to hold on to the good that remains, to resist a little the
stream of history as it seems to flow toward its cataract.

For the past decade and a half, John Paul II was a good leader. He
had his failures: losing the fight for recognition of Christianity in the
European constitution, watching the democratic energy he generated
during his 1998 visit to Cuba dissipate without much apparent damage to Castro's dictatorship, seeing his efforts to influence China's anti-
religious regime peter out. But he had his successes as well: convincing even his bitterest opponents in the Church to join in at least the
verbal rejection of abortion, regularizing Vatican relations with Israel
to allow his millennial visit to the Holy Land, inspiring the defeat of
the Mafia in Sicily.

With the drama of his final illness and death, he offered a lesson
about the fullness, the arc, of human life. With the prophetic voice he
used in his later writings, he pointed to spiritual possibilities that
were being closed by what he once called the "disease of superficiality."

Always he was present, one of the world's conspicuous figures, pushing on history where he could, guiding the Church as much as it would be guided, choosing the best among the available options—doing all that a good leader should.

But before that—for over a decade at the beginning of his pontificate, from his installation as pope in 1978 through the final collapse of Soviet communism in 1991—John Paul II was something more, something different, something beyond mere possibility. He wasn't simply a good leader. He was inspired, and he seemed to walk through walls.

Certain images remain indelibly fixed—the skeptical Roman crowd, for instance, falling in love with the new Polish pope in the first seconds of his pontificate as he gave his lopsided smile and called out, not in Latin, but Italian, from the papal balcony: "I don't know if I can make myself clear in your . . . *our* Italian language. If I make a mistake, you will correct me." He had a perfect sense of timing, as the actor John Gielgud observed after watching him, and in the whirlwind of those early years he seemed incapable of doing anything that wasn't news: skiing, mountain-climbing, gathering crowds of millions to pray with him everywhere from Poland to Australia, performing the marriage of a Roman street-sweeper's daughter because she'd had the pluck to ask him—snapping the Lilliputian threads of courtly precedent and royal decorum with which the Vatican curia traditionally tied down popes as though he didn't even notice.

The "postmodern pope," American magazines dubbed him, caught up in the media circus of his superstar status, the John Paul II magical mystery tour that swept across the globe through the 1980s. Certainly he had, all his life, the elements of stardom—the whole package of good looks, and charm, and curiosity, and intelligence, and physical presence, and, especially, an obvious and easily triggered sort of *joy*: the ability to please and the ability to be pleased that combine to make a man seem radiantly alive.

As a young priest, he was a polished, careful subordinate, clearly destined for high office in the Church—but he was also a recognized minor poet during a period when Polish poetry was the most flourishing in the world. As archbishop of Krakow, he was a full-time political player in the complex dance of Soviet-dominated Poland—but he was also an important philosophical interpreter of Thomistic metaphysics and Husserlian phenomenology, teaching courses at the Catholic University of Lublin, the only non-state university in the Communist world. As pope, he was a mystic who spent hours a day in solitary prayer—but he was also a natural for television. He seemed perfectly at home receiving the stately bows of ambassadors in the Clementine throne room—but when thousands of teenagers in Madison Square Garden chanted at him, "John Paul II, we love you," he was equally comfortable winning their hearts by shouting back: "Woo-hoo-woo, John Paul II, he loves you!"

And yet, to call all this "postmodern"—to imagine these elements are simple contradictions, absurdly juxtaposed in a characteristically postmodern way—is to believe something about John Paul II that he himself never did. It is to imagine that helicopters are ridiculous beside devotion to the Blessed Virgin, or that prayer gainsays philosophy, or that faith ought not to go with modern times.

This is another form of the poverty of the possible, the thinness of the choices and narratives that seem available at any particular time. Every step John Paul II took in those early years was a denial that our options were as limited as they appeared—in the political life of the world, in the religious life of the Church, and in the intellectual life of our cultures. For the impoverished imagination of the time, he seemed both far behind and far ahead of the rest of the world. But he never saw his medievalism as a reactionary antimodernism, or his modernism as an enlightened anti-medievalism. Christianity always seemed to him simultaneously an ancient faith and the newest hope for the world. He prayed constantly that he would live long enough to see the Jubilee of 2000, for he thought he was called to shepherd humankind into the third millennium that he claimed would be a "springtime of evangelization."

Toward the end of his pontificate, the tyranny of available options may have begun to close in on him. Certainly, in the first days after his death on April 2, the media have proved incapable of picturing him in any way other than caught in the clash of accepted political categories. John Paul II was a voice for peace—but he hated abortion! He was a radical critic of materialism—but he rejected women's ordination! He was one of the architects of the great opening of the Church at the Second Vatican Council—but he disciplined heterodox Catholic theologians!

The *New York Times* oddly and disturbingly used the pope's death as an occasion to editorialize in favor of euthanasia: "Terri Schiavo was a stark contrast to the passing of this pontiff, whose own mind was keenly aware of the gradual failure of his body. The pope would certainly never have wanted his own end to be a lesson in the transcendent importance of allowing humans to choose their own manner of death. But to some of us, that was the exact message of his dignified departure." It's hard to imagine a more egregious use of the word "transcendent" or a more grotesque inversion of the legacy of a man who always insisted that life wasn't a choice but a gift. In truth, however, the *Times* was merely one among many publications that saw the pope only through the lens of current social politics. In all the thousands of obituaries that have appeared in the past week, hardly one failed to speak of the pope's "contradictions" somewhere along the way.

There's a reason. John Paul II's work in the Church must seem a hodgepodge when explained with the old narrative of Vatican II as entirely a struggle between liberal reformers and conservative traditionalists. His theology of the body, laid out in four years of addresses he began in 1979, must appear a mess when encountered with the view that libertines and reactionaries divide between them the only possible ways to think about human sexuality. And his politics of rightly ordered freedom must be unintelligible in a world that thinks itself limited to the alternatives of tyranny and radical license.

For the man himself, there was no contradiction at all, and he spent his pontificate trying to create new possibilities for history. You can see it perhaps most clearly in the defeat of communism—when he showed his ability to open doors where the rest of the world saw only walls.

After an unscheduled discussion during the 1979 papal tour of the United States, national security adviser Zbigniew Brzezinski joked that when he met with President Carter, he had the impression of speaking to a religious leader, and when he met with John Paul II, he felt he was talking to a world statesman. It was a joke with a bite. Of all American presidents, Jimmy Carter may have been the one most constrained by his thin conception of the available options, and all he could do was complain—in that failing voice of the would-be prophet he always seemed to end up using—that things ought to be different than they seemed to be.

John Paul II *made* them different. There's a temptation to overestimate the pope's role in the demise of Soviet communism. The labor unions, the anti-Stalinist intellectuals, and the churches all contributed enormously. The United States' long resistance during the Cold War, through presidents from Truman to Reagan, held Soviet expansion at bay while the Marxist economies ground toward their collapse: "We pretend to work, and they pretend to pay us" ran a factory workers' joke at the time in the Russian satellites behind the Iron Curtain.

And yet, however weak the Communist edifice may have been in actuality, it still seemed formidable, and the pope was at the center of the cyclone that blew it down. The KGB's Yuri Andropov foresaw what John Paul II would be, warning the Politburo in Moscow of impending disaster in the first months after the Polish cardinal became pope. Figures from Mikhail Gorbachev to Henry Kissinger have looked back on their careers and judged that the nonviolent dissolution of the Communist dictatorships would not have happened without John Paul II.

"How many divisions has the pope?" Stalin famously sneered. As it happens, with John Paul II, we have an answer. At the end of 1980, worried by the Polish government's inability to control the independent labor union Solidarity, the Russians prepared an invasion "to save socialist Poland." Fifteen divisions—twelve Soviet, two Czech, and one East German—were to cross the border in an initial attack, with nine more Soviet divisions following the next day. On December 7, Brzezinski called from the White House to tell John Paul II what American satellite photos showed about troop movements along the Polish border, and on December 16 the pope wrote Leonid Brezhnev a stern letter, invoking against the Soviets the guarantees of sovereignty that the Soviets themselves had inserted in the Helsinki Final Act (as a

way, they thought, of ensuring the Communists' permanent domina-
tion of Eastern Europe). Already caught in the Afghanistan debacle
and fearing an even greater loss of international prestige and good will,
Brezhnev ordered the troops home. Twenty-four divisions, and John
Paul II faced them down.

When President Carter urged Americans in 1977 to overcome
their "inordinate fear of communism," he clearly thought the only
path out of the Cold War was agreement to the continuing existence of
Communist regimes. This was the lie John Paul II was never willing
to tell. It remains a mystery what the organizers of the annual "World
Day of Peace" were hoping for when they asked the pope to contribute
a reflection in 1982, but what they got from the apostle of peace was a
letter denouncing the "false peace" of totalitarianism. In the end, the
path out of the Cold War was neither Henry Kissinger's hard realpoli-
tik nor Jimmy Carter's soft détente. It was instead John Paul II's insistence that communism could not survive among a people who had
heard—and learned to speak—the truth about human beings' freedom, dignity, and absolute moral worth.

Think of the number of regimes based on lies that gave way without
violent revolution during his pontificate. The flowering of democracy
was unprecedented, and he seemed always to be present as it bloomed.
There was Brazil, where the ruling colonels allowed the free elections that
replaced them. There was the Philippines, where Ferdinand and Imelda
Marcos fled from marchers in the street. There were Nicaragua, and
Chile, and Paraguay, and Mexico. And looming over them all was the
impending disintegration of the Soviet empire in Poland, East Germany,
Czechoslovakia, Lithuania, Latvia—on and on, people after people who
learned from the pope a new possibility for history, born from an ability
to hear and speak the truth about the regimes under which they lived.

For John Paul II, the possibility of political truth was a philosophically obvious *fact*, demanded by the theory of personalism he developed as he used the modern phenomenology of Edmund Husserl to
move intellectually beyond the dry versions of neo-Thomistic philosophy he had studied in seminary. It was a theological fact, as well,
derived from—and pointing back toward—the awareness that human
beings are created in the image of their free Creator. It was even a his-

torical fact, learned during the long humiliation of Poland first by Nazi Germany and then by Soviet Russia while he was young. And it became, in the end, a mystical fact for John Paul II, joined—through Mary and the secrets of Fatima—to God's direct providence in history.

The mystical unity begins, for the pope, in what the papal biographer George Weigel calls the "shadowlands." Yuri Andropov's grim predictions about the impact of the Polish pope did not fall on deaf ears. On November 13, 1979, the Central Committee in Moscow approved a KGB plan entitled "Decision to Work Against the Policies of the Vatican in Relation with Socialist States." Much of the document dealt with issuing anti-Catholic "propaganda" in the Soviet bloc and the use of "special channels" in the West to spread disinformation about the pope. But another section ordered the KGB to "improve the quality of the struggle" against the Vatican.

What this meant became completely clear only with evidence released just last month. Elements of the Soviet security forces, working through the Bulgarian secret service, made contact with a Turkish assassin named Mehmet Ali Agca and aimed him at the pope. And on May 13, 1981, Agca shot John Paul II in St. Peter's Square with a Browning 9-mm semiautomatic pistol, striking him in the belly to perforate his colon and small intestine multiple times. In the pope's last book, *Memory and Identity*—a collection of philosophical conversations that appeared in Italy this February—he shows that he always knew the origin of Agca's attempt on his life: "Someone else masterminded it and someone else commissioned it." The assassination attempt was a "last convulsion" of communism, trying to reverse the historical tide that had turned against it.

But it was also something more. "One hand fired, and another guided the bullet," he tried to explain after he left the hospital. On May 13, 1991, Pope John Paul II traveled to Portugal and placed the bullet with which he had been shot ten years before in the crown of the statue of Mary at the site of her original apparitions at Fatima. It wasn't till 2000 that the Vatican offered an explanation—and, along the way, revealed what had been called "the third secret of Fatima," a prophecy about a pope gunned down, hidden since it was given by the Blessed Virgin to three Portuguese children on July 13, 1917.

For John Paul II, the pieces all came together: the endless rosaries prayed since 1917 for the "conversion of Godless Russia" as the Blessed Virgin had asked, the "secret" vision of a shot pope she had further revealed at Fatima, the thirteens repeated in the dates, the special devotion to Mary that he marked with the large "M" on his coat of arms—and the truth of human freedom, asserted against the Communist lie.

He had sophisticated philosophical, theological, and historical reasons to see chances for political change where even the good leaders of his time saw only the poverty of the possible. He had poetic and aesthetic reasons, as well, to suppose it all somehow made sense: If "the word did not convert, blood will convert," he said of martyrdom in "Stanislaw," the last poem he wrote before becoming pope. But we cannot understand the man—we cannot grasp how, for him, history was always open to new possibilities—unless we also understand that it was, most of all, a *mystical* truth: the unity of things seen and unseen, the coherence of the spirit and the flesh.

"The intellect of man is forced to choose perfection of the life, or of the work," William Butler Yeats once sadly observed. But this was yet another thinness of possibility—that we can make either our lives or our works beautiful and whole, but not both—which John Paul II refused to admit. In his magisterial biography *Witness to Hope* (first published in 1999, and soon to appear in a third and updated edition that carries the story through the pope's death), George Weigel reports innumerable telling facts about the life of Karol Wojtyla before he became John Paul II at age 58, the youngest pope in more than a hundred years. His mother died when he was 8, for instance—and then his only brother when he was 12, and his father eight years later: "At the age of 20," he would look back to say, "I had already lost all the people I loved."

But though Weigel reports such facts and the pope's own occasional reflections upon them, he hardly ever draws a psychological conclusion—and he never offers a picture of what Wojtyla's subjective life was like or makes a guess about the interior monologue of his emotional life. On a first reading, this resolute refusal to psychologize may seem odd: People *are* their psyches, after all. We read biographies to understand their subjects, which we do only as we learn who they are and the psychological causes that shaped them into those particular people.

Indeed, John Paul II himself told Weigel in a 1996 interview, "They try to understand me from outside. But I can only be understood from inside." To the general reader of biographies, there is something absurd when Weigel quotes this line—and immediately goes on to describe the "inside" of Karol Wojtyla by mentioning the history of the Nazi and Soviet occupations of Poland, the philosophical necessity for grounding humanism and freedom in the truth about human existence, and the theological centrality of the virtue of hope.

And yet, over the last few years—as the world watched John Paul II teach, even with his death, one last lesson about the shape of human life—it has become clear that Weigel was right to think of the pope in this way. We have millions of words from the man: the 14 major encyclicals, 15 apostolic exhortations, 11 apostolic constitutions, and 45 apostolic letters; the popular books like *Crossing the Threshold of Hope*, scribbled on yellow pads during long plane flights; the scholarly works he wrote as a young theologian; the thousands of prayers and exhortations he delivered during the innumerable audiences he tirelessly gave as pope. And in all those words, there is hardly a hint of what a psychologist would demand: a persona that somehow stands apart from the history through which he lived and the intellectual growth he experienced.

It is not that he was a private person, in the usual way we speak of such people: refusing to discuss themselves and burying their psyches in their public work. It is, rather, that the center of the man—the focal point of his unified life—was the narrative arc of his story: *what* he was and *how* he got that way.

The closest John Paul II came to explaining himself may have been *Roman Triptych: Meditations*, a collection of three poems written during the 2003 papal trip to Poland, which he foresaw would be his last visit home. The only new poetry the pope published during his pontificate, the book is not first-class verse. But it remains fascinating autobiography, for each poem of the triptych shows a man considering human existence in one of its apparently divided aspects: as the life of the artist, as the life of the intellectual, and as the life of the believer. And through all three of the linked poems, the author seeks to express the unity he could always sense was drawing it all together.

It was a unity that derives, finally, from God's providential purpose in history. But history and Karol Wojtyla's biography grew together, more and more as the years went by, and the divine presence he felt in history joined the divine presence he could feel in the arc of his life. The man *was* his story—and in that story, he could seek perfection of both the life and the work.

Consider just one scene from his early life. When the Nazis began their occupation of Poland in 1939, they were determined to do more than conquer the country. "A major goal of our plan is to finish off as speedily as possible all troublemaking politicians, priests, and leaders who fall into our hands," the German governor, Hans Frank, wrote to his subordinates from the office he established in Krakow's old royal residence, Wawel Castle. "I openly admit that some thousands of so-called important Poles will have to pay with their lives, but . . . every vestige of Polish culture is to be eliminated."

The Nazis' destruction of Poland's Jews was more deliberate and systematic than their slaughter of Poland's Catholics, but the unified goal was clear from the beginning: By the time he was done overseeing the murder of thousands of priests and hundreds of college professors, and the deaths of millions of ordinary citizens along the way, Frank boasted, "There will never again be a Poland."

Among the schemes for the elimination of Polish culture was the closing of all secondary schools and universities—including seminaries. When the 22-year-old Karol Wojtyla entered studies for the priesthood in 1942, the entire Catholic educational system was underground and illegal. The first years of his priestly formation were snatched in secret, usually at night and always while waiting for death to find him as it found so many others in Poland.

When Frank closed the seminary in Krakow, up the hill near Wawel Castle, the German S.S. took over the building and used it for the next five years as an administrative headquarters. And when the Nazis abandoned Krakow on January 17, 1945, as the Red Army's 1st Ukrainian Front closed in on the city, the archbishop—Adam Stefan

Sapieha, the "uncrowned king of Poland" who had dared to mock Frank openly—quickly moved to reclaim the seminary before the Russians seized it.

It turned out that, by the end of the war, the S.S. had begun using the building as a makeshift jail, and Sapieha found the seminary with its roof collapsed, its windows shattered, and its rooms scarred from the open fires the inmates had built to keep from freezing. Worst of all was the failed plumbing, and in the hurry to save the building, young Wojtyla and another seminarian were sent in with trowels to clear out the cold, hard feces left by the prisoners.

Picture, for a moment, that scene: the brilliant 24-year-old—already known among his contemporaries as an actor and a playwright, already clearly destined for great things, already arrived at the fullness of his intellectual powers—chipping away for days in rooms full of frozen excrement.

And contrast it with another scene, 34 years later, when Karol Wojtyla made his first trip to Communist Poland as Pope John Paul II. He arrived on June 2, 1979, and by the time he left eight days later, 13 million Poles—more than one-third of the country's population—had seen him in person as he traveled from Warsaw, to Gniezno, to the shrine at Czestochowa, and ended in Krakow. Nearly everyone else in the nation saw him on television or heard him on the radio. The government was frightened to a hair trigger, and outside observers all had the sense that the Communist regime was doomed, one way or another, from the first moment the pope knelt down and kissed his native soil.

The enormous crowds could sense it, too. On the night of Friday, June 8, tens of thousands of young people gathered outside St. Michael's Church in Krakow for a promised "youth meeting" with the pope. *"Sto lat! Sto lat!"* they shouted over and over: "Live for a hundred years!" Abandoning his prepared speech, John Paul II joked with them—"How can the pope live to be a hundred when you shout him down?"—in an effort to calm the situation. But by 10:30 the emotions of the young crowd had reached a fever pitch.

The temptation for demagoguery must have been enormous: tens of thousands of young Poles—children, really—waving crosses above their heads, chanting in their ecstatic madness for this man to lead them, hungry for martyrdom, ready to trample down the government troops that waited nervously to meet them. A single hint, a single gesture, and the city could have been his—the whole of Poland, perhaps, for the emotion was electric across the country. But all that blood would have been his, too, and he knew the time was not yet right. "It's late, my friends. Let's go home quietly," was all he said, and inside the car that carried him away, John Paul II wept and wept, covering his face with his hands.

For anyone else, these two scenes would stand in contradiction: Once this man was so powerless that he was forced in the middle of a frozen January to clean open rooms that had been used as toilets, but later he was so powerful that thousands of people would have gladly died if he had but lifted his hand. For Karol Wojtyla, however, there seemed no contradiction at all. They were both demanded by the vocation to which God had called him. They were both involved with service and obedience. They were both the next thing that needed to be done.

This is the only way to make sense of John Paul II. He spent his life refusing the poverty of the possible, the worldly notion that our choices and explanations are limited to contemporary political categories—and all the apparent contradictions in his thought melt away when we realize he was perceiving options that no one else could see.

With his 1991 encyclical on democratic freedom and economics, *Centesimus Annus*, he issued what is by any objective measure the most pro-Western—pro-American, for that matter—document ever to come from Rome. And then, with the denunciations of the "culture of death" in the 1995 encyclical *Evangelium Vitae*, he issued Rome's most anti-Western and anti-American document. It looks like an impossible combination, until we remember that between them came the 1993 encyclical *Veritatis Splendor*—"The Splendor of Truth," joined with *Centesimus Annus* and *Evangelium Vitae* as the three-part message that formed the central theological achievement of his pontificate. The unity of truth—the only sustainable ground for a healthy society—is what lets us grasp both the rightness of democracy and the murderousness of abortion.

That's not to say his pontificate was an unbroken string of successes. He felt the Christian schism deeply, but his many overtures to the Eastern Orthodox Churches were mostly unrequited, and the healing of Christianity is still far away. He never understood the Middle East with the same clarity that he grasped Eastern Europe, and after the fall of Soviet communism he didn't have the same direct impact on world history. When he opposed the first Gulf War in 1991—and allowed Iraq's murderous deputy prime minister, Tariq Aziz, to pay a state visit to Rome and Assisi before the second Gulf War in 2003—he seemed to have become locked into a single model for democratic reform, as though Saddam Hussein could be overcome with the same nonviolent, soft-power techniques that had worked in Catholic countries from Poland to the Philippines.

Similarly, he often appeared to have greater success evangelizing the rest of the world than he did evangelizing his own Church. The orthodoxy of the new Catechism he issued in the 1990s and the example of his personal spirituality stopped the slide of post-Vatican II Catholicism into a theological simulacrum of liberal Protestantism. His unique connection to young people—manifested at the huge outpourings for World Youth Day events—created a new generation of "John Paul II Catholics" among young people who have never known another pope. But on the older generations formed before his pontificate, particularly in America and Western Europe, he found little purchase. The liberal Catholic establishment never forgave him for either his failures or his successes, and they blamed him when the American priest scandals became public in 2002—though the priests involved were of the generation formed before John Paul II became pope.

But along the way, he refused to falter. He seems never to have been frightened of anything in his life, and he expected everyone else to share his confident courage: "Be not afraid," he began his pontificate by echoing from the gospel. The 1981 bullet wound slowed him down a little, a 1994 fall in his bath slowed him more, and by the time he reached his 82nd birthday in 2002, he was showing the signs of his impending death. But even at the end he was "a body pulled by a soul" to remain active, as the Vatican official Joaquin Navarro-Valls put it, and his constant motion throughout his life seems breathtaking.

In the 27 years of his pontificate he was seen in the flesh more often than anyone else in history—by over 150 million people, according to one estimate. He traveled to more than 130 countries, created 232 cardinals, and never slowed in the Vatican's endless schedule of audiences, consistories, synods, and meetings. He named 482 saints and beatified another 1,338 people, more than all his predecessors, in his confident belief that the possibility of sanctity was still alive in the world. He produced the first universal Catechism since Vatican II, revised canon law, reorganized the Curia, and made huge advances in Jewish-Christian and Catholic-Protestant relations.

That set of features—his complete courage and his boundless energy—gave him enormous freedom, particularly when combined with his certain conviction that there must exist a way to living in truth no matter how thin the merely possible seemed. He was, in fact, the freest man in the twentieth century. As a measure of his greatness, think of him this way: He could have been a Napoleon. He could have been a Lenin. Instead, he was the vicar of Christ, the heir of St. Peter, steward of a gospel recorded long ago.

History labors down its worn tracks, and the poverty of human possibilities leaves us few choices. Or so it often seems.

But not always. Not while we remember that living in truth is always possible. Not while we remind ourselves of the message of hope preached ceaselessly by Karol Wojtyla. Not while we recall John Paul the Great.

II

POLITICS

AND SOCIETY

# MRS. CLINTON'S VERY, VERY BAD BOOK

## By P. J. O'Rourke

*(Originally published in the February 19, 1996, issue)*

> When Chelsea needed permission at school to get aspirin,
> she told the nurse to 'Call my dad, my mom's too busy.'
>
> —Eleanor Clift and Mark Miller,
> *Newsweek*, March 1, 1993

IT TAKES A VILLAGE TO RAISE A CHILD. The village is Washington. You are the child. There, I've spared you from reading the worst book to come out of the Clinton administration since—let's be fair—whatever the last one was.

Nearly everything about *It Takes a Village* is objectionable, from the title—an ancient African proverb that seems to have its origins in the ancient African kingdom of Hallmarkcardia—to the acknowledgments page, where Mrs. Clinton fails to acknowledge that some poor journalism professor named Barbara Feinman did a lot of the work. Mrs. Clinton thereby unwisely violates the first rule of literary collaboration: Blame the coauthor. And let us avert our eyes from the Kim Il-Sung–type dust-jacket photograph showing Mrs. Clinton surrounded by joyous-youth-of-many-nations.

The writing style is that familiar modern one so often adopted by harried public figures speaking into a tape recorder. The narrative

voice is, I believe, intended to be that of an old family friend, an old
family friend who is, perhaps, showing the first signs of Alzheimer's
disease:

> On summer nights, our parents sat together in one another's yards or
> on porches, chatting while the kids played. Sometimes a few of the
> fathers dressed up in sheets and told us ghost stories. We marched with
> our Scout troops or school groups or rode bikes in holiday parades
> through our town's small downtown, to a park where all the kids were
> given Popsicles.

Elsewhere the tone is xeroxed family newsletter, the kind enclosed
in a Christmas card from people you hardly know:

> One memorable night, Chelsea wanted us to go buy a coconut. . . . We
> walked to our neighborhood store, brought the coconut home, and tried
> to open it, even pounding on it with a hammer, to no avail. Finally
> we went out to the parking lot of the governor's mansion, where we
> took turns throwing it on the ground until it cracked. The guards
> could not figure out what we were up to, and we laughed for hours
> afterwards.

Hours?

However that may be, let us understand that we have here a
Christmas card with ideas, "a reflection of my continuing meditation
on children," as Mrs. Clinton puts it. And we need only turn to the
contents page to reap the benefits of her many lonely hours spent in
philosophical contemplation of puerile ontology: "Kids Don't Come
with Instructions," "Security Takes More Than a Blanket," "Child
Care Is Not a Spectator Sport," "Children Are Citizens Too."

Bold thoughts. Brave insights. "It is often said that children are
our last and best hope for the future," claims Mrs. Clinton. "Children,"
she ventures, "need to hear from authoritative voices that kindness and
caring matter." And she flatly states, "The teenage years, we all know,
pose a special challenge for parents."

"Children," says Mrs. Clinton, "are like the tiny figures at the center of the nesting dolls for which Russian folk artists are famous. The children are cradled in the family, which is primarily responsible for their passage from infancy to adulthood. But around the family are the larger settings *of paid informers, secret police, corrupt bureaucracy, and a prison gulag.*" I added the part in italics for comic relief, something *It Takes a Village* doesn't provide. Intentionally.

The profound cogitations of Mrs. Clinton cannot help but result in a treasure trove of useful advice on child rearing. "[T]he village needs a town crier—and a town prodder," she says. I shall be certain to propose the creation of this novel office at the next Town Meeting in Sharon, New Hampshire. I'm sure my fellow residents will be as pleased as I am at the notion of a public servant going from door to door at convenient hours announcing, as Mrs. Clinton does, "We can encourage girls to be active and dress them in comfortable, durable clothes that let them move freely."

Some of this needful counsel is gleaned from Mrs. Clinton's own experience of partly raising one child with only a legion of household help courtesy the taxpayers. Not that Mrs. Clinton always had it easy:

> *But for two years when Bill was not governor (and Chelsea was still very young) our only help was a woman who came during work hours on weekdays. . . . My own version of every woman's worst nightmare happened one morning when I was due in court at nine-thirty for a trial. It was already seven-thirty, and two-year-old Chelsea was running a fever and throwing up after a sleepless night for both of us. My husband was out of town. The woman who normally took care of Chelsea called in sick with the same symptoms. No relatives lived nearby. My neighbors were not at home. Frantic, I called a trusted friend who came to my rescue.*

Whew, that was a close call.

Anyway, Mrs. Clinton has swell tips on everything from entertaining toddlers ("Often . . . a sock turned into a hand puppet is enough to fascinate them for hours") to keeping older kiddies fit ("If your children need to lose weight, help them to set a reasonable goal and make a sensible plan for getting there"). She is determined that every child should reach his or her full potential in mind and body ("One of my pet theories is that learning to tie shoelaces is a good way of developing hand-to-eye coordination"). And what parent will not applaud Mrs. Clinton's hint "to explain to the child in advance what the shots do, perhaps by illustrating it with her favorite dolls and stuffed animals"? This is also an excellent method of educating offspring about sexual abuse and, perhaps, capital punishment. Don't call the White House if the kid refuses to be left alone in the room with Fuzzy the Bunny.

Furthermore, Mrs. Clinton taps the expertise of—what else to call them?—experts. "The Child Care Action Campaign . . . advises that 'jigsaw puzzles and crayons may be fine for preschoolers but are inappropriate for infants.'" And Ann Brown, the chairhuman of the Consumer Product Safety Commission, is cited for suggesting that "baby showers with a safety theme are a great way to help new and expectant mothers childproof every room in their homes." Oh, Honey, look what Mom brought—a huge bouquet of rubber bands to put around all the knobs on our kitchen cabinets.

But *It Takes a Village* is so much more than just a self-help book for idiots. Mrs. Clinton also shares her many virtuous thoughts with us. "From the time I was a child I loved being around children." And she lets us in on her deep personal sorrows. "Watching one parent browbeat the other over child support or property division by threatening to fight for custody or withhold visitation, I often wished I could call in King Solomon to arbitrate." Though one shudders to think of the lawsuits the Children's Defense Fund would have brought against old Sol for endangering the welfare of a minor, bigamy, and what Mrs. Clinton calls "the misuse of religion to further political, personal, and even commercial agendas."

Mrs. Clinton explains, however, that church is good. "Our spiritual life as a family was spirited and constant. We talked with God, walked with God, ate, studied, and argued with God." And won, I'll warrant. "My father came from a long line of Methodists, while my mother, who had not been raised in any church, taught Sunday school." Interesting lessons they must have been. I myself am a Methodist. But Mrs. Clinton apparently belongs to the synod from

Mars. "Churches," she says, "are among the few places in the village where today's teenagers can let down their guard and let off steam." She says that in her Methodist youth group, "we argued over the meaning of war to a Christian after seeing for the first time works of art like Picasso's *Guernica*, and the words of poets like T. S. Eliot and e. e. cummings inspired us to debate other moral issues." I can only wonder if any of those words were from *one times one* by cummings:

> *a politician is an arse upon*
> *which everything has sat except a man*

Until now the First Lady has had two media aspects or avatars. There was Hillary the zealous and committed, ideological wide-load, antithesis to that temporizing flibbertigibbet and political round-heels her husband. Then there was Hillary guile incarnate, swindling the widows and orphans of Arkansas in bank stock, real estate, and cattle trading deals, sending her minions to rifle the office of Vince Foster before his body had cooled and loudly touting the virtues of feminism while acquiring her own wealth and prestige by marriage to a promising lunk. But *It Takes a Village* contains plentiful evidence that we members of the press do not know the true woman. We have failed to penetrate the various masks of the public persona. We have neglected to learn who the real Hillary Rodham Clinton is. She's a nitwit.

But Mrs. Clinton really can't be stupid. Can she? She has a big, long résumé. She's been to college. Several times. Very important intellectuals like Garry Wills consider her a very important intellectual like Garry Wills. Surely the imbecility of *It Takes a Village* is calculated, cynical, an attempt to soften the First Lady's image with ordinary Americans. Mrs. Clinton chooses a thesis that can hardly be refuted, "Resolved: Kids—Aren't They Great?" Then she patronizes her audience, talks down to them, lowers the level of discourse to where it may be understood by the average—let's be frank—Democrat. This is an interesting public-relations gambit, repositioning the Dragon Lady to show how much she cares about all the little dragon eggs. But if the purpose of *It Takes a Village* is to get in good with the masses, then explain this sentence on page 182: "I had never before known people who lived in trailers."

Is the First Lady a dunce? Let us marshal the evidence:

| ARGUMENTS CONTRA STUPIDITY | ARGUMENTS PRO |
|---|---|
| President of her class at Wellesley | It was the 60s, decade without quality control |
| Involved in Watergate investigation | So was Martha Mitchell |
| Partner in most prestigious law firm in Arkansas | Examine phrase "most prestigious law firm in Arkansas" |
| Went to Yale | Went to Yale |
| Married Bill | Married Bill |
| Is good on television | Not as good as Tori Spelling |

The jury seems to be out. We will have to rely for our answer on old-fashioned textual analysis.

In *It Takes a Village*, Mrs. Clinton is highly critical of *The Bell Curve* by Richard J. Herrnstein and Charles Murray. One whole chapter of her book is titled "The Bell Curve is a Curve Ball." Mrs. Clinton shows no evidence of having read even the dust jacket of *The Bell Curve*, but never mind; let us take her underlying point that innate intelligence is hard to measure. Then let us postulate something we might call a "bell trough" and draw a conclusion from this that innate stupidity is hard to measure too. "Smart is not something you simply are, but something you can become," says Mrs. Clinton. And ditto, my dear, for dumb.

There are times in *It Takes a Village* when Mrs. Clinton seems to play at being a horse's ass, when she makes statements such as "some of the best theologians I have ever met were five-year-olds." Mommy, did they put Jesus on the cross before or after he came down the chimney and brought all the children toys?

But some kinds of stupidity cannot be faked. Says Mrs. Clinton: "Less developed nations will be our best models for the home doctoring we will then need to master." And she tells us that in Bangladesh she met a Louisiana doctor "who was there to learn about low-cost techniques he could use back home to treat some of his state's more than 240,000 uninsured children." A poultice of buffalo dung is helpful in many cases.

Mrs. Clinton seems to possess the highly developed, finely attuned stupidity usually found in the upper reaches of academia. Hear her on the subject of nurseries and preschools: "From what experts tell us, there is a link between the cost and the quality of care." And then there is Mrs. Clinton's introduction to the chapter titled "Kids Don't Come with Instructions":

> *There I was, lying in my hospital bed, trying desperately to figure out how to breast-feed. . . . As I looked on in horror, Chelsea started to foam at the nose. I thought she was strangling or having convulsions. Frantically, I pushed every buzzer there was to push.*

> *A nurse appeared promptly. She assessed the situation calmly. . . . Chelsea was taking in my milk, but because of the awkward way I held her, she was breathing it out of her nose!*

The woman was holding her baby *upside down.*

But let us not confuse stupid with feeble or pointless. Stupidity is an excellent medium for the vigorous conveyance of certain political ideas. Mrs. Clinton is, for instance, doggedly pro-Clinton. Anyone who makes the least demur to the Clinton administration agenda (whatever it may be this week) is an extremist: "As soon as Goals 2000 passed, it was attacked by extremists." And she says the "extreme case against government, *often including intense personal attacks on government officials and political leaders* [italics added by an extremist, me], is designed not just to restrain government but to advance narrow religious, political, and economic agendas." That crabbed, restrictive screed the Bill of Rights comes to mind. Mrs. Clinton claims to have once been a Goldwater Republican. Perhaps she just muffed her note-taking during his 1964 nomination acceptance speech. I suppose that looking back at her diaries, rediscovered in the East Wing book room, she found the following entry: " 'Extremism in the defense of liberty is (illegible).' Remind myself to ask that nice girl in PoliSci class who's president of SDS what the Senator said."

Nor does Mrs. Clinton miss a chance to swipe at family values, often putting the phrase in quotation marks to signify ironic scorn. Clever device. "This is real 'family values' legislation," she says of the Family and Medical Leave Act, a law she calls "a major step toward a national commitment to allowing good workers to be good family members"—something workers never were, of course, until the government made them so.

Poverty, injustice, the need to take a couple of days off work—in the Mrs. Clinton world view there is no social problem that's not an occasion for increased political involvement in private life.

> *Imagine hearing this kind of "news you can use" sandwiched in the middle of the Top Ten countdown: "So you've got a new baby in the house? Don't let her cry herself red in the face. Just think how you'd feel if you were hungry, wet, or just plain out of sorts and nobody paid any attention to you. Well, don't do that to a little kid. She just got here. Give her a break, and give her some attention now!"*

> *Videos with scenes of commonsense baby care—how to burp an infant, what to do when soap gets in his eyes, how to make a baby with an earache comfortable—could be running continuously in doctors' offices, clinics, hospitals, motor vehicle offices, or any place where people gather and have to wait.*

You think getting your driver's license renewed is a pain now? Just wait until the second Clinton administration.

There is no form of social spending that Mrs. Clinton won't buy into (with your money). "I can't understand the political opposition to programs like 'midnight' basketball," she says. And no doubt the Swiss and Japanese, who owe their low crime rates to keeping their kids awake till all hours shooting hoops, would agree.

Mrs. Clinton has no thought for the infinite growth of cost and dependency inherent in entitlement programs. She blithely speaks of "a single mother in Illinois who . . . described herself as falling into the childcare netherworld because she makes too much to qualify for state programs but finds that the price of daycare 'is well out of reach.'" Why should every taxpayer in the nation become Miss Illinois's husband when there is one particular taxpayer honestly obliged to do so?

And Mrs. Clinton is oblivious to the idea that the government programs she advocates may have caused the problems the government programs she advocates are supposed to solve. "Whatever the reasons for the apparent increase in physical and sexual abuse of children, it demands our intervention," she says. But what if the reason *is* our intervention?

Only the lamest arguments are summoned to support Mrs. Clinton's call for enormous expansion of state power. She uses a few statistics of the kind that come in smudgy faxes from minor Naderite organizations: "135,000 children bring guns to school each day." She recollects past do-goodery: "In Arkansas we enlisted the services of local merchants to create a book of coupons that could be distributed to pregnant women. . . . After every month's pre- or postnatal exam, the attending health care provider validates a coupon, which can be redeemed for free or reduced-price goods such as milk or diapers." (In 1980, Arkansas had an infant mortality rate of 12.7 per 1,000 live births, almost identical to the national average of 12.6. As of 1992, the Arkansas rate was 10.3 vs. a national average of 8.5.) And Mrs. Clinton offers pat little anecdotes of this ilk:

> I will never forget the woman from Vermont whom I met at a health care forum in Boston. She ran a dairy farm with her husband, which meant that she was required by law to immunize her cattle against disease. But she could not afford to get her preschoolers inoculated as well. "The cattle on my dairy farm right now," she said, "are receiving better health care than my children."

Of course the dairy farmers could have, I don't know, sold a cow or something, but that would have been playing into the hands of anti-government extremists. Clucks the First Lady: "The influence of profit-driven medicine continues to grow."

Indeed the profit motive is to blame for many, many of America's problems. Mrs. Clinton talks long and often about the "harsh consequences of a more open economy." So unlike the lovely time people are having in North Korea. Mrs. Clinton opines that "one of the conditions of the consumer culture is that it relies upon human insecurities to create aspirations that can be satisfied only by the purchase of some product or service." Such as vaccinations for kids, maybe.

Yet, at bottom, Mrs. Clinton cannot really be called a commie or a pinko or even a liberal in the contemporary hold-your-nose sense of the word. She spends too much time arguing both sides of the social, if not political, issues—a thing done deftly by her husband and rather less so by her. Says Mrs. Clinton, "It would be great if we could get kids to postpone any decision about sex until they are over twenty-one." Though perhaps they may be allowed to decide what sex they are, since adolescents, says Mrs. Clinton, "need straight talk about contraception and sexually transmitted diseases to help them deal with the consequences of their decisions." But, she says, "After many years of working with and listening to American adolescents, I don't believe they are ready for sex or its potential consequences."

"I share my husband's belief that 'nothing in the First Amendment converts our public schools into religion-free zones,' " says Mrs. Clinton, and on the next page she endorses the joint Justice Department/Department of Education guidelines on religious activities in the public schools, which state: "Schools may not provide religious instruction, but they may teach about the Bible." It's real old. It's real long. There are Jews in it. Quiz tomorrow.

And on the subject of pedagogics in general, Mrs. Clinton crafts this jewel of equivocation: "I strongly favor promoting choice *among* public schools" (italics *not* in the original).

Mrs. Clinton does her best to steal conservative thunder or, anyway, troglodyte rumblings. She frames herself as wife, mother, and Christian, favors making divorces harder to get, mentions responsibility about every third page, and goes as far as to tell this bald-faced lie: "We reject the utopian view that government can or should protect people from the consequences of personal decisions." CC: Miss Illinois.

Mrs. Clinton doesn't even dislike business, as long as business is done her way. She gives examples of corporate activities that statists can cozy up to. For instance, "A number of our most powerful telecommunications and computer companies have joined forces with the government in a project to connect every classroom in America to the Internet." And she vapors: "Socially minded corporate philosophies are the avenue to future prosperity and social stability."

If a name must be put to these stupid politics, we can consult the *Columbia Encyclopedia* under the heading of that enormous stupidity, fascism: "totalitarian philosophy of government that glorifies state and nation and assigns to the state control over every aspect of national life." Admittedly, the fascism in *It Takes a Village* is of a namby-pamby, eat-your-vegetables kind that doesn't so much glorify the state and nation as pester the dickens out of them. Ethnic groups do not suffer persecution except insofar as a positive self-image is required among women and minorities at all times. And there will be no uniforms other than comfortable, durable clothes on girls. And no concentration camps either, just lots and lots of day care.

Nonetheless, the similitude exists. The *Columbia* article points out that fascism "is obliged to be antitheoretical and frankly opportunistic in order to appeal to many diverse groups." "Elitism" is noted, as is "Fascism's rejection of reason and intelligence and its emphatic emphasis on vision." Featured prominently in the fascist paradigm is "an authoritarian leader who embodies in his [or her!] person the highest ideals of the nation." The only classical fascist element missing from *It Takes a Village* is "social Darwinism." It's been replaced by "social creationism," expressed in such Mrs. Clinton statements as "I have never met a stupid child."

Lest the reader think I exaggerate the First Lady's brown-shirt (though from a New York designer and with nice ruffles) tendencies, let me leave you with a few vignettes from Mrs. Clinton's ideal world:

> *At the Washington Beech Community Preschool in Roslindale, Massachusetts, director Ellen Wolpert has children play games like Go Fish and Concentration with a deck of cards adorned with images—men holding babies, women pounding nails, elderly men on ladders, gray-haired women on skateboards . . .*

> *Journalists and news executives have responsibilities too. When violence*

*is newsworthy they should report it, but they should balance it with
stories that provide children and adults with positive images of themselves and those around them, taking care not to exacerbate negative
stereotypes.*

*I tried some rice pilaf with lentils, beans, and chick peas with a group
of fifth and sixth graders, who not only ate what was served but said
they liked it.*

"Children have many lessons to share with us," says Mrs. Clinton.
And on page 153 of *It Takes a Village* we share a good one:

*When my family moved to Park Ridge, I was four years old and
eager to make new friends. Every time I walked out the door, with a
bow in my hair and a hopeful look on my face, the neighborhood kids
would torment me, pushing me, knocking me down, and teasing me
until I burst into tears and ran back in the house.*

# STATUS-INCOME DISEQUILIBRIUM

## By David Brooks

*(Originally published, as "The Tragedy of SID:*
*Status-Income Disequilibrium," in the May 6, 1996, issue)*

THE EDITOR HAD TRIUMPHED. All through a long New York spring evening, it had been John Updike this and Norman Mailer that. He'd kept his tablemates at the Freedom Forum's annual Free Expression Dinner in a state of conversational bliss, and when the meal was over everybody at his table was in such high spirits they decided to go down to the lounge for a few drinks. The Regency Hotel has a little room called The Library, where the martinis are $11. The editor was joined by an investment banker from Morgan Stanley and a lawyer from Wachtel Lipton and his wife. And he was just as amusing in the bar, filling the night with publishing tales. Feeling expansive, he decided to pick up the tab, putting it on his expense account, and when the whole group stumbled outside to the corner of 61st Street and Park Avenue, he was seized by his high spirits and called out, "Does anybody want to share a cab?"

The lawyer looked uncomfortably at his wife. "Actually, we're walking distance, just up on 65th," he said, motioning up Park. The investment banker said she lived just a block and a half away, toward 5th Avenue.

The editor decided not to splurge on a cab after all. He caught a cross-town bus at 57th Street and then waited nervously near the token booth for the number 1 subway train at Columbus Circle. A foul-smelling homeless person shouted something at him until the train finally came, taking him up to 103rd Street and Broadway. He walked over to his apartment building, which had a check-cashing place downstairs and a storefront operation offering low phone rates to El Salvador. The elevator (with a bare lightbulb flickering overhead) took him upstairs to his scratched steel door. He opened it and was in his dining room. The people who live on Park and Madison have foyers, foyers so long you're tired by the time you reach the living room. But the editor couldn't afford an apartment with a foyer. He stepped over the threshold and found himself looking across his cluttered table into the kitchen and wondering where he'd left the cockroach spray. Suddenly he was feeling miserable.

Our editor, a composite, was suffering from Status-Income Disequilibrium (SID). The sufferers of this malady have jobs that give them high status but low income. They lunch on an expense account at The Palm, but dine at home on macaroni. All day long the phone-message slips pile up on their desks—calls from famous people seeking favors—but at night they realize the tub needs scrubbing, so it's down on the hands and knees with the Ajax. At work they are aristocrats, Kings of the Meritocracy, schmoozing with Felix Rohatyn. At home they are peasants, wondering if they can really afford to have orange juice every morning.

Status-Income-Disequilibrium sufferers include journalists at important media outlets, editors at publishing houses, TV news producers, foundation officers, museum curators, moderately successful classical-music performers, White House aides, military brass, politicians who aren't independently wealthy, and many others. Consider the plight of the army general, who can command the movements of 100,000 men during the week but stretches to afford a Honda Accord for weekend outings. Or of poor John Sununu, who ruled the world when he was White House chief of staff but had to feed, educate, and house eight children on $125,000 a year. The disparity is not to be borne.

There are two sides to the status-income equation. On one end is the Monied Class, those with plenty of dough who can use it to acquire status. But I am concerned with the Titled Class. Historically, when we think of the Grand Titles, we think of Prince, Duke, Earl, and Baron. But in the age of meritocracy, the Grand Titles are Senior Fellow, Editor in Chief, Assistant to the Secretary. Or titles that include an employer's name—the *New York Times*, the White House, Knopf—in which case it scarcely matters which position the individual holds.

The Titled Class has always resented and secretly envied the Monied Class. But for journalists, writers, and politicos, the pain now is acute. Until recently, a person who went into, say, the media understood that he or she would forever live a middle-class life. But now one need only look at Cokie Roberts or David Gergen to see that vast wealth is possible. Once it becomes plausible to imagine yourself pulling in $800,000 a year, the lack of that money begins to hurt.

Furthermore, the rich used to be remote. An investment banker went to Andover and Princeton, and a radio producer went to Central High and Rutgers. But in the new media age, the radio producer also went to Andover and Princeton. The schlumps she wouldn't even talk to in gym class are bond traders on Wall Street with summer houses in East Hampton. The student who graduated from Harvard *cum laude* makes $85,000 as a *New York Times* reporter covering the movie business. The loser who flunked out of Harvard because he spent all his time watching TV makes $1.2 million selling a single movie script.

Consider the situation of our composite editor. He's earning $110,000 a year as a top editor at, say, *Time* magazine. His wife, whom he met while they were studying at the Yale drama school, is a program officer at a boutique foundation that offers scholarships to Brooklyn high-school students. She makes $65,000. In their wildest imagining they never dreamed they'd someday pull in $175,000 a year.

Or that they'd be so poor. Their daughter turned 10 last year and needed a separate bedroom from her brother. They were lucky to get a fairly bright three-bedroom for $2,750 a month, even allowing for the dingy neighborhood and the cockroach-infested building. Jessica's tuition at Dalton is about $18,000, once you throw in the extras, and it costs at least $16,000 to send Max to the Ethical Culture School. The parking spot for the 1988 Camry is $275 a month, the part-time nanny who picks up Max from school costs about $12,000 a year (off-the-books cash; there goes any chance of serving as Attorney General), and after throwing in the costs of various ballet (Max) and rock-climbing (Jessica) lessons, the family is left with an after-tax disposable income for food, laundry, subway tokens, clothes, and leisure of about $600 a month. Which explains why the editor hasn't bought a new tie in three years and why he wakes up at 4 in the morning wondering where next year's tuitions are going to come from. It explains why he can't face his accountant, who knows that out of his $175,000 annual income, he gave a grand total of $450 to charity.

Members of the Titled Class are good at worrying about their reputations. All their lives they've mastered the art of having other people think them smart and wonderful. But the person who suffers from Status-Income Disequilibrium can scarcely spare an hour worrying about his reputation because he has to spend all his time worrying about money (when in fact all he wants from money is to have enough so he doesn't have to worry about it).

And it is not as if the Titleholder these days fills his mind with thoughts about truth and beauty, or poetic evocations of Spring. It is not as if he is compensated for his meager $110,000 salary with the knowledge that he can spend his days amidst the Higher Things. If he's in publishing, say, he spends his days thinking about market niches, the same thing those summer-house-owning executives at AT&T think about. When a book comes in, he wonders first which market it will serve: the Jewish market, the gay market, the depressed women's market? If every day he could publish a memoir by a neurotic lesbian Holocaust survivor with her own syndicated radio program, he'd have his own imprint in a year.

And it's not as if he is less ambitious than the partners at Skadden Arps, or that he does less schmoozing than the muni-bond traders at Kidder Peabody. The media person is in business just like $600,000-a-year smoothies in Fortune 500 executive suites. It's just that they are working in big-money industries and he's in a small-money industry.

The Titleholder is at the tail end of the upper class. Our composite editor is rich enough to send his kids to Dalton and Ethical Culture, but all the other parents make as much in a month as he makes in a year. When his wife wasn't working, she used to pick up Jessica from Dalton. She'd wait outside on the sidewalk, she and 150 nannies. She'd try to arrange play dates with the other kids, but their nannies weren't willing to travel all the way uptown to 103rd Street, so they'd end up going to the playgrounds off Central Park West. And she'd sit, a little uncomfortably, with the nannies on the benches that ring the playgrounds, trying to find common conversational ground. If a Martian were to land in a Manhattan playground, he would conclude that human beings are white as children and grow up to be black with Trinidadian accents.

Eventually, the kids of a SID sufferer begin to notice the income difference between their family and all their classmates' families. It happens around birthday time. The other kids in the class have birthday parties at Yankee Stadium (they've rented out a skybox) or at FAO Schwartz (they rented out the whole store for a Sunday morning). The SID kid has his party in his living room, with a picture of a donkey on the wall and a 69-cent blindfold you can peek through if you really want to.

Often, the child of a SID victim will get invited for play dates by classmates who live in the Dakota or on Central Park South, big, high-ceilinged places with servants' wings and dining rooms the size of tennis courts. These are the apartments of those who live in the forest canopy, where everything is light and clear and odorless and most of all uncluttered. People who live in the canopy enjoy wide-open spaces. Their apartments are filled with long expanses of counter space, wall space, settee space, table space, and floor space, all of it luxuriously spare. And it's just the same in their offices. People in the Monied Class have big offices and luxurious wood surfaces. And they have secretaries to route the paper flow, and their secretaries have secretaries to file things away, so there is nothing left stacked up to cover the wide-open expanse of a Monied person's desk. The briefcases of the Monied Class are wafer thin, with barely enough space to squeeze in a legal pad—because their lives are so totally in control they don't have to schlep things around. They can travel luggage-free to London because, after all, they've got another wardrobe waiting for them in the flat there.

The life of a SID sufferer, by contrast, is cluttered. He's got a little cubicle at his newspaper or magazine, or a little office at his publishing house or his foundation. And there are papers everywhere: manuscripts, memos, yellowed newspapers, magazine clippings. And at home, the kitchen of the SID sufferer has jars and coffeemakers jamming the available counter space, and pots hanging loosely from a rack on the wall. The SID sufferer has books jammed all around the living room, some dating back from college (*The Marx-Engels Reader*), and there are magazines and frayed copies of the *New York Review of Books* lying on the bed-stands.

The contrast is clear when it comes time for the annual class dinner. One pair of parents take it upon themselves to throw a dinner for all of the other parents of the kids in their daughter's second-grade class. The host parents are inevitably executives at Goldman Sachs or CFOs at some media conglomerate. The affair is catered (Little Dorothy Caterers—with the slogan "We're not in Kansas anymore"). And everybody else gets to come admire a dining-room table that can seat 26.

When a Titleholder with a household income of $175,000 a year enters a room filled with Monied persons who earn $1.75 million a year, a few social rules will be observed. First, everyone will act as if money does not exist. Everyone, including the Titled person near bankruptcy, will pretend it is possible to jet off to Paris for a weekend and the only barrier is finding the time. Everyone will praise the Marais district, and it will not be mentioned that the Monied person has an apartment in the Marais, while the Titled person stayed in a one-star hotel somewhere in the suburbs. The Titled person will notice that the Monied Class spends a lot of time planning and talking about vacations, whereas all the Titled person wants to talk about is work.

These conversations between those who are Titled and those with Money are fraught with peril. For example, a person who has made $10 million in the garbage-collection business has to defer in conversation to an editor at *Esquire*. On the other hand, an editor at *Esquire* has to defer to a person who has made $300 million in the garbage-collection business. A TV producer who went to Yale and Oxford is higher than an apartment-building owner who went to SUNY-

Binghamton but lower than the owner of a hot restaurant who went to Brooklyn Community College. You've got to be sensitive to the invisible social hierarchies.

And at the back of the Titled person's mind there is the doubt: Do they really like me, or am I just another form of servant, one who provides amusement or publicity instead of making the beds? The sad fact is, the rich tend not to think this way. The millionaires think it would be neat to be a think-tank fellow and appear on the *NewsHour with Jim Lehrer*. Look at Mortimer Zuckerman, who owns the *New York Daily News*, the *Atlantic, U.S. News & World Report*, and a goodly chunk of Manhattan. He'll drive out to Fort Lee, New Jersey, so he can do a taping for the cable channel CNBC. It's not enough to have more money than most countries. He wants to be a pundit.

I don't know about Zuckerman, but most people in the Monied Class who fantasize about becoming a public intellectual can't actually fathom what it would be like to make less than $300,000 a year. They, like everybody else, suffer from Bracket Amnesia. As soon as you reach one income bracket, you forget what life is like in the lower brackets (in the way women forget about the pain of childbirth). The Monied know that the middle classes can't afford any dress they fancy, or ski when they please, but this knowledge is an abstraction.

Still, the rich feel a lack. First of all, they have to pay for all the foundation dinners they attend, while the Titled people go free. The rich are the johns of the foundation dinner-party circuit. Second, the Titled people are, in effect, paid to be interesting. They are paid to read and think and come up with interesting things to say (it's astonishing that so many do this job so badly). And the rich feel vulnerable because despite their vast resources they still rely on the publicity machine for their good reputations, which these professional dinner-party ironists control.

For their part, members of the Titled Class react in diverse ways to the pressures of Status-Income Disequilibrium. Some try to pass for members of the Monied Class. First, they dress the part. They buy those blue shirts with white collars and, to go with them, bright paisley ties that make it seem like the wearer has 100 electrified sperm crawling up his chest. Or, if women, they'll scrounge together enough dough for a Chanel suit. They keep their shoes polished daily, so that the sheen almost matches that of their hard briefcases. They buy glasses with largish frames, in contrast to the tiny "artsy" frames of the rest of their media friends. In this way, they believe, they can walk into a society restaurant like Mortimer's and nobody will think they are just a bunch of editors trying to pass as moguls.

And they use their expense account to the max. Like an asthma sufferer taking the cure at an Arizona resort, a SID sufferer can find temporary relief from his affliction while traveling on business. He can stay at the Ritz-Carlton for $370 a night, with phones and televisions in every room in his suite. Hotel dry-cleaning will be as nothing; a room-service omelet will arrive every morning at 7:30 sharp. He will rent a Mercedes, or hire a car and driver, and for once he will be able to slide through life like one of the elite, in the clean, elegant world he so richly deserves.

But then the business trip ends and it is back to earth.

Which explains why other members of the Titled Class go the other way and aggressively demonstrate that they reject the luxuries the Monied Class enjoy. You will see them wearing Timberland boots with their suits, a signal that they haven't joined the Money culture. Their taste in ties and socks will tend toward the ironic; you might see them wearing a tie adorned with the logo of a local sanitation department, a garbage truck driving over a rainbow.

At home this sort of SID sufferer will luxuriate in his poverty. He will congratulate himself for the fact that he lives in an integrated neighborhood, though he couldn't afford the pearly-white neighborhoods along Park Avenue. He'll note proudly that he is in touch with normal Americans, since he, unlike all the elites he works with, still cleans his own dishes, still scrubs his own toilet. (In fact, the distinction between normal Americans and SID sufferers is that it never occurs to the former to congratulate themselves on their populism every time they do the dishes.) Most of all, he will congratulate himself on choosing a profession that doesn't offer the big financial rewards, for his decision not to devote his life to money grubbing. He does not mention to himself that in fact he lacks the quantitative skills it takes to be, say, an investment banker, and he is unable to focus on things that bore him, the way lawyers can. There never was any great opportunity to go into a more lucrative field.

How can we alleviate the suffering of those who suffer from Status-Income Disequilibrium? For SID sufferers who are politicians or leading public officials, the answer is LEEP, the Lifetime Earnings Equalization Plan. The big lobbying firms, which hire politicians and top officials when they retire, could simply begin paying the politicians a decade or two before they actually go to work for the firms. That is, instead of making $125,000 a year for 20 years in public life and then $1.1 million for 10 years in private industry, the public figure would have his income equalized at $600,000 a year for the entire 30-year period.

For journalists, media types, and other SID sufferers, there is no easy solution at hand. One can envision the rare high-income/high-status people—William F. Buckley, Martin Peretz, Lewis Lapham—getting together to form charitable organizations to benefit their deprived brethren. These organizations could give out prestigious awards to low-status billionaires. Or they could give six-bedroom homes to high-status/low-income types.

But the needs are so great, I fear that only the federal government has sufficient resources to address them. In most cities, people are perpetually $1,500 a month away from happiness. Whatever their income, they imagine that an extra $1,500 a month would give them everything they need. But in New York, Washington, and Los Angeles, where SID is found in its greatest concentrations, people are $250,000 a year away from happiness. It will take a lot of money to bring these people's incomes into line with their status. Only the federal government has that kind of money.

Under the federal plan I envision, anybody who could prove that five of his reasonably close friends earned seven times more than he would be eligible for federal aid. This aid would not come in the form of a cash grant. Under a cash program, some SID sufferers would lose the work ethic and simply try to scrape by on the federally provided $250,000 a year. But a targeted in-kind benefit—mortgage stamps—would have the right effect. The government would send out monthly mortgage stamps to pay the cost of any newly bought home valued at more than $1.1 million. The recipient would still be responsible for paying tuition costs, ski-trip costs, wardrobe costs, and other essentials. He would preserve his high-status career, but he would not feel ashamed when he returned home at night.

Ultimately, such a program would benefit the entire nation. Because SID sufferers control the American media, government, and the terms of civic discourse, their anxieties dominate the national culture. Their bad mood depresses everybody. If they were richer, the entire country would feel better about itself. And this would have a positive impact on the lives of American children everywhere. This would once again be a country in which little boys and girls could dream of becoming the literary editor at *Elle* and still be secure in the knowledge that they will be able to do their work from a six-bedroom apartment overlooking Central Park.

# A THOUGHT EXPERIMENT IN SUPPORT OF ESTATE TAXES

## By Irwin M. Stelzer

*(Originally published, as "Inherit Only the Wind,"*
*in the May 26, 1997, issue)*

CONSERVATIVES, WHO SEEM INTENT on tearing their movement apart, have come together on two seemingly unrelated issues: affirmative action and inheritance taxes. Both, it seems, are bad. Affirmative action is bad because it *gives* certain groups (blacks, Hispanics, women) an unfair advantage in life's race for success. And inheritance taxes are bad because . . . well, because they *deny* certain groups (inheritors of the product of someone else's success) what Trent Lott calls "a little jump start . . . so that they can be successful."

It seems that if your parents were successful, you are entitled to special treatment by government—a reduced tax (many conservatives prefer zero tax) on income received from your ancestral benefactors. If the Republicans have their way and the new budget accord passes the Congress, if you work hard and have taxable income of $1.2 million, you will pay almost $500,000 in income taxes (married couple, joint return). But if you receive an inheritance of $1.2 million, you pay no taxes. (Technically, it's not you, but the estate that pays, which amounts to the same thing.)

The argument I am about to pose in favor of a draconian inheritance tax is based in part on economics and in part on broader considerations of equity and social policy. I am a firm believer in tax cuts and consider myself very much a libertarian. Given these facts, my position might seem a little quixotic, not to say quirky. So, in the hope of retaining what little affection my fellow conservative friends may have for me, let me classify the discussion I am about to undertake as a "thought experiment"—an attempt to face the implications of a conservative worldview as fearlessly as possible.

Like any thought experiment worthy of the name, it is an effort designed to lead my friends to a conclusion far different from the one they now hold. But the opposite is also true—it might result in a rebuttal sufficiently persuasive to convince me of the errors of my ways. In that event, I will willingly accept the notion that some persuasive purpose as yet unrevealed to me is served by lowering the inheritance tax rather than by raising it to a level that is close to confiscatory.

My concern is to preserve and strengthen the American culture of entrepreneurship. So let me make it clear that I do not wish to argue in favor of the current system of estate taxes, because the current system does almost nothing to further the entrepreneurial culture (unless you are a lawyer looking to hang out an estate-tax shingle). And so, for the purposes of this argument, I will bend to the most public and tear-stained conservative objection to the estate tax and say that estates in the form of family farms and small businesses should be taxed at a rate of 0 percent. Economically, it doesn't make much difference. Taxes on such assets constitute less than 7 percent of all estate taxes collected in this country. Only one in 25 farmers leaves a taxable estate; the median estate tax paid by farmers is only $5,000; and right now inheritors of small businesses may take over 14 years to pay their estate taxes, with interest charged at a bit below market rates.

Nevertheless, we should eliminate any possibility that our new system of inheritance taxes might result in forced liquidations, and thus go with the zero rate. I say this only to clear this bit of underbrush from our path. But in truth, if a business or farm is taxed at fair market value, the inheritor should still be able to borrow against that asset to raise the necessary taxes and retain a substantial equity position in the business he or she has come into by chance of birth.

But for purposes of this thought experiment (and despite the possibility that it will result in the establishment of bogus businesses as vehicles for tax avoidance) let's pander a bit to conservatives' desire to preserve the balance sheets of inherited farms and businesses in their pristine condition, and exempt them from our inheritance tax.

Now, on to the confiscation.

To begin with, remember that the only wealth transfer that an inheritance tax can interfere with is the part that involves financial assets. An inheritance tax cannot deny children the most important inheritances they receive from their parents. For one thing, it is no longer in substantial dispute that intelligence is substantially heritable; we simply do not know what portion. Nor can tax policy deny children what economist Gary Becker, in his *Treatise on the Family*, calls "endowments of family reputation and connection; [and] knowledge, skills, and goals provided by their family environment." Nor can children be denied the advantages that inhere in what Glenn Loury, in a recent issue of the *Public Interest*, calls "networks of social affiliation"—education, the "parenting skills" of one's mother and father, acculturation, nutrition, and socialization in one's formative years.

A study by Thomas Dunn and Douglas Holtz-Eakin for the National Bureau of Economic Research, for example, suggests that parents' human capital, far more than any financial assets with which they may endow their offspring, affects "the propensity to become self-employed." In other words, the entrepreneurial spirit and the tools needed for success are the best things that parents can bequeath children—and it would be difficult to demonstrate that the fact children know they may be showered with money they did not earn later in life enhances that spirit. The principal objections to our inheritance tax are that it would give the government more money to waste, and that it could easily be avoided, either with the help of specialists in that art or simply by working less so as to accumulate less. Neither objection withstands scrutiny.

The revenues from such a tax could be used to reduce the marginal rates of income tax, producing all the wondrous benefits conservative supply-siders confidently predict for such a reduction. By such a move we would have reduced the tax on work by increasing the tax on the less productive activity of being around when someone dies. The government gets no more revenue, and the economy grows faster. And—

libertarians take note—our new inheritance tax would not represent an increase in the government's power over the citizenry. The compulsion to pay income taxes would be reduced, dollar for dollar, to the extent that the compulsion to pay inheritance taxes was increased. Put slightly differently, the amount of national income subject to government seizure is not increased.

None of this is to deny that the higher the inheritance tax rate, the greater the incentive to devise schemes to avoid the tax. Or that there will be many a slip between taxpayer cup and tax-collector lip. But that is true of most taxes: We all know how sales taxes on big-ticket items are avoided by shipping purchases to out-of-state addresses, and how capital-gains taxes are avoided by a variety of techniques such as selling short against the box. The possibility of some tax avoidance, in short, is not a compelling reason to fail to levy a tax that is otherwise justifiable on economic and policy grounds.

No doubt, every effort will be made by those planning their estates to evade a 100 percent inheritance tax. And, given their ability to enlist the finest minds the nation is wasting in the accounting and legal professions—what the Heritage Foundation calls "a substantial cottage industry devoted to estate tax avoidance"—there will be evasion. But such evasion is unlikely to have major antisocial consequences.

Gifts during one's lifetime—$10,000 a year can be transferred tax-free—already provide a bit of an escape. And more can be done in one's children's interests with the money that would otherwise be left behind: Instead of leaving financial or other assets to be taxed, parents might spend large sums enriching the educations of their offspring while the parents are living. To the extent that the inability to leave financial assets to the next generation encourages parents to spend still more on their children's education, parents' incentives to work to meet the costs of such education are increased, and the nation's stock of intellectual capital is enriched.

The other possibility is that oldsters might simply retire sooner in order to accumulate less money. If they elect this option of early leisure, the supply of labor will be reduced, a bad thing from society's point of view. But the now not-so-rich heirs will have to work harder, which would increase the available supply of labor (and younger labor at that). No precise computation of the net effect on labor supply and costs is possible, but there is no reason to believe that society will be the loser by trading some years of oldies' labor for some years of more intense labor by younger folk.

Finally, those about to depart this mortal coil, anticipating that terminal event, might donate their money to charity. No social loss there. Or they might simply spend their money on pleasure binges, a final thumb-in-the-eye to the waiting tax collector. If a 100 percent inheritance tax did indeed induce such consumption, it would replace the eventual consumption of the heirs to whom those who earned the money would in other circumstances have directed it. It is difficult to see what large-scale social loss is involved, unless we want to argue that older people derive less satisfaction per dollar spent than do younger people. This seems highly unlikely, given seniors' superior experience in separating the truly pleasurable from the merely fashionable.

It is important to note at this point that no inheritance tax should apply to transfers of wealth between spouses. Inheritances of spouses are not now taxed, and should not be, since the inheritance consists of accumulated income earned by both, whatever the distribution of work between office and home happened to be.

So there you have it: A policy toward inheritance taxes that is consistent with opposition to affirmative action and other government preference programs, that encourages young people to work, that induces seniors to invest in the intellectual capital of their offspring and to increase their charitable donations, and that promises to lower income taxes and thereby stimulate growth.

These of course are in the main economic arguments against favored tax treatment of inherited wealth. Such arguments can never be dispositive; public policy must be based on more than economics. But it also must be informed by economics, and those who choose to override economic arguments have the burden of rebutting them by showing that the economic costs of their proposals are exceeded by some higher values. And they must, it seems to me, also be quite explicit in describing the social values that should be given precedence over the long-held and very American ideal of equality of opportunity, a level playing field, or whatever term might best describe a fair field with no favors.

Nothing here suggests that parents be denied the opportunity of passing on to their children any social advantages inherent in their birth or upbringing. Nothing here suggests that parents be in any way limited in what they can spend on adding to their children's intellectual capital, or reduces their incentive to work hard to provide their offspring with the best the world's educational institutions have to offer. I only wonder what economic or social purposes are served by preferential tax treatment that gilds the lily of birth.

What am I missing?

# THE CLINTON LEGACY

## By Noemie Emery

*(Originally published in the August 10–17, 1998, issue)*

LINDA TRIPP, SAYS MARGARET CARLSON, when she pressed the "on" button of her little tape recorder, "lost membership in the family of man." Read herself out of the human community. Lost contact with the whole human race. And for what crime? Not murder, not larceny, not even lying; but for recording and spreading truths others wanted kept secret. By most standards, this is not wholly lovely, but as grounds for damnation, it appears rather thin.

Not so, it seems, in the Clintons' America, where Linda Tripp's offense and others like it have become mortal sins. And as this goes on, something still stranger is happening: Real sins—sins in the Bible, like adultery and bearing false witness, two of the activities captured on Tripp's tapes—are being defined down to meaningless pranks. Adultery is "just sex" and nothing to bother with. Likewise, lying about it is just "lying about sex" and also trivial. Even lying under oath about sex is no big deal. From all of this, the true dimensions of the Clinton Project—the Clinton legacy, one might venture to call it—have begun to emerge.

The Clinton Project is not really about politics. It is about values. That is, it is about an inversion of values. Many have wondered whether the Clintons and their friends are truly immoral—engaged in knowing wrongdoing—or merely amoral, unable to tell right from wrong. Now, it appears neither is accurate. In the strange p.c. terms of their culture, the Clintons appear to be "differently moraled"—that is, they have morals, even quite strong ones, but ones of which no church or state has ever heard. This is the Church of Bill, in the State of Bill, with its own mores and standards. There is the Bible, with its boring old Ten Commandments, where certain acts are simple no-no's. Then there is the Bible of Bill, in which Thou-shalt-nots are downsized to glitches, and trendy new sins are invoked in their place. We are at the verge of a meaningful moment. Let us pause for a look at what the new morality has wrought.

It was back in January, when Bill Clinton was alleged to have said that, according to his interpretation of Scripture, certain forms of sex are not adultery and possibly not even sex, that we began to realize we might have a moral thinker of rare imagination on our hands. And sure enough, his policies have embodied his unique point of view.

Thus abortion, an issue groaning with grave value questions— What is a life? When does it start? When is taking it justified?—is drained of its moral dimension and becomes a mere medical matter, a personal choice. On the other hand, smoking, which *is* a choice and a health matter, acquires solemn moral overtones. Is drawing smoke through your lungs, which one day may hurt you, *morally* wrong, while ripping a life from a womb and ending it *isn't*? In the Clinton code, yes.

Thus Clinton proclaims he will carefully monitor the ads taken out by tobacco companies, because parents "have the right to know" who is luring their children into smoking. On the other hand, Clinton doesn't think parents have the right to know if their children are supplied with abortions or transported for abortions out of state. His surgeon general didn't even think parents had a right to know when schools gave young teenagers condoms. Traditionally, moral codes have sought to discipline and regulate—to moralize—sexual conduct, not out of stuffiness, but because unregulated sex can cause havoc. As this is the traditional view, it must now be uprooted. So sex becomes the one behavior one must never, ever judge.

Indeed, long before Bill Clinton appeared, trailing his fragrant scandals behind him, the left had already marked out sex as the one great exception to its general political enterprise: the island of license in its sea of restriction, in its ocean of meddling and interference. Socialist in all else, the left here believes in unrestricted markets.

Communitarian to a fault in economics, health, education, welfare, you name it, it is libertarian in this to an extreme. "Privacy" here is the watchword. Government must safeguard the right not to tell a partner one may be giving him or her a fatal illness, along with the right to kill a human being inches or days from being born. Smoking and fat tend to kill over time, but they kill older people, and many people survive them. AIDS kills younger people and always is fatal. Antismoking crusaders justify their campaign by citing the high cost of treating lung cancer. Per patient, AIDS costs much more, yet the causes of AIDS are never mentioned, much less condemned.

While the administration works itself up to near hysteria over the harm done to young people by secondhand smoke, an epidemic of sexually transmitted diseases (STDs) goes unremarked. The office of Republican congressman Tom Coburn, a physician, cites data showing that 12 million Americans, two-thirds of them under 25, acquire new STD infections every year: That's 12 times as many as start smoking. Five of the 10 most frequently reported infectious diseases in the country are STDs. Some of these ailments cause cancer in women; some cannot be checked by condom use. Like most epidemics, this one feeds on ignorance: Many people don't know that these dangers exist. Why no crusades to save the young from the peril of sexually transmitted diseases, in this most caring and safety-conscious of administrations?

The reason is obvious: the fear of even seeming to censure promiscuity *for any reason whatsoever*. Such censure would violate the code of the strange new religion. People might think you were reading the Bible. How out of step can you be?

Smoking, of course, is a dumb thing to do, and a White House might plausibly use its moral authority to discourage the practice. It is only in the context of other ills considerably more deadly that the intensity of the campaign against smoking must be seen as perverse. Likewise, taping a phone call that one party thinks private is a betrayal. But given the context in which it occurred, its singling out for special condemnation seems odd.

Linda Tripp betrayed a young woman who trusted her. But that offense did not occur in a vacuum. Around, before, and after it, there is good evidence that these other things have happened too:

1. The president of the United States sexually exploited a young woman in his employ.

2. The president of the United States sexually assaulted an aide in his office who had come seeking a job.

3. The governor of Arkansas exposed himself to a state employee of low rank and no power.

4. The most powerful man in the world sent his flacks on missions to destroy the reputations of several women, whose only crime was that he had approached *them*.

5. The president of the United States put an airheaded intern in the position of lying under oath to protect him.

6. The airheaded intern tried to get another government employee to lie under oath.

7. The Pentagon revealed confidential information about an employee to a hostile reporter, who used it to damage her.

Surrounding Linda Tripp's act were many betrayals, of which hers must appear the most innocent. She herself was betrayed by the Pentagon and by Lewinsky, who urged her to lie. Like Tripp, Lewinsky is betrayed and betrayer, used by the president as she tried to use Tripp. But the source of these acts is always Bill Clinton, truly the root of all evil in this sordid case. Tripp made the tapes because she was being pressured to lie under oath about Kathleen Willey, the woman who claims Clinton harassed her, in the Paula Jones lawsuit. Tripp was also afraid of Clinton's lawyers and fixers and the dirt they had dumped on these women and others. In all of this, devoted disciples of the First Church of Clinton apparently see nothing amiss.

Of all these sins, Linda Tripp's would appear the least deadly— just as smoking seems less perilous for young people than promiscuous sex, binge drinking, or hard drugs. Thus, of course, by Clintonian standards, smoking and taping become the all-important sins. It's an inversion we have seen before. Six years ago, Daniel Patrick Moynihan wrote of the social equivalent and called it "defining deviancy down." Back then, faced with obvious social dysfunction—violent crime, aggressive panhandlers, deranged people sleeping in the streets and subways—some people coped by calling the situation "normal" instead of aberrant and dangerous. So, too, defenders of the moral swamp that is the Clinton administration deal with its obstruction of justice over sex and fund-raising scandals by defining these down, either as commonplace—"Everyone does it"—or as too trivial to mention—"So what?"

But this downsizing of sins into glitches is only part of the story. As Charles Krauthammer has noted, when some forms of deviancy are defined down, there is always a parallel movement in the opposite direction: "defining deviancy up." As real crimes are downgraded to background street noise, fascinating new crimes, like date rape, hate speech, and insensitivity, are invented and pushed up in their place. In fact, it is the decay of the real that *requires* the creation of the fraudulent: People need rules, no matter how ludicrous, to supply a sense of order to their world and a sense of their own effectiveness. "Helpless in the face of the explosion of real criminality, . . . we satisfy our crime-fighting needs with a crusade against date rape," says Krauthammer. "Like looking for your lost wallet under the street lamp even though you lost it elsewhere, this job is easier, even if not terribly relevant to the problem at hand." Unable to say much about AIDS—mustn't condemn promiscuity—or about infants in dumpsters—too much like late-term abortion—liberals vent through their jihad against tobacco. Hillary Clinton can't make her husband keep his hands off the help, but she sure can ban smoking. Her White House may have high rollers in the Lincoln Bedroom and sex in the pantry, but you can't say it isn't smoke-free.

This also explains the feminist rage over Tripp. For six years, all the members of the Nina Burleigh school of presidential assessment who treasure the Clintons for abortion and quotas have worked hard to reconcile their political theories (and their dreams about Hillary) with the unbuttoned urges of Bill. Thanks to Linda Tripp and her tapes, this is no longer possible: Bill stands exposed as a lech and a liar, his wife as a very old kind of feminine victim, and their whole model life as a sham. Thanks to the tapes, Clinton's agenda is dead in the water; his party in trouble; his heir losing altitude. As his legacy is, too. Because of the tapes, the Clintons will go down in history as the second coming of the Warren G. Hardings, not the Franklin D.

Roosevelts. Because of the tapes, the feminist groups have been forced to surrender their pretense of caring for women, driven to explain that (a) assault isn't assault when a liberal does it or (b) any woman should be thrilled to be harassed or assaulted by such a strong supporter of abortion rights. Of course, they want Linda Tripp disemboweled. They could say, "She blew up our charade." But somehow it sounds so much better to say, "She betrayed that poor girl." It allows them to vent *and* feel righteous. Or so they can tell themselves.

Somewhere in his meandering through history—Truman today, Reagan tomorrow; TR and FDR on weekends; JFK in between—Clinton caught the idea that presidents get remembered when they identify a threat to human freedom and dignity—the Axis, Jim Crow, the Evil Empire—and mobilize the country against it. But what to fight when the "health-care crisis" failed and all the other good stuff is taken? Besides, both Clintons seem so hemmed in by the scandals that there is precious little they can rail against. Greed? Lust? Gluttony? Buck-passing? Cowardice? Dissimulation? And they face another small problem: On what ground can they appeal to people? Duty? Honor? Self-restraint? Courage? Self-discipline? TR's bully pulpit, from which his successors have rallied the nation, has shrunk to the size of a pinhead, on which the Clintons are trying to dance. They were forced to embrace smoking as a last resort—one sin or vice in which neither has been tempted to indulge. Feeling called upon to moralize about something, they have cloaked their little cause in grand moral language, railing against a cluster of dubious admen as FDR railed against the Axis during World War II. Meanwhile, when called upon to account for their own actions, they continue to stonewall, obstruct, dissimulate, and to trash and use other people. It's a new kind of morality, but their own.

In the 1970s, Jeb Stuart Magruder, a figure out of our last major debacle, admitted he had "lost [his] moral compass" in the course of the Watergate scandal. The Clintons have done something different: Their compass is not lost, just reset with north and south reversed. When Clinton and company went to Washington in 1993, talking about new ways of seeing and doing, who could have known what they meant? "Reinventing government" is tame by comparison. Six years later, government is much as it was, but the moral traditions of millennia are under assault. With a president likened to Zeus (by Nina Burleigh) and personally keen on Biblical allusion, his administration sounds more and more like a religious cult. He has his disciples (Carville and Blumenthal) eager to serve him. He has his vestals (Carlson and Clift) eager to tend him. Who then can blame him for using his "mandate" to try to make life anew? But this is a prophet who should be without honor, for his is a devilish work. It is a work of confusion, inversion, and chaos that ruins perspective and sets all our standards adrift. A crime is a choice and a choice is a crime. Convicted felons like Webb Hubbell and Susan McDougal are innocent victims, while people engaged in self-preservation are expelled from mankind.

Should friends tape friends? No, they shouldn't. But the questions don't end there. Should friends try to talk friends into crime? Should adult men present ditzy interns with the kind of dilemma that propelled "that woman . . . Miss Lewinsky" to fame? Should employers make such use of those in their power? Is this the way presidents act? Under the present administration, apparently yes.

In fact, the whole sorry train of betrayals was set in motion by one person, Bill Clinton, president of the United States, who betrayed his wife, his employees, and his office repeatedly. A walking source of moral contagion, he is not only corrupt in himself, but the source of corruption in others.

Before they knew him, Linda Tripp and Monica Lewinsky were unlikely candidates for legal entanglements. Bill Clinton has flouted the time-honored standards—but by his standards, he is righteous. The question for the rest of us is whether we will let him seduce us into accepting his transvalued morality as ours.

# I DISONESTI

## By David Tell

*(Originally published, as that week's featured
"Parody," in the March 1, 1999, issue)*

### Giuseppe Conason's

### I Disonesti

*Opera in three acts, libretto by the composer,
first performed in Washington during the 1998–99 season*

## ACT I

SCENE ONE. The Gingricci, an alliance of clerics and enraged albinos led by Salamandro, have shut down most of the city and surrounded the Doge's palace. Don Guglielmo, leader of the Disonesti noblemen, confers with his deputy, Podesta, his bodyguard, Carvillo, and his valet, Blumentalio. Don Guglielmo blames the emergency on masculine pride, and despairs that palace life has withheld from him the worldly counsel of high-born women (*"Che orrore! Mia moglie è frigida e i miei testicoli sono blu"*). Don Guglielmo departs, and, alarmed by his gloom, his associates ponder their fate. Podesta declares that the Disonesti must confidently adhere to principle (*"Dobbiamo commissionare un fuocogruppo di Penn e Schoen"*). Blumentalio replies that he, a mere servant, can be confident only of his lowly station (*"Sono l'uomo più bello e più intelligente del mondo"*). Carvillo calls for moderation and magnanimity (*"Distruggiamo! Massacriamo! Assassiniamo!"*).

SCENE TWO. Later that evening. Don Guglielmo half-heartedly pastes military dispatches to a ledger on the desk of his adjutant, Giorgio di Stefanopoli. Donna Monica, disguised as a palace intern, arrives with a pizza. Don Guglielmo is made suspicious by his visitor's appearance (*"Ha gambe enormi e porta un beretto"*), but he accepts the food. While he eats, the lady reveals that she is not a simple intern at all, but a woman of royal birth (*"Mi chiamo Monica, principessa ebraica-americana"*) who has come to help him defeat the Gingricci. Impossible, Don Guglielmo responds; his power is spent. Nonsense, Donna Monica cries, and she urges him to inspect the clever plan she has hidden in her pants (*"Molto piccante, no? Venti dollari dal Segreto di Vittoria"*). Don Guglielmo is revivified by Donna Monica's support and exults over their meeting (*"Buongiorno, un pompino fantastico! Upsidesi—il mio sperma. Scusa"*). In his excitement, Don Guglielmo accidentally spills a small drop of Giorgio's paste on Donna Monica's dress.

## ACT II

Pentagonia, two years later. Out of misguided pity, Donna Monica has befriended an ugly, fat peasant named Linda. She recounts to Linda the story of her profound intellectual partnership with Don Guglielmo, and explains how she has exiled herself to this remote territory purely out of concern that he was eating too much pizza (*"Desidero che Nancy Hernreich manga merda"*). Linda reassures Donna Monica that no serious person could find scandal in the paste stain on her dress (*"Sento l'odore di un gran contratto di libro"*). Donna Monica responds that, though Salamandro is now dead, she still fears a resurgence by the Gingricci and worries that Don Guglielmo is not getting the feminine advice he needs (*"Ha tette minuscole, Kathleen Willey, ma Eleanor Mondale—vacca!—è provocantissima"*). After Donna Monica departs, Linda signals out the window, and the Inquisitore Indipendente enters the room. The Inquisitore reminds Linda that he is acting as a secret agent of the Gingricci, who have promised him a share of their tobacco profits should he succeed in destroying Don Guglielmo (*"Ohimè, preferivo veramente una nomina al Tribunale*

*Supremo"*). Linda informs the Inquisitore of Donna Monica's soiled gar-
ment, and to celebrate the mischief they will make, the two schemers
pull the head off a parakeet and drink its blood (*"Ha, ha, ha, ha, ha,
ha, ha!"*).

## ACT III

SCENE ONE. The Doge's palace, a short time later. Egged on by
Druggio and Isicoffo, two Gingricci provocateurs, the citizenry is in
an uproar over the Inquisitore's false accusation of a liaison between
Don Guglielmo and Donna Monica. The Disonesti consider how
best to proceed. The Doge's consiglieri, Ruffo and Chendalli, recommend a response of transparent candor and simplicity (*"Primo, il
privilegio esecutivo, e poi la definizione della parola «solo»"*). Because she
knows him to be a man of scrupulous rectitude, Don Guglielmo's
wife, Rodhama, agrees (*"Dirò «è una conspirazione vasta di destra» e
spererò di non vomitare"*). Don Guglielmo's best friend, Vernono
Giordano, offers to arrange for the Duke of Revlona to provide
Donna Monica a secure redoubt from the mob (*"Parlerò a Don
Renaldo di MacAndrews e Forbes, e presto! Una missione completa"*). The
group's initial optimism is dashed when Podesta reports that
Giorgio has defected and denounced Don Guglielmo in the public
square (*"Ha fatto menzione di «imputazione» in Piazza di Sam e Cokie"*).
His colleagues depart, and the Doge, crestfallen and distracted, asks
his scribe, Bettina, to remind him of their advice, but she has
already forgotten what they said.

SCENE TWO. The Grand Council. The oligarchs are debating a
Gingricci proposal to expel the council's Moorish members, to whom
Don Guglielmo has bravely extended full voting rights. The better to
achieve this and other sinister aims (*"Niente fluoruro nell'acqua!"*), one of
the Gingricci, Signor Barro, proposes that Don Guglielmo be replaced
as Doge by the ineffectual Don Alberto of Perezzia. Supporting Barro,
two of his confederates, Rogano and Lindsigrammo, produce the
Inquisitore's fraudulent alchemical analysis of Donna Monica's clothing (*"Siamo innamorati dei dettagli pornografichi"*). All seems lost until
Don Guglielmo dramatically appears and makes clear the truth (*"Non*

*ho avuto rapporti sessuali con quella donna"*). Moved by his obvious sincerity, the spectators disembowel the Inquisitore and set fire to his liver. A chorus of citizens reaffirm their love for the Doge (*"Noi, il popolo, siamo stupidi e immorali"*). As the curtain falls, Don Guglielmo rededicates himself to public service (*"Mi portate un tamburo di bongo e un sigaro"*).

# NOTES ON THE HAIRLESS MAN

## By David Skinner

*(Originally published in the June 21, 1999, issue)*

MEN WITHOUT CHESTS—that was C. S. Lewis's striking description of graduates of the postwar English schools, with their faculties trained to dismiss the virtues of patriotism and piety. These Englishmen, Lewis worried, would become lifelong enemies of the sublime, unable and unwilling, when push came to shove, to defend themselves or their countrymen. American men, I am happy to report—even the sensitive new age guys—still have something of a chest, thanks to our enduring fitness mania. But have you noticed how bare those chests are? Late twentieth-century America is increasingly a land not of men without chests but of men without chest hair.

I first realized this a couple of years ago while watching the otherwise forgettable B-movie romance *Picture Perfect*, starring Jennifer Aniston and Kevin Bacon, who was then just under 40. As the two characters get ready to, well, put the 13 in PG-13, Bacon takes off his shirt and his chest is completely hairless. Okay, Kevin Bacon may not prove much. But not long after that, I saw Al Pacino, whose fuzzy talent has graced the screen in such classics as *Scarface* and *Serpico*, appearing as hairless as an angel in *The Devil's Advocate*. (Once you're aware of the hairless-man phenomenon, by the way, you can no longer see a movie without noticing it. Sorry.)

Now, only 20 percent or so of adult white males are totally without what's technically referred to as "terminal pigmented chest hair." And yet, in the last few years, practically every Hollywood male sex symbol, when standing half-dressed for his more intimate scenes, looks as if he has absolutely no chest hair. Tom Cruise. Matt Damon. Keanu Reeves. Brad Pitt. All of them look like boys. One even sees older actors depilated to look like the boy-man stars who now capture every significant romantic role. The traditional Hollywood aesthetic in which old was never sexy has been carried to a new extreme: Now only the immature is sexy. Forget heroin chic, the hip aesthetic of the early '90s; say hello to permanent adolescence. And this new look trickles down. A big-city cop of my acquaintance confided not long ago that he shaves his chest. For several years now waxing salons have not been for women only. One of these days, no doubt, a cosmetic surgeon will come up with the philosopher's stone of our age— how to transplant hair from men's chests to their heads—and make a fortune.

But the hairless man represents more than just a simple change in cosmetic fashions, like the widening and narrowing of ties. These men without chest hair carry on, knowingly or not, quite a tradition. In ancient Rome and Greece, the romantic associations of men and boys were jeopardized by the appearance of facial and bodily hair. It meant the boy was now a man, which meant he was no longer available. The Roman epigrammatist Martial lampooned men who plucked their hair to stay boyish.

> *Why pluck the hairs from your gray fanny?*
> *That's a chic touch which men admire*
> *In girls, not in a flagrant granny.*

Martial also took issue with a man who insisted on calling him brother (*frater*), a term that also meant lover:

> *I'm shaggy-legged and bristle-cheeked*
> *Daily you depilate*
> *Your silky skin. Your voice is light;*
> *You lisp in a charming way—*
> *My voice, as my loins can testify,*
> *Is gruff, and so, I'll say:*

*We're less alike than eagles and doves*
*Or lions and does, so Mister*
*Don't call me "brother," or*
*I'll have to call you sister.*

Obviously today, as ever, the phenomenon cannot be disentangled from the romantic ideals of male homosexuals. As *Salon* columnist Camille Paglia authoritatively noted in a recent piece, "depilation has become highly fashionable in the gay male world. . . . Not since Greek athletes scraped their oiled, sandy bodies with the *strigil* . . . have men had such a fetish for girl-smooth skin." But what is most interesting about the hairless man is that he is no longer exclusively gay; he is, rather, the American male ideal.

Last decade's gay "clone" has become this decade's hetero stud. The subject of countless overwrought academic "queer theory" treatises, the gay "clone" was usually defined as an archetypal boy cruising men on the streetcorners and in the clubs of big cities. Boyish and neatly dressed (jeans and T-shirt ironed), he displayed a vanity and sense of style that were a "perfect" representation of manliness. And then, some-where along the line, the straight male began to imitate him. To see the gay clone today, one need only flip through magazines like *Men's Fitness* or *Men's Health*, two glossies that have made vanity a lifestyle.

A typical article from *Men's Health* tells readers how to decrease calories and stress ("Assign numerical values to the major parts of your life, such as work, marriage, and family; this can help you better appor-tion your time") while increasing earnings, physical strength, and sex drive. And "if Jane Goodall's research assistants have been creeping around your backyard, perhaps it's time to ask a dermatologist about hair removal with lasers. . . . A typical back treatment takes four hours and costs $500 to $2,000. Nose and ear procedures cost around $200. Backside denuding is at the doctor's discretion." After which, you can turn over and be made to look like the hairless man on the magazine cover. These magazines are an education in how to look exactly like a '90s man without having to think about what it means to be one.

Nor is this simply another case of gay fashion being a trendsetter for straights. The newly prominent hairless man is a sign of the convergence of gay and straight culture. Male vanity and the desire to prolong adolescence are becoming mainstream traits, no longer the markers of a subculture. Just two years after Ellen DeGeneres's "coming out" scored a ratings bonanza for her then-declining, now-off-the-air TV show, the arguments between gay activists and their critics over how visible homosexuality should be on prime time TV are already seeming quaint. Such arguments presume that there is a dominant, hostile majority culture. But there isn't. There are only tiny protest groups that get laughed at when they count the number of gay characters in TV shows and movies. The mainstream culture is the culture of the hairless man, at best indifferent to old-fashioned, grown-up male traits.

Here is a mainstream cultural moment. Cinematic stud Mark Wahlberg was interviewed earlier this year by Matt Lauer on the *Today* show. By the admittedly bland standards of morning television, the contrast in personalities should have made it an interesting conversation: Strong silent type who recently played an outsized porn star in the movie *Boogie Nights* confronts Sensitive New Age Guy, the kind of softy Americans want to see first thing in the morning. Instead, the only contrast in the interview was that of a regular SNAG versus a postmacho SNAG. It took Wahlberg, the postmacho SNAG, only seconds to reveal his vulnerable side: "It's kind of hard, you know, because the whole macho thing, you know, it's—coming from Boston, it's—it's also an—an athletic place, you know, and there's not too much opportunity there. So being the tough guy is the thing to do. . . . It was—it was difficult to—to accept the role in *Boogie Nights* only because I was—and it's stupid now to think about it, but I was worried about what my friends would think, you know, and—and stuff like that . . ."

Machismo is never so talked about as when it is absent. But there was a worthwhile question answered by the interview: What do you get when you put two SNAGS together? Answer: a conversation about being gay.

LAUER: You said in an article in *Premiere* magazine that when you were growing up, it was tough to repress the fact that you were . . . creative. It was a little bit like being gay and not being able to tell your parents.

WAHLBERG: Yeah.

LAUER: How does it feel to be in a place right now where it's cool to be gay—sorry, it's cool to be creative? You know what, it could be either way.

WAHLBERG: It's cool to be gay, too. It's cool to be gay.

LAUER: I loved your look when I said it. You kind of looked at me and said 'What?'

WAHLBERG: It's cool to be gay, too.

In fact, there is nothing ironic in Wahlberg's playing spokesman for the gay community. The rapper formerly known as Marky Mark was central to one of the most important sightings of the hairless man. Before his success in *Boogie Nights* had him making appearances on Charlie Rose and other talk shows, Wahlberg was a model for the famous Calvin Klein ad that appeared in countless magazines, but nowhere more prominently than on that humongous billboard above Times Square. Striking a pose in his skivvies, Wahlberg looked like a bit of rough trade freshly showered for a special occasion. But more important, he was, except for a butchy hairdo, as smooth skinned as the day he was born.

Not only has the mainstream gone gay—remember the quaintly controversial IKEA commercial featuring two thoroughly domesticated gay men picking out items for their home?—but gay life has gone mainstream. The course of this change can be seen in Hollywood movies. It was just a few years ago that the gay hit *The Adventures of Priscilla, Queen of the Desert,* a bawdy and occasionally hilarious movie from Australia, inspired a mediocre American imitation starring tough-guys Wesley Snipes and Patrick Swayze—hairless men both. In 1997's *My Best Friend's Wedding*, Julia Roberts's gay pal, who is her cover date for the wedding, seems to be the only character capable of romance; a faux-heterosexual tic has him stealing all the scenes he is in. Amidst so many public displays of friendship to support the comedy's bland premise (the possibility of good friends' getting married) the gay character refreshes the movie by leading the rehearsal dinner in a round of "I Say a Little Prayer." It's a weird throwback moment in which the movie's greatest display of devotion—a scene that could have been stolen from an old Gene Kelly musical—is romantically meaningless. It's also a reflection of the question at the center of the movie: Is love just an intense form of friendship?

Well, yes, according to various pop-culture trends of the '90s. The super-successful girl pop band the Spice Girls were practically a propaganda squad detached by the friends of friendship to demote eros to the status of a lower passion. Two of the most popular sitcoms of the decade, *Seinfeld* and *Friends*, were both predicated on the elevation of platonic love, one cynically and the other in a way that was painfully cute. Ross from *Friends*, the show's one regular male character of serious romantic intent, doesn't even merit being called a SNAG. His whiny boyish mannerisms suggest he can barely live up to the guy part. Men who really do love women have been, if not written out of television and Hollywood, playing second fiddle to their emasculated brothers.

In her famous 1964 essay "Notes on Camp," Susan Sontag, the voice of New York's then cultural vanguard, felt compelled to explain the obvious overlap between the self-consciously theatrical style described in her essay and homosexual taste, which, she wrote, constitutes "the vanguard of camp": "The camp insistence on not being 'serious,' on playing, also connects with the homosexual's desire to remain youthful." In September 1996, *New York* magazine published a prescient article describing the decline of the other defining characteristic of gay life: militancy. Referring to the " 'Hallmarkization' of gay sensibility," the author, Daniel Mendelsohn, argued, "If you take away the edge and the kitsch, there's not much left—and what remains isn't all that different from what you find in straight culture."

This seems to already overstate the difference between the sometimes campy and sometimes edgy singles culture of gays and the less campy and less edgy singles culture of straights. Traditionally, big cities are magnets for both gays and young people who are looking for careers first and spouses later. In places like New York, the romantic lives of a young straight and a young gay—both divisible into units of temporary attachments—aren't really that different. The difference between young married people and young unmarried people is far greater. If an icon of gay sexuality like the hairless male has gone mainstream over the last decade, it is because mainstream America wasn't intrinsically hostile to gay visibility to begin with.

What has been lost as the hairless man, an eternal boy, has become our male ideal? Real romance, for one significant thing. The hairless man is perhaps searching for romance, but only insofar as it supplies self-fulfillment and steers him clear of the burdens of love and family. Which is a pity. In order for real romance to occur, there must be some connection with matrimony. The hairless man would have to be robbed of his adolescent affectations and forced to mature. Defenders of a traditional culture have been overly fixated on gay characters, openly gay actors, and gay love stories. Such entertainment will succeed or fail on its merits as entertainment. Yet, it is the embarrassment of heterosexual love that should concern us.

Manliness cannot, after all, be reduced to a hard body, high income, and regular exercise. And yet, a pretty boy, the hairless man, has become the signature of American romance, thus mistaking the acorn for the tree, potential for the final product, leaving us with too many suitors and too few fathers, and stories about sex and love that never end in marriage and family. The problem, to paraphrase C. S. Lewis, is that we cannot raise geldings and expect them to be fruitful. We cannot turn middle-aged men back into boys and expect them to be leaders, elders, the carriers of what wisdom comes with age. We cannot erase general notions of manliness from popular culture and expect today's boys to be tomorrow's protectors and providers. Where can one find reflections of manliness, if everywhere you turn, the American male seems boyish, hairless, shorn of any sign that he is an adult?

# THE INNER HISTORY OF THE McCAIN CAMPAIGN

## By Tucker Carlson

———⬥———

*(Originally published, as "On the Road: From*
*New Hampshire to California, a Diary of the Real*
*McCain Campaign," in the March 27, 2000, issue)*

*Franklin, New Hampshire—January 30, 2000*

It's Super Bowl Sunday and John McCain is sitting on his campaign bus finishing off the second of two hamburgers. McCain has just given a rousing speech to a packed VFW hall, and he's hungry. An aide has arrived with an appliance-sized cardboard box of McDonald's food. As McCain eats, dripping ketchup liberally on his tie, the aide tosses burgers over his head to the outstretched hands of reporters. One of the burgers comes close to beaning George "Bud" Day, a 70-ish retired Air Force colonel who has been traveling with McCain. Around his neck Day wears the Congressional Medal of Honor, which he won for heroism during the years he spent with McCain in a North Vietnamese prison camp. "Where's the booze?" Day growls. Someone gestures to the back of the bus, and Day soon disappears to rejoin a group of fellow former POWs who, by the sound of it, have already located the bar.

"Senator," says a reporter who came on for the first time at the previous stop, "can I ask you a couple of questions?" McCain laughs. "We answer all questions on this bus. And sometimes we lie. Mike Murphy is one of the greatest liars anywhere." McCain points what's left of his hamburger at Murphy. "Aren't you Mike?" Murphy, a 37-year-old political consultant who is both McCain's message guru and his comic foil, nods solemnly. "Murphy has spent his life trying to destroy people's political careers," McCain says. "I'll have yours done on Tuesday," Murphy replies.

The reporter looks a little confused, but goes ahead and asks his question, which is about McCain's strategy for winning the New Hampshire primary. Before McCain can answer, Murphy jumps in with an insult. "The problem with the media," he says, "is you're obsessed with process, with how many left-handed, Independent soccer moms are going to vote." McCain translates: "You're assholes, in other words," he says, chortling and grinning so wide you can see the gold in his molars. About this time, one of the POWs sticks his head into the compartment where McCain is sitting. Sounds of clinking glasses and raspy old-guy laughter follow him from the back of the bus. "We're picking your cabinet back there, John," he says.

It takes only a day or two of this sort of thing for the average political reporter to decide that John McCain is about the coolest guy who ever ran for president. A candidate who offers total access all the time, doesn't seem to use a script, *and* puts on a genuinely amusing show? If you're used to covering campaigns from behind a rope line—and virtually every reporter who doesn't cover McCain full time is—it's almost too good to believe. The Bush campaign complains that McCain's style and personality have caused many reporters to lose their objectivity about him. The Bush campaign is onto something.

There are reporters who call McCain "John," sometimes even to his face and in public. And then there are the employees of major news organizations who, usually at night in the hotel bar, slip into the habit of referring to the McCain campaign as "we"—as in, "I hope we kill Bush."

*Nashua, New Hampshire—February 1*

Primary day has arrived, and the final distinctions between McCain's mobile primary campaign and your average sophomore road trip to Vegas are breaking down. By 8:00 a.m., the last of the coffee, bottled water, Diet Coke, and candy have disappeared from the bus. All that remains is beer and donuts. McCain is eating the donuts. He's in a sentimental mood. Late polls have shown him likely to beat Bush today, but he doesn't seem particularly jubilant about it. Instead McCain mentions three times how much he will miss rolling through New Hampshire in a bus. He seems to mean it. With McCain you get the feeling that the pleasure is in the process—that he considers the actual election a signal that the fun part is over. "It's been the great experience of my life," he says. "I'm feeling a little wistful."

McCain returns to his hotel suite and spends most of the afternoon chatting with his POW friends. At 7:00 the networks declare him the winner. The room erupts in cheers. All except McCain, who stands by himself, arms folded in front of him, unsmiling and not saying a word.

After his speech a few hours later, McCain and his wife are hustled into a conference room in the hotel for their first round of postvictory television interviews. Outside, the scene in the lobby looks like the end stages of a particularly rowdy wedding reception. The campaign has hired a couple of heavily tattooed Manhattan nightclub DJs to run the sound and lights. One of them—the guy with five earrings and control of the CD player—recently came off tour with the Foo Fighters and Nine Inch Nails. He's blasting a tune by Fatboy Slim. Hundreds of people are dancing and cheering and yelling.

Inside, where McCain is, the room is dark and still. Cameramen and sound technicians are fiddling with coils of wires on the floor. A photographer, exhausted from days on the road, has taken off his boots and is lying flat on his back asleep surrounded by camera bags. A CNN crew works to dial up the satellite link to *Larry King Live.*

McCain seems oblivious to it all. He has his eyes locked, unblinking, on the blank camera in front of him. His teeth are set, his chin thrust forward in go-ahead-I-dare-you position. Between interviews, he maintains the pose. McCain looks on edge and unhappy, not at all like a man who has just achieved the greatest political triumph of his life. There is no relief on his face.

It's a dramatic change from a week or two before. Back then, before he had seriously considered the possibility that he could become president, McCain seemed determined to run the most amusing and least conventional campaign possible. His style became more free-form by the moment. In the final days before the New Hampshire primary,

McCain took to pulling wackos out of the crowd at his town meetings and giving them air time. "Anyone who makes the effort to show up in costume deserves the microphone," McCain explained when a reporter asked what he was doing. At one point he handed the mike to a man dressed like a shark. A few days later he turned the stage over to a guy with a boot on his head and a pair of swim fins glued to his shoulders like epaulets.

For a politician it was risky, almost lunatic behavior—imagine if the shark man had started raving about Satanism, or the pleasures of child pornography. McCain appeared to thrive on it. Now, sitting in the dark waiting for Larry King, he seems burdened, or at least bewildered. Something unexpected has happened to John McCain: He won. He is the dog who caught the car.

It's close to midnight when the staff bus leaves the hotel for the Manchester airport. There's a case of champagne on the floor near the driver, but everyone is drinking beer. The whole thing is so amusingly improbable—the joke that came true. A few minutes later, Mike Murphy scans the AP wire and learns that McCain's lead has grown to 19 points. He chuckles. "What a caper," he says.

The bus finally pulls onto the tarmac and comes to a stop beside an elderly-looking jet with Pan Am markings. Rep. Lindsey Graham of South Carolina, who has spent all week stumping for McCain, peers out the window and spots it. He looks slightly concerned. I think I can tell what he's thinking: Didn't Pan Am go out of business years ago? "What kind of plane is that?" he asks Murphy. "It's a Russian copy of a 727," Murphy says. "It was decommissioned from Air Flug in the '70s. The Bulgarian mechanics checked it out and said it runs fine. We're not wasting precious campaign dollars on expensive American-made, quality aircraft. A minivan full of vodka and a sack of potatoes and we got it for the whole week."

Murphy seems to be joking, though over the next month, as the campaign travels from coast to coast and back again and again, the plane does take on a certain Eastern European feel. The flight attendants speak in hard-to-pin-down foreign accents. The paint around the entryway is peeling. The bathrooms are scarred with cigarette burns. The right engine periodically makes loud, unexplained thumping noises. Occasionally, in flight, the plane lists dramatically to one side for no apparent reason. Almost every landing ends with at least three bounces along the runway. As the plane touches down at a private airstrip in rural Ohio one afternoon, a voice comes over the intercom with a disconcerting announcement: "Ladies and gentlemen, welcome to Indianapolis."

None of this bothers McCain, who has successfully bailed out of four airplanes and knows he's not going to die in one. (Nervous reporters joke that if the plane does start to go down everyone on board will try to hop into his lap.) He spends most of his time in the air asleep. Presidential candidates traditionally sit at the front of the plane, behind a curtain where they can confer privately with their staffs. McCain does very little in private. After each event he reboards the plane like any other commuter, opens and closes a series of over-head bins in search of a place to store his coat, then finds a seat in economy class and sprawls out, head back and mouth open. Before long he is snoring quietly.

If it's after four in the afternoon, just about everyone else has a drink. Cocktails are a recurring motif on the McCain campaign. The candidate himself rarely drinks more than a single chilled vodka, and then only in private. Members of his staff are almost always in the bar till closing. (When the bar at the Copley Plaza in Boston finally stopped serving one night, one of the campaign's traveling press secretaries went to his room, emptied the contents of the minibar into a pillow case and returned to keep the festivities going.) At the front of the plane, right outside the cockpit and across from the cigarette-burned lavatory, are coolers of beer and wine, surrounded by baskets of candy bars and plates of cheese cubes. At the back is a bar—not a rack of miniature airplane bottles, but a table laid out with quarts of booze, ice, and mixers. Minutes after takeoff a crowd gathers near the rear galley.

A cable news producer works to wrench the cap off a beer bottle with a cigarette lighter as a group of cameramen sit nearby chatting and drinking horrible airplane champagne out of two-piece plastic cups. John Weaver, McCain's taciturn political director, stands at the bar pouring himself an unusually large drink. In the row next to him is the campaign's advance team, which is busy stuffing confetti guns—

thick plastic pipes with $CO^2$ canisters at the bottom—with orange streamers in preparation for the next rally. They're drinking, too. Cindy McCain, the candidate's wife, approaches, a glass of wine in hand, only to be intercepted by an MTV correspondent who looks about 15. "Could I get a quick interview?" asks the MTV girl. "Sure," says Cindy. Sitting off to the side, watching it all, is Greg Price, the guy who will drive the bus when the plane lands.

Price has been with McCain since the beginning of the New Hampshire campaign, when he was hired from a charter bus company in Ohio. He is 30, a laid-back, chain-smoking Navy veteran with no previous interest in politics. Price initially expected to be back home within a couple of weeks. That was in August. In December, he returned to Columbus briefly, got married, then left to rejoin McCain two days later. He has seen his wife for a total of 24 hours since. She is seven months pregnant. The New Hampshire primary changed Price's life.

Like a lot of former fighter pilots, John McCain is superstitious. He wears lucky shoes, eats lucky food, makes certain to get out on the correct side of the bed. His pockets are filled with talismans, including a flattened penny, a compass, a feather, and a pouch of sacred stones given to him by an Indian tribe in Arizona. He jokes about all of this, but he's not really kidding. At some point, McCain began to suspect that Price was a lucky bus driver. The campaign's rising poll numbers seemed to bolster this theory; the subsequent 19-point New Hampshire blowout proved it.

In the weeks since, Price has gone everywhere with McCain. Campaigns typically hire new bus drivers in each city. Those who travel stay in inexpensive hotels near the rest of the campaign staff. Price has stayed in McCain's hotel every night, sometimes in a suite. On some trips he has been a passenger rather than a driver. He has come to know McCain's family; on the night of the Arizona and Michigan primaries he sipped cocktails in the candidate's living room in Phoenix. ("You're never going home again," Cindy McCain told him when CNN announced that her husband had won both states.) And despite a long night at the bar in the Dearborn Hyatt, he is at the wheel of the bus at 8:00 a.m. Sunday morning to take McCain over to *Meet the Press*.

*Detroit, Michigan—February 20*

McCain lost the South Carolina primary last night, but you'd never know it from the way he's acting. He's in a great mood. As the bus rolls past miles of rubble-strewn vacant lots on the way to the television studio, McCain is laughing and telling story after story—about the late Rep. Mo Udall, about the Naval Academy, about the time he watched an Indian woman give birth in the corner of a bar in New Mexico. He doesn't seem upset about South Carolina. He hasn't come up with any talking points to explain his loss there. He doesn't appear to be preparing for *Meet the Press* in any way. McCain's aides aren't even sure how long he's going to be on the show this morning. Half an hour? Fifteen minutes? No one seems to know. (The full hour, McCain discovers when he gets to the studio.) It's obvious that no one really cares, least of all McCain.

McCain has never had a reputation as much of a detail guy. He can do a pretty good campaign-finance-reform rap. He can talk forever about the need to open up Reagan National Airport to long-haul flights to the West Coast. He seems to know everything about American Indian tribes in Arizona. Venture far beyond those topics and the fine print gets blurry. As he explained one morning a few weeks ago, there's no reason to get sucked into "Talmudian" debates over policy. "I won't bother you with the details," McCain often says when a member of the audience at one of his speeches asks about a specific piece of legislation. "That's a very good question," he'll respond, and then neglect to answer it.

It's an effective technique on the stump. Most people don't really want to know the details. But it is also a reflection of the candidate's personality. McCain can be kind of reckless. In fact, he enjoys being kind of reckless, and so does his staff.

Not surprisingly, McCain is having a pretty rough time on *Meet the Press*. One of his most prominent supporters in South Carolina, it turns out, is affiliated with a magazine that has been hostile to the organized civil rights movement. Tim Russert is hammering McCain on the subject. McCain looks like he isn't sure what to say. In the next room, McCain's aides are watching the show by remote. John Weaver is eating a piece of melon and chuckling about the campaign's unofficial slogan, "Burn it Down." "It's like Stokely Carmichael," Weaver says. "Power to the people!" He throws his fist into the air. "Burn it down—I love that." A few days later, at the bar on the plane, Weaver comes up with a new slogan: "Eradicate Evil." "We're going to have T-shirts printed," Weaver says. "They're going to have 'E$^2$' above crossed light sabers."

*Saginaw, Michigan—February 21*

McCain seems to be taking his own slogans to heart. At a rally this morning in Traverse City, he spent more time than usual beating up on the Republican party. "My friends," he said gravely, "my party has lost its way. My party has become captive to special interests." In conversations with reporters, he has begun to make disparaging references to the "Christian right," the "extreme right," and the "bunch of idiots" who run Bob Jones University. On the bus from Saginaw to Ypsilanti, he goes all the way, recalling with a smile "that old bumper sticker: The Christian Right is Neither."

Part of this is calculated rhetoric: McCain knows most evangelicals aren't planning to vote for him anyway. Bashing them might bring him more votes from moderates. But part of it is heartfelt. During the race in South Carolina, leaflets were distributed at political events that savaged Cindy McCain for her early-'90s addiction to prescription painkillers. McCain blames conservative Christian groups (and to some extent, the Bush campaign) for the flyers, as well as for a series of ugly push polls. For the first time, he talks about his opponents in a way that seems bitter. "They're going around saying Cindy's a drug addict who's not fit to be in the White House," McCain says, his fists clenched. "What am I supposed to do? Come out and make a statement that my wife is not a drug addict?"

*St. Louis, Missouri—March 2*

He is still mulling the question a couple of weeks later when the campaign plane touches down in St. Louis. McCain is in town for a few hours to participate by remote in a televised forum with Bush and Alan Keyes. It is the last scheduled debate. McCain knows he must do well. He and half a dozen advisers gather in the conference room of a television station downtown to eat barbecue and prepare. McCain is resigned to appearing tonight with Alan Keyes ("If we tried to keep him out of the debate, he might chain himself to my front door"), but it is clear that the very thought of George W. Bush makes him agitated. McCain is angry at Bush. Very angry.

I happen to be standing next to the coffee maker when McCain walks over to pour his ninth cup of the day. He's thinking about what he needs to do in the debate, and about mistakes he has made in weeks past. "I've got to try not to get down into the weeds tonight," he says, to himself as much as to me. Bush may be a dishonest candidate running a vicious campaign, but in the end . . . McCain looks up from his coffee. "Nobody gives a shit."

It's a good point, and absolutely true. Voters say they dislike attacks ads, but they generally believe them. They may feel sorry for a candidate who is being bashed over the head, but they tend to assume he must have done *something* wrong. And no matter how they feel about the accuracy of an attack, voters almost always perceive complaints about negative campaigning as whining. McCain knows all this. He also knows that the public doesn't believe that his campaign has behaved any more honorably than Bush's—particularly after McCain was caught lying last month about calls his campaign was making to voters in Michigan. Still, he is finding it hard to choke back how he feels. And how he feels is aggrieved.

McCain feels aggrieved fairly often, but for some reason his aides hate to admit it. One morning in New Hampshire, a reporter asked McCain what he would do if his 15-year-old daughter Meghan were raped and became pregnant. Would he allow her to have an abortion? McCain's face reddened as he listened to the question. After a family discussion, he replied slowly, "the final decision would be made by Meghan." Reporters pounced. But isn't that a pro-choice position? No it's not, barked McCain. He looked furious.

Except he wasn't, it was explained later. Moments after McCain got off the bus, Todd Harris, the campaign's traveling press secretary, loped to the back where half a dozen reporters were still sitting, replaying their tapes and checking their notes. Harris had heard that someone, probably a wire-service reporter, was planning to describe McCain's response to the pro-choice question as "angry." Harris was determined to stop the adjective in its tracks. "Who's calling him 'angry'?" he demanded. No one confessed. McCain wasn't angry at all, Harris explained. He was merely "tense."

An hour and a half later, McCain's mood was upgraded. A friend and I were sitting in a diner in downtown Manchester having breakfast when Todd Harris walked up to our booth carrying a statement from McCain on the abortion question. "I misspoke," it began, and went on to explain that if Meghan McCain were to get pregnant, the entire family, not Meghan alone, would decide what to do next. Dutifully retrieving our notebooks, my friend and I took this down. What about McCain's state of mind on the bus this morning? I asked. If he wasn't angry, is it fair to say he was irritated? That's acceptable, said Harris, nodding. "The AP's going with 'irritated.'"

With three minutes to go before air time in St. Louis, McCain is standing in the makeup room with a small group of advisers practicing his final comments. Rick Davis, his campaign manager, is humming "Ode to Joy" and pacing in the corner. McCain is using a thick blue marker to jot down some final revisions on a piece of scrap paper. His arm hooks in the shape of a sickle when he writes. His script is terrible. Looking out across an imaginary audience, McCain tries to recite what he has written. "I am a proud Reagan conservative," he says. "I am . . ." He stumbles, stops, then closes his eyes. For an instant he looks defeated, like he may not be able to continue. "I'm drawing a blank," he says. Mike Murphy leans forward until he is inches from McCain's face. "It's okay," he says softly.

And in seconds, it is. Soothing McCain is a large part of Murphy's job. McCain loves funny stories, and during lulls in the conversation on the bus he often asks Murphy to tell the one about the candidate he worked for who seemed to have Alzheimer's. Or about the campaign ad he claims he once made that accused an opponent of selling liquor to children. As Murphy tells the story, no matter how old it is, McCain breaks into hysterical, chair-pounding, hard-to-breathe laughter. McCain is genuinely amused by Murphy—he calls him "Murphistopheles," "The Swami," or simply "008," James Bond's little-known political consultant brother—but he is also calmed by his presence. A minute later, McCain grabs a final cup of coffee and heads into the studio.

The debate goes fairly smoothly for McCain, despite the obvious disadvantage of appearing by remote. Afterward, as he sits in a chair having his makeup removed, Murphy renders the verdict. "You were better than last time," he says. "You were good." "Do you think so?" asks McCain. It's not a rhetorical question. McCain honestly wants to know. "You were better and he was better," replies Murphy, "so it was sort of a blur."

*San Jose, California—March 5*

It soon becomes clear that a blur was not good enough. Two days before the California primary, it is obvious to virtually everyone that McCain will not win the nomination. His poll numbers have stopped rising. On the bus McCain seems, by turns, happier and more frustrated than ever. He is probably both. McCain prefers a righteous fight to almost anything, and Bush has given him new reason for outrage. A pair of rich Bush supporters in Texas have paid for an ad that attacks McCain's record on environmental issues. The ad is nasty and misleading, but what really incenses McCain is the idea of it. *Billionaire Texans attacking my integrity? Outrageous.* McCain gets hotter with every campaign stop.

"Tell Governor Bush to tell his cronies in Texas to stop destroying the American political system!" he shouts to a crowd in Ohio the Sunday before the primary. "If they get away with it," McCain tells reporters on the bus in California that night, "then I think it will change the nature of American politics forever. It will destroy it." The following morning, Bush's Texas Cronies have become "Governor Bush's sleazy Texas buddies." By afternoon, McCain is accusing Bush and his supporters of trying "to steal this election." Stopping them, he says, "is a race against time." Finally, on what turns out to be one of the campaign's final bus rides, from LAX to the hotel, McCain's rhetoric reaches the boiling point. "If this is allowed to go unchecked," he says, "there's never going to be another young American who's ever going to vote again, over time."

McCain sounded about as angry as a presidential candidate can, or for that matter ever has. Except that in real life, he didn't. McCain is one of those people who have to be seen to be properly understood. On paper he can come off as a red-faced blowhard. In person the effect is far more complicated. McCain can accuse a person of subverting democracy and grin as he says it, all without being phony or disingenuous. He can rant about the evils of the special interests as he cheerfully attempts to eat an éclair with a plastic spoon. I've seen him do it. John McCain is a happy warrior, maybe the only real one in American politics.

*Los Angeles, California—March 6*

With defeat a day away, McCain is becoming even looser. He no longer seems mad about losing. He seems to feel vindicated. To McCain, a loss to the massive Bush machine is proof that everything he has been saying for the past year is true: That money is the decisive factor in politics. That the system is rigged to exclude outsiders and mavericks. That the Establishment felt so threatened by his honesty that it mobilized to crush him. Most of all, McCain considers his defeat evidence that he ran an honorable campaign—he lost because he would not do anything to win.

In speeches, he continues to swing wildly at Bush. On the bus, his jokes are getting more outrageous. ("We ought to call this The Bullshit Express," he says to Murphy. "Get someone to paint 1-800-BULLSHIT on the side.") Members of his staff are taking pictures of each other, presumably to capture a moment that is about to end. There is no longer much reason to pretend. Or for that matter to be polite about the opponent. Murphy has taken to wearing a pin that says "W stands for Wuss."

*Beverly Hills, California—March 7*

By quarter to eight on the night of the California primary John McCain's presidential campaign has minutes to live. Tim Russert has just told McCain's guys that the latest round of exit polls from California looks bad. McCain is going to lose. He has already lost New York and Ohio and a couple of other states. The networks haven't called the race yet, but the official pronouncement is imminent. McCain isn't one to drag things out. "All right, Johnny," he says, looking around the Beverly Hilton Hotel suite for John Weaver, the campaign's political director. It is Weaver's job to arrange concession calls to the Bush campaign. Weaver hates doing it, and for the moment he has disappeared.

"Johnny," McCain calls again.

Weaver's voice floats out of an adjoining bedroom. "Do I have to?" he asks. "Yep," says McCain.

A few minutes later, Weaver appears with a cell phone. His mouth is puckered, like he just took a shot of something sour. Bush is on the line. McCain takes the phone without hesitating. Then he leans back in his chair, feet on the coffee table in front of him, chilled vodka in hand, and congratulates the man he has come to despise. "My best to your family," McCain says. The conversation is over in less than 30 seconds.

And that's it—the end of John McCain's run for president. Now it's time to face the reporters waiting in the lobby, and from there on to the concession speech. For a moment the room is silent. A few of McCain's aides look like they might cry. Not McCain. He is buzzing with energy. "Let's go," he says, bouncing out of his chair. "Onward."

# THE BEAUTY OF SUBURBAN SPRAWL

## By Fred Barnes

—◆—

*(Originally published, as "Suburban Beauty:
Why Sprawl Works," In the May 22, 2000, issue)*

THE OLD TOWN SECTION OF ALEXANDRIA, Virginia, just across the Potomac River from Washington, D.C., is about as close to utopia as it gets for devotees of traditional communities and critics of suburban sprawl. It's a lovely example of "mixed use" zoning—shops, offices, homes are interspersed—and a monument to late eighteenth-century American architecture. Its streets are tree-lined and narrow, forcing cars to move slowly. The buildings are low-rise and close to the street. It's both pedestrian-friendly and accessible to mass transit. Old Town Alexandria is "a unique place to visit to engage in civilized activity," insist Andres Duany, Elizabeth Plater-Zyberk, and Jeff Speck, three widely respected urban planners whose new book, *Suburban Nation: The Rise of Sprawl and the Decline of the American Dream*, is the most coherent and important attack on American sprawl to appear so far.

And yet there's a problem. When I moved a half dozen years ago to an Alexandria neighborhood not far from Old Town, my neighbors turned out to be immigrants—from Old Town. The family on one side has four kids and wanted a bigger house and a yard large enough for football and lacrosse games. Old Town was too cramped. The family of four on the other side also needed more room and the father was eager to landscape his back and front yards with dogwoods and azaleas and cherry trees. He couldn't plant a big garden in Old Town.

So what kind of neighborhood did they move to? One with many of the distinguishing characteristics of suburban sprawl: a cul-de-sac, single-use zoning, McMansions, decks behind the houses and no front porches, two-car garages and four-car families, five minutes from a mall and nearer to an interstate highway than to mass transit.

These refugees from Old Town were followed by two more families I know. One lived on the edge of Old Town adjacent to a swath of public housing. They were tired of worrying about the crack dealer who arranged his business deals on the pay phone across the street. And they wanted a spacious house with a yard. The other family had always wanted to live in Old Town. So when their kids grew up and left home, they moved to a townhouse there. They didn't stay long. Old Town was too noisy and parking spaces were too few. They moved to a quieter, lower-density neighborhood miles farther from Washington. They have no trouble parking their two cars now.

All this is merely anecdotal evidence, but it's consistent with an irrefutable fact of American life. For all the scorn that's heaped on the suburbs—and especially on subdivisions of nearly identical houses on the fringe of metropolitan areas—people like living there. And not just middle-class drones either. My friends who left Old Town are upper-middle class, highly educated, and reasonably well-to-do. Like millions of others, they prefer a big house with a yard and plenty of room, plus a place to park their fleet of cars. Old-fashioned towns crammed with stores and homes and apartments or new imitations of them like Seaside, Florida (the town in the movie *The Truman Show*) have enormous curb appeal, but they're too crowded and expensive for most people. They just aren't where most Americans want to live. And neither are dense city neighborhoods, even ritzy ones like Georgetown in Washington.

This is hard for those with an urban sensibility or a bias for college towns to believe, given their aversion to suburban America. Much of suburbia, after all, is grotesquely ugly, with ubiquitous strip malls and streets lined with fast-food joints. As often as not, neighborhoods in the inner ring of suburbs are decaying. In the exurbs, many homes are newer, but poorly designed and cheaply built. Then there's the traffic congestion that lengthens daily commutes. It's unavoidable because suburbanites are hopelessly car dependent. Yet the truth is they're mostly contented. They've come face to face with sprawl and they like it. And who can blame them?

A bevy of people, it turns out, from the heavyweight authors of *Suburban Nation* to Rosalyn Baxandall and Elizabeth Ewen, two professors from the State University of New York, whose *Picture Windows: How the Suburbs Happened* is a much more lightweight entry in the sprawl wars. Planners don't like suburban communities because much of the planning is done by real estate developers. Intellectuals have always looked askance at a suburban lifestyle they believe to be culturally barren: I can't think of a single novel or play that treats the suburbs kindly. Transportation specialists resent the refusal of suburbanites to abandon their cars and use mass transit. Environmentalists are mad at them for gobbling up open space. Liberals look down on them because the farther one gets from the city center, the more likely residents are to be conservatives. And Hollywood thinks suburbia is crass and soulless. Thus, the Oscar-winning movie of 1999, *American Beauty*, depicted every suburbanite as repellant for one reason or another.

The loathing of the suburbs has now morphed into a potent political movement that ostensibly targets sprawl as a specific type of suburban development, but is actually aimed at suburbia itself. *Suburban Nation* is likely to become this movement's bible. One of its authors, Andres Duany, is already the intellectual leader of the antisprawl cause. He's a Miami architect and the creator, along with his coauthor, Elizabeth Plater-Zyberk, of Seaside, Florida, and Kentlands, Maryland, and other eye-catching "new urbanist" towns. More than anyone else, Duany has made "sprawl" a buzzword and a growing issue in community after community. Videotapes of Duany's lecture and slide show on sprawl have been circulating like *samizdat* for several years, and the power of his argument against the current state of America's suburbs has been fully captured in *Suburban Nation*.

The case is flawed, but not easily dismissed. "The dominant characteristic of sprawl," Duany and his coauthors write, "is that each component [of a community] is strictly segregated." Housing, shopping centers, office parks, and civic institutions are physically separated, causing the residents of suburbia to "spend an unprecedented amount of time and money moving from one place to the next." And since nearly everyone drives alone, "even a sparsely populated area can generate the traffic of a much larger traditional town." As bad as the congestion is, life is worse for those who don't drive. Kids, the poor, and the elderly are isolated. Teenagers become bored and sometimes violent. Old people "know the minute they lose their license, they will revert from adulthood to infancy and be warehoused in an institution where their only source of freedom is the van that takes them to the mall on Monday and Thursday afternoons." And what jobs sprawl does provide, the poor can't get to.

What strengthens the case against sprawl by Duany and company is that it has a conservative ring to it. They propose to replace sprawl with communities designed and built "in the traditional manner of the country's most successful older neighborhoods." Their models are the cheerful suburbs that sprouted up in the first third of the twentieth century. All the elements of community life were integrated. Stores and offices were nearby and people walked to work and to shop, or they rode trains or trolleys. Their lives weren't dominated by their automobiles. They weren't trapped in traffic congestion for hours every day and had more time for family life. Streets were designed to make neighborhoods peaceful, not to rush cars through as fast as possible. The suburban communities were more densely populated and closer to downtown. There were few highways, not many cars, no exurbs, and no sprawl. People sat on their front porches. They rode bicycles.

This is a pleasant vision and perhaps appropriate for a country of a hundred million people. But it's utterly impractical for a postindustrial nation with 270 million people. Two-thirds of American families own their own homes, a phenomenal achievement. Without suburbs that extend far into the countryside, there would be millions fewer homeowners.

The authors raise the familiar cry about the lack of affordable housing. But where is housing least affordable? In fancy suburbs like Old Town Alexandria, Seaside, and Kentlands—the places extolled by Duany as models for the modern implementation of his old-fashioned vision. Housing is far less expensive the farther you get from a city. In other words, the more sprawl, the more affordable housing. The homes may look alike and be miles from offices or stores, but average working families can afford them.

The authors of *Suburban Nation* are candid enough to concede that not all suburban sprawl is bad. "In truth, a lot of sprawl—primarily affluent areas—could be considered beautiful." Even a McMansion— an enormous house that's bigger than the authors think it ought to be—"provides excellent value for its price." Inside, American houses are roomy and functional, but outside, "our public realm is brutal."

Cars and highways are the chief culprits. But as much as Duany, Plater-Zyberk, and Speck hate the automobile, they admit they've yet to give up their own cars. They have an alibi or at least an explanation: "The problem with cars is not the cars themselves but that they have produced an environment of dependence."

Of course, people always depend on some form of transportation to get where they want to go. In the compact, close-in suburbs that critics of sprawl propose to build, folks would be dependent on their feet or bikes or trolleys. For most people, however, driving a car makes more sense. It gives them freedom and mobility and saves time that would be lost if they used mass transit. *Suburban Nation* admits as much. The book's opposition to the automobile is largely aesthetic and sociological. The plethora of parking lots outside suburban shopping centers irritates the authors. "Such excess is inevitable," they write. No parking space, no peace: "Anyone who has shopped in suburbia knows that the inability to find a parking space makes the entire proposition unworkable." A car is an isolation chamber: "a potentially sociopathic device." And it's never the answer to any problem.

"The only long term solutions to traffic are public transit and coordinated land use," Duany and his coauthors assert. They promote the idea of "induced traffic." Building more highways will cause more traffic congestion, not less. This is nonsense. As Steven Hayward of the Pacific Research Institute has pointed out, long lines at a grocery store would not prompt anyone to say, "Well, we can't build more grocery stores. That would only bring out more customers." Building more highways wouldn't lure more cars. The cars come anyway. What foes of sprawl won't accept is the inescapable fact that most Americans would rather suffer in daily traffic jams than use mass transit. Trains, trolleys, and light rail aren't a viable option.

*Suburban Nation* argues that the preference for cars didn't come about naturally. The authors repeat the canard that the auto industry bought up streetcar companies across the country and tore up the tracks in a conspiracy to promote the use of cars. To a large degree, they claim, "the atomization of our society into suburban clusters was the result of specific government and industry policies rather than some popular mandate." Government-guaranteed loans at low interest paid for suburban homes. The federal interstate highway system provided roads. But these programs were not imposed on reluctant Americans. Next to Medicare, they're the most popular government programs ever.

It's the antisprawl movement that wants to force a lifestyle and a housing pattern on unwilling Americans. For activists like Duany, democracy is often an impediment. Listen to this complaint in *Suburban Nation*:

> *It is painful but necessary to acknowledge that the public process does not guarantee the best results. In fact, on certain issues, such as transit, population density, affordable housing, and facilities for special-needs populations, the public process seems to produce the wrong results. Acting selfishly, neighbors will typically reject a LULU (locally undesirable land use) even if its proposed location has been determined based on regional, social, or even ethical considerations.*

What should officials do in such cases? Go forward anyway, the authors suggest, because they know what's best: "Decision-makers must rely on something above and beyond process, something that may be called *principles.* Affordable housing must be fairly distributed. Homeless shelters must be provided in accessible locations. Transit must be allowed through. The environment must be protected." The authors express the hope that *Suburban Nation* "will provide a foundation upon which to make difficult decisions on behalf of the public good." With or without public support.

*Picture Windows* doesn't add much to the sprawl debate. The authors are professors who've heard from students that the Long Island suburbs aren't that bad a place. They agree, but only because those suburbs aren't like suburbs anymore. Gays and lesbians and immigrants live there. Women who've wised up and become feminists live there. Lots of unhappy people who've gotten a raw deal in life live there. The biggest problems are too many gated communities and not enough government-built housing and rental units. To its credit, however, *Picture Windows* does catalogue the "anti-suburban snobbery" of America's intelligentsia.

That snobbery is shared even by those who live in the planned communities and affluent, mixed-use inner suburbs beloved by the authors of *Suburban Nation*. Return to Old Town Alexandria for a moment. A favorite solution of antisprawl activists is something called "in-fill," which involves developing vacant spaces in cities and inner suburbs. Well, Alexandria has a large vacant area on the outskirts of Old Town where the federal government wants to build a headquarters for the Patent Office and seven thousand of its employees. A subway stop is nearby, and so is a train station. But residents of Old Town aren't pleased with this opportunity to fight the spread of suburbia with in-fill. They'd like the Patent Office to be built elsewhere, somewhere far over the horizon in the outer reaches of suburbia. Somewhere in the land of sprawl.

# GORE'S SPOILED BALLOT

## By David Tell and William Kristol

*(Originally published in the November 20, 2000, issue)*

THE PRESIDENTIAL ELECTION OF 2000 is the impeachment drama of 1998–99 all over again. And Al Gore is Bill Clinton. Only Gore's behavior is worse—worse because Clinton's misdeeds were of a gravity about which people might at least plausibly disagree. What Gore has done is directly challenge something explicitly articulated in the Constitution and therefore indisputable—and indisputably central to our system of government: the mechanism by which we have selected our chief executives for more than 200 years. This is rather a big deal, is it not?

No good can come of the massive confusion Gore's designated lieutenants have deliberately sown, in his name and at his behest, since Election Day last week. They have publicized unsubstantiated—indeed, altogether baseless—accusations of illegality against the popular-vote canvass conducted in Palm Beach County, Florida. They have loudly insisted that this purported illegality will be corrected only when Gore is finally awarded Florida's 25 decisive electoral votes—*whether or not* it can ever be shown that his name was checked on a plurality of valid ballots originally cast in that state. Worst, perhaps, they have done violence to civic understanding in America by repeatedly suggesting that because Gore appears to have won a plurality of the nationwide popular vote, he somehow *deserves* Florida's electoral votes—and thus the presidency.

It is a scandal that any major-party presidential candidate should ever authorize such a claim to be made on his behalf. As a matter of constitutional law, the nationwide popular vote is an entirely irrelevant consideration here. No man has ever campaigned for the nationwide popular vote, and no man has ever been elected president because he's won it. Like it or not, the Electoral College is *everything*. Intimating otherwise, and in the same breath circulating fictions about polling-place irregularities, the Gore camp has done its best to ensure that should George W. Bush eventually be elected president, some faint whiff of illegitimacy will hang over his administration. It will be unfair and corrosive. We hope that doesn't happen.

But if it does, it will still be better than either of the two alternatives Team Gore prefers. It remains possible that Gore's campaign will yet succeed by more or less legal and ordinary means—that the ongoing review of Florida's Election Day ballot will ultimately secure him the votes he needs to overtake Bush. In which case it will be proper and necessary for Gore to be inaugurated come January. Trouble is, our president will then be a man who has in the meantime proved himself wholly unconscious of, even hostile to, the most fundamental obligations of his office. The same will be true in the unlikely event that Gore captures the White House by the bizarrely extralegal means he and his lawyers are now proposing to the Florida courts: that Palm Beach County's November 7 ballot be invalidated and replaced by a full-scale, do-over election in that lone jurisdiction. In which case Gore will have become president by instigating a *genuine* crisis of governmental legitimacy from which the country—for reasons we will come back to—might have difficulty recovering.

No one should be surprised by what's already transpired, really. Not long ago, after all, Bill Clinton made systematic assault on essential elements of our democracy's republican character. That the president must consistently accept and respond to questions about his conduct; that his subordinates must never become a personal palace guard; that he must always obey the law—all these traditional doctrines of constitutional formalism Clinton defied. Democratic partisans, nearly the whole of the party, sustained him in this defiance. They thereby signaled their rejection of constitutional formalism—its organization of government around impartially administered rules and procedures—in favor of a politics devoted first and foremost to the business of winning this week's fight.

Then these same Democrats nominated one of their own, Clinton's unflaggingly loyal vice president, to succeed him. And now Al Gore has made war, for the convenience of his ambition, on *the* rule and *the* procedure around which the nation's entire public life quadrennially revolves: the election of the president.

We should all of us clearly understand the precise nature of this war. In late October, when suspicions emerged that the Democratic ticket might triumph on Election Day without a popular plurality, Gore spokesmen were quick to broadcast a preemptive demand that no one dare question the legitimacy of such a result. And they were right to do so. Hours after the polls closed last Tuesday, however, when it seemed clear that Gore and his running mate had won the popular vote—but might actually *lose* the Electoral College by a hair in Florida—Democratic campaign representatives and associated party leaders wasted no time at all executing a total *volte face* of spin.

By 4:00 a.m. on Wednesday, Gore talkers had begun ritually asserting that of "first" importance was the fact that Gore had won the popular vote—and that this fact was somehow inextricably related to the "will of the people" the election was meant to express. By Wednesday afternoon, Senate minority leader Tom Daschle had declined to promise that his Democratic caucus would accept the "legitimacy" of a Bush presidency. Democratic National Committee chairman Joe Andrew had announced that George W. Bush was absent from the election's "big picture"—that Gore alone had "earned and won the support of the American people." In New York, Hillary Rodham Clinton had wished aloud that Gore should be given all the votes she knew people "intended for him to have."

And in Nashville, Gore himself had popped briefly into view to share his concern that developments he left unspecified had called into question "the fundamental fairness of the process as a whole." And incidentally, he offered, "Joe Lieberman and I have won the popular vote."

Then Gore retreated, Bill Clinton–style, into silence. And soon enough his lawyers were litigating, David Kendall–style, all those purported "illegalities" in Palm Beach County. And his fund-raisers, Terry McAuliffe–style, were ponying up the cash the lawyers would need to litigate some more. And Jesse Jackson, Jesse Jackson–style, was in Florida collecting—but not revealing—evidence that Gore-supporting minority voters had been subjected to "intimidation" at polling stations across the state. And the usual know-nothing celebrities and cynical law professors were taking out another full-page ad in the *New York Times*, this one decrying the fact that while Al Gore had been elected president by "a clear constitutional majority of the popular vote and the Electoral College" (whatever that is), that result had so far been "nullified" in a manner that threatens "our entire political process." Maybe we should have "new elections in Palm Beach County," this Emergency Committee of Concerned Citizens suggested.

During the Lewinsky scandal, the last time a leading political figure so spectacularly violated some taboo, the nation listened in passive astonishment as the malefactor's allies constructed a similarly ridiculous set of excuses for him—and launched heedless attacks on anyone or anything that might stand in the way of his victory. The nation listened and listened and listened. Until the arguments seemed no longer bogus but comfortably familiar. And we lost all collective capacity for effective resistance.

This time, this year, as the order and integrity of a presidential election hangs in the balance, it is important that Americans stay focused and alert to the end. It is important that they know and remember two things in particular.

First, it is a *lie* that Palm Beach's presidential ballot last week was "patently illegal," as Gore partisans charge. True, as you have no doubt heard, Florida election law requires standard paper ballots to list candidates in a specified order, with the check box to the right of each name. True, too, Palm Beach observed neither of these strictures.

But that is because Palm Beach County employs machine-readable ballot cards, to which the rules for paper ballots do not apply. A separate provision of the Florida Code governs the use of such cards. The arrangement of their printed text is supposed to conform to that of paper ballots, but only "as far as practicable." And the placement of their check boxes need not conform to paper ballot requirements at all: The boxes may appear "in front of or in back of the names of the candidates."

Palm Beach's ballot was approved by representatives of both major parties in advance of the election. It was then published in the newspaper and distributed to the voters by mail. And it was used successfully, without complaint or incident, by upwards of 95 percent of those voters on Election Day. Yes, it does seem likely that some number of Palm Beach voters were confused by the ballot and failed to cast the votes for Al Gore they had intended. There may even have been enough of them to give Gore a statewide plurality—had they cast valid ballots.

But they didn't. And as a narrow legal matter, there really isn't much more to say than that. Two thousand confused voters cannot render invalid several hundred thousand ballots cast by Palm Beach voters who managed to follow the rules. And nothing in Florida statute or precedent says otherwise. The Palm Beach ballot was legal.

And yet, say Gore's men, the confused Palm Beachers *wanted* to vote for Gore, which means that Florida as a whole wanted to vote for Gore, which means that Gore really should have won the state's electoral votes and really should be declared our president-elect. The "will of the people," as reflected in the nationwide popular vote, must be effected, or last week's entire election was a fraud.

Ah. Here's the second and broader point Americans must remember as they listen to this complaint in the coming days. It is not true, as Gore campaign chairman Bill Daley has contended, that our national elections are designed to ensure that "the candidate who the voters preferred becomes our president." Our national elections instead are designed to ensure that the candidate the voters *voted for* becomes our president. And it is only from such votes, filtered through the Electoral College, that any meaningful "will of the people" can be determined. Any effort to impute such a national will from some other source and use the imputation to delegitimize an election whose results seem vaguely inconsistent is an effort to overthrow the constitutional system and replace it with banana republic–level chaos. In the United States, we do not conduct mulligan ballots whenever some losing candidate's supporters claim they were somehow prevented from getting it right the first time.

If, when Florida concludes its recount, it turns out Al Gore won Florida on Election Day, he should be president. If Bush proves the winner, the same should be true. No other outcome is acceptable. And none should be tolerated.

# KARL ROVE, WHITE HOUSE IMPRESARIO

## By Fred Barnes

━◆━

*(Originally published, as "The Impresario: Karl Rove, Orchestrator of the Bush White House," in the August 20–27, 2001, issue)*

IN LATE JULY, BILL BENNETT, the former education secretary and drug czar, got a telephone call from the White House. Would he be interested in serving as special presidential envoy on Sudan, where Christians are persecuted and slavery thrives? The caller wasn't Clay Johnson, President Bush's personnel director, or a State Department official. It was Karl Rove, senior counselor to Bush and political adviser. Bennett thought about the offer, then said no.

Weeks earlier, a senior Republican congressman recommended to the White House a nominee to serve on the part-time oversight board of a quasi-governmental corporation. The job paid $20,000. Johnson said there would be no problem. But the nomination never came about, and the congressman later discovered what had happened. Rove had substituted another choice for the post.

That Rove plays a major role in staffing the Bush administration—every appointment, even the most insignificant, crosses his desk—is startling enough. He's a campaign consultant by trade, and his line authority at the White House is limited to political operations, strategic planning, and public liaison. What's more startling is that personnel matters and his official duties are only a tiny part of what Rove does.

Cocksure, decisive, feared in Washington and inside the national political community, Rove is first among supposed equals in advising Bush, cabinet members included. His ideas animate the Bush presidency. His political maneuvering propels Bush's agenda. Rarely has a president's success depended so much on the skill of a single adviser. It's only a slight exaggeration to say: As Rove goes, so goes Bush.

Rove is the conceptualizer of Bush as a "different kind of Republican," whose presidency transforms the GOP into a majority party by adding new constituencies (Latinos, Catholics, wired workers) to a conservative base. Rove charts the long-term (90-day) White House schedule, including which issues Bush will stress. This, in effect, makes him both Bush's chief congressional strategist and the man behind Bush's message. For the fall, Rove's scheme calls for Bush to play up his "compassionate conservative" side, emphasizing education and conservative values. The aim is to counteract Bush's image as a conventional Republican, which Rove believes was created by the president's stress on tax cuts during his first six months in the White House.

There's still more, much more, to Rove's vast portfolio. He's both policy adviser and policy implementer. He took over the simmering issue of U.S. Navy bombing practice on the Puerto Rican island of Vieques and engineered the decision to terminate it (against the Navy's wishes). He became the leading White House expert on stem cell research and arranged for a stream of outsiders to meet with Bush, including Leon Kass, the University of Chicago professor whom the president tapped last week to head his council on the ethics of biomedical research. When Bush's faith-based initiative stalled this spring, Rove stepped in at the president's behest and, along with faith-based director John DiIulio, rejuvenated the effort and won House approval. He's a major force behind the president's plan to reform Social Security with personal investment accounts. He lobbied critical Republican House members from New Jersey to back Bush on a patients' bill of rights (most did).

Then there are Rove's more mundane political chores. He picks out prospective Republican candidates and encourages them to run. "That's my job," Rove says. The latest: congressman John Thune of South Dakota, who now appears likely to challenge Democratic senator Tim Johnson. When Tom Davis, head of the House GOP campaign committee, told Rove that Randy Forbes, not the candidate favored by governor Jim Gilmore, offered the best chance to pick up a Democratic House seat in a special election in Virginia in June, Rove responded, "I know." Rove dispatched a spate of Bush administration officials to stump for Forbes, who won.

A balding 50-year-old with glasses, Rove has become the hottest speaker on the Republican circuit. When he addressed the Midwest Republican Leadership Conference in Minneapolis in July, he drew a more enthusiastic response than Vice President Dick Cheney. "He's a hero to Republicans," says former congressman Vin Weber, who attended the conference. That same weekend he spoke at a fund-raiser for Kentucky representative Ann Northup in Louisville and a Republican National Committee event in San Francisco. In Virginia in June, he addressed both the state party convention and a gathering of well-heeled Republican donors hosted by Gilmore. Rove, by the way, negotiated the selection of Gilmore as RNC chairman last winter with the governor's chief of staff, Boyd Marcus. Gilmore had balked at being "general chairman" with little authority. He got the full chairman's job, but Rove assigned Bush loyalist Jack Oliver to the committee in the newly created post of deputy chairman.

Rove assigned himself one of the most important tasks at the White House: keeping the Republican party's conservative base solidly behind Bush. This is virtually a full-time job. He stays in almost daily contact, by phone or e-mail, with important conservative players in Washington, like National Rifle Association lobbyist Chuck Cunningham. He meets regularly with a group of conservative intellectuals in Washington, listening to their ideas and saying little himself. He talks to conservative journalists. He attends conservative gatherings. When attorney general John Ashcroft balked at addressing the Conservative Political Action Conference last February, Rove volunteered, though his family was moving from Austin to Washington that weekend. On August 1, he briefed the weekly meeting of Washington activists hosted by Grover Norquist of Americans for Tax Reform on congressional reapportionment. As Rove spoke, House GOP whip Tom DeLay entered the meeting, and Rove gently poked fun at him. Rove's appearance was warmly received.

All this activity, plus Rove's long and trusting relationship with Bush, has made him not only the most influential adviser to Bush, but one of the most powerful presidential aides since the advent of the modern White House under President Franklin D. Roosevelt. The media, however, tend to treat Rove as a top adviser whose duties are purely about gaining popularity and winning elections. As reporters see it, to use an analogy from the Clinton era, it's as if campaign consultant James Carville had joined the president's top staff and begun to throw his weight around. When Rove gets involved substantively in an issue, reporters treat that as proof the issue has become tainted with politics. But in truth, Rove is not Bush's Carville. He has always advised Bush on substance—while Bush was governor, during the campaign, and now. It was Rove who organized the teams of policy advisers who prepped Bush in the campaign and now fill high-level jobs in his administration. "Rove's a generalist," says Weber. "He's one of those rare people who operate at the intersection of policy and politics. When you get someone who's really good at both, that's the indispensable person."

Rove's official title is "senior counselor," but he refuses to spell out all that entails. David Keene of the American Conservative Union says Rove is the "central point" in an otherwise compartmentalized White House. Norquist calls him the "Grand Central Station where everything switches through." Marshall Wittmann of the Hudson Institute, an ally of senator John McCain and critic of Bush, says Rove is "perceived as the nerve center of the administration." Roy Blunt of Missouri, the deputy GOP whip and Rove's chief contact in the House, says of him: "He's everywhere."

The aides from earlier White Houses who rivaled Rove in influence had a distinct advantage: They served as chiefs of staff. But neither John Sununu in Bush's father's White House nor James Baker in Ronald Reagan's had the long personal relationship with the president that Rove has with Bush. And neither had devised the themes and masterminded the campaign of the president he served, as Rove has. "We're used to a White House that's not built on a long-term relationship," says Blunt. One like Clinton's or Richard Nixon's. In the Nixon White House, only the combination of H. R. Haldeman, the chief of staff, and domestic adviser John Ehrlichman matched Rove's clout. Perhaps Harry Hopkins in FDR's White House was more influential. And Sherman Adams, Dwight Eisenhower's chief, probably was.

The case of Adams is instructive. He exemplifies the peril of being a highly visible White House aide in a partisan environment. With Eisenhower immune from attacks as a war hero, Adams became the target for political foes and reporters and was forced to resign for improperly accepting gifts. And now Rove is under attack from Democrats and the media. "It's dawned on people he's the leading conservative in the administration and he's the leading policy adviser to Bush," says Republican consultant Jeffrey Bell. "The press and the non-Republican institutions in this town have found out how important he is to Bush's success," says Charles Black, a Washington lobbyist and Bush campaign adviser. That alone makes him subject to scrutiny, and he's all the more a target because criticism of Bush as a lightweight or a radical conservative hasn't caught on. Since foes of Bush view Rove as the president's brain, their strategy is decapitation: Cut off the head (Rove) to kill the body (Bush).

From all appearances, Rove doesn't take the attacks very seriously. Some he shouldn't, such as the barbs of Democratic national chairman Terry McAuliffe, who routinely zings Rove in his speeches and TV interviews. At a Los Angeles fund-raiser in July, he indicated that Rove was getting away with unethical conduct and that Democrats would increasingly go after him. He cited a meeting Rove had at the White House with corporate officials of Intel, who were seeking approval of the merger of a supplier and a Dutch company. "Isn't it a shame that's come to light," McAuliffe said sarcastically.

McAuliffe has no credibility, especially on ethical issues, but Henry Waxman, ranking Democrat on the House Government Reform Committee, does. Waxman is a fierce partisan, but he's also smart, relentless, and taken seriously by the press. On the basis of media accounts of Rove's meetings at the White House with executives of companies in which Rove owned stock, Waxman has sought a congressional inquiry. White House counsel Alberto Gonzales informed Waxman that conflict-of-interest rules don't apply to those meetings. Waxman responded that even if there's merely an appearance of conflict, the question must be turned over to the Justice Department for investigation.

There the matter stands, but only for now. Waxman has been stymied by the White House in seeking related documents and a full list of those with whom Rove has conferred. And Dan Burton, the Republican who chairs the House Government Reform Committee, has refused to conduct an investigation. Waxman, however, does not give up easily. His recourse is the Senate, controlled by Democrats. Majority leader Tom Daschle has said he doesn't favor a Rove probe. But Waxman aides insist, after talking to Daschle's office, that he was referring only to an investigation to retaliate against Republicans for badgering the Clinton White House, not a legitimate inquiry into Rove's dealings.

Should the Senate Governmental Affairs Committee, headed by Joe Lieberman, take up the matter, that could be trouble for Rove. He could be interviewed under oath by committee investigators, forced to turn over documents, and pressed into testifying at a public hearing. All that may sound farfetched, but it's not implausible. Democrats have always been good at "oversight" hearings that turn into gotcha sessions with a partisan payoff. And for the moment, Rove is the biggest game in town.

The outside advisers who talk to Rove every other week— Washington veterans Weber, Black, Ed Gillespie, Haley Barbour, and Bill Paxon—are worried about the attacks. Rove, who says the attacks are "part of the political game" in Washington, may be more concerned than he lets on. He says he's finicky on ethical matters. He told me he walked out of a session with New York governor George Pataki when the topic of dredging the Hudson River came up. That issue specifically involved General Electric, another company in which he held stock. The other meetings, including several with John Chambers, the CEO of Cisco Systems, consisted only of general policy discussions or friendly chats, not matters that directly affected any company. Thus, he and Gonzales insist no conflict of interest arose.

Rove has not been accused of exploiting his office to boost his stocks. In a June 15 letter to Rove, Waxman said: "I am writing not to make accusations about your conduct but to seek more information about your involvement in policy matters that may involve your hold-ings." In fact, Rove sacrificed millions in earnings by selling his polit-ical consulting firm and joining the Bush campaign in 1999 and now the White House staff. (Carville *made* millions by *not* joining Clinton's staff.) As an outside adviser, he could have collected lucrative fees for placing Bush campaign ads. With Bush as president, he could have signed a consulting contract with the RNC and worked for other clients, political and corporate, as well. Instead, Rove makes $140,000 a year as a government employee.

Absent White House dawdling, the trouble over Rove's stock would have been avoided. Rove says he offered to sell all his stock (worth $1.6 million at the time) before joining the administration but was told to wait for a certificate of divestiture to be issued by the coun-sel's office. He badgered White House lawyers, Rove says, but they didn't produce the document until June 6. He sold his stock the next day. In the interim, he'd met with Intel and other corporate executives. The delay in selling his stock proved costly to Rove. A Bloomberg News analysis found his stocks dropped 8.6 percent from January 20 to June 6, a loss of roughly $138,000.

Besides Democrats, Rove has the press gunning for him. When James Jeffords of Vermont quit the GOP in May and Democrats took control of the Senate, Rove was widely criticized for heavy-handedness in dealing with Jeffords. Howard Fineman of *Newsweek* said Bush would have to "rein in Rove" to recover politically. Actually, Rove had little to do with Jeffords's defection. Rove's attachment to conserva-tives is particularly annoying to the Washington press corps, which believes Bush must move to the center ideologically. Meanwhile, the *Washington Post* has gotten on Rove's case, hyping his minimal role in a bid by the Salvation Army for an exemption from antidiscrimination laws, then reporting he'd become the focus of critical attention.

There's another potential trouble spot for Rove: the White House staff. Rove says he was leery of signing up because internal feuds are chronic in Washington. "I'm not good at internecine warfare," he says. As things have turned out, Bush's staff is famously collegial. But the organizational structure is a recipe for competition, envy, and backbit-ing. At the top are four generalists—chief of staff Andy Card, deputy

Josh Bolten, communications chief Karen Hughes, and Rove—plus an active vice president. Rove dwarfs Card in influence. He and Hughes worked together for Bush in Texas and during the presidential campaign, and are close. But they also compete for Bush's favor—with a lot at stake. Rove urged Bush to vow to veto a liberal patients' bill of rights. Hughes argued against the use of the word "veto." She lost and Rove was vindicated, as the veto threat aided Bush in getting a patients' bill more to his liking through the House. Rove and Hughes also disagreed on embryonic stem cell research. She was for it. He made sure that Bush heard the concerns of pro-lifers and social conservatives. In the end, the compromise Bush announced last week was one Rove had floated months before.

On the Navy's bombing of Vieques, Rove took control of an issue that initially had been under Card's supervision. A binding referendum loomed, in which Vieques residents were likely to bar the Navy. Bush was already irritated at protests over the bombing. Rove persuaded him to call a halt to bombing runs. Rove has insisted he didn't force the Navy to go along, but what a participant in Vieques deliberations calls the "ultimate-decision meeting" was held in his office. Rove, of course, has as a top priority luring Latino voters. Bombing a Puerto Rican island wasn't helping.

Rove didn't have to grab the faith-based initiative. Bush handed it to him—and not to Card or Bolten or a White House aide with less on his plate. The president had chatted with Michael Joyce, the ex-president of the Bradley Foundation, about it during a White House ceremony in May. Bush was fearful the issue was languishing. He called Rove, instructed him to talk to Joyce, and told him to get the issue moving again. Joyce, on his own, was ready to start an outside lobbying effort to assist John DiIulio, the college professor who runs the program. Rove helped energize GOP leaders in Congress. The initiative, watered down, defied expectations and passed the House in July, beginning a winning streak for Bush proposals.

So what's the problem in all this? Nothing yet, and maybe nothing ever. But Rove's remarkable ascendancy in Washington brings expectations. If they aren't met, Rove will be held accountable inside the administration, on Capitol Hill, and by the media. White House aides won't blame the president. They'll finger Rove. Some congressional Republicans are squeamish about Bush's insistence on pressing ahead with Social Security reform. Rove thinks the issue is no longer an effective club for beating up Republican candidates. Tom Davis, the House campaign chief, isn't so sure. Young voters like the idea of investment accounts funded by payroll taxes, Davis says, but "the intensity is with older voters." If the issue polarizes seniors against Republicans, "it kills us." Davis frets this could occur in congressional elections next year.

The 2002 race is the next big test of Rove's skill. He is the man with the plan. It calls for a "compassionate conservative" president who holds his conservative base while attracting a wave of new voters to his party. One of Rove's specific duties is outreach—to Latinos, new economy workers, Catholics, suburban women, union households, and what he dubs "resource dependent communities," where coal mining or farming is dominant. His goal is to reproduce what President William McKinley and his adviser Mark Hanna achieved at the turn of the 20th century, namely a broadly based, majority party.

It's a dazzling vision, more appealing and perhaps more realistic than anyone else's. The first test was whether Bush could emerge as a successful president. He has. Another is to shape Bush's image to woo nontraditional Republicans. "I think he is viewed as being more conventionally conservative than he is," says Rove. So Bush will now stress education and values, not taxes and defense, and hope to be seen as an unconventional conservative. If Republicans hold their own in the 2002 elections, Rove will deserve at least a small measure of credit. If they suffer badly, he'll face cries for his ouster.

Finally, there's the reelection test in 2004. Never before have a president and a party had so much riding on a single person whose name won't be on any ballot. Rove could wind up as one of the greatest political strategists in the past century. But it's a risky business and there's little margin for error.

# THE GREATNESS
# THAT WAS GIULIANI

## By John Podhoretz

*(Originally published in the September 24, 2001, issue)*

WHAT RUDY GIULIANI DID last week in the aftermath of the attack on New York was not all that different from what he has done in the midst of any crisis during his mayoralty. In each case—a terrible fire, a water-main break, the crash of TWA Flight 800, a neighborhood blackout—he dons the garb of an emergency worker, tours the scene, gathers the heads of his agencies, designs a plan of action, and then makes public appearances during which he informs the city about what is going on in exhaustive detail. And he does all this within a couple of hours of the incident at hand.

New Yorkers have grown so used to Giuliani's omnipresence during a crisis that we were probably far less impressed by his performance last week than the rest of the world was. We take it for granted—so much so that in the past few years, Giuliani's crisis-management style has often been the subject of grumbling and partisan attack. When his administration closed off public access to City Hall, citing terrorism concerns, conventional wisdom in New York held this to be a reflection of the mayor's paranoid grandiosity. When Giuliani announced plans in 1998 for a multimillion-dollar center to coordinate city and federal agencies in the event of a major terrorist disruption, he was accused of wanting to set up a "bunker." The choice of the word "bunker," with its Hitlerish associations, was all too intentional.

Now the bunker itself, located in one of the ancillary World Trade Center buildings, has been destroyed as a result of a terrorist plot that the most paranoid among us could never have dreamed up. There is no chance that the fences Giuliani built around City Hall will ever be removed—despite the fact that all of the mayoral candidates seeking to succeed him had promised to tear them down. And even the mayor's enemies stand mute before the evidence of his astounding conduct on Tuesday, September 11. Trapped in a building near the World Trade Center, Giuliani had to keep his head about him in order to find his way to safety, helping others along the way. He had to collect himself to do his job, only to learn that three of the four most senior members of the city's fire department had been killed—heartbreaking news that would soon be followed by word that they had been joined in certain death by another 300 firefighters and scores of police officers as well.

And yet there he was, only hours later, standing before cameras and offering a kind of strong reassurance the country had yet to receive from any public official. "We will strive now to save as many people as possible and to send a message that the city of New York and the United States of America is much stronger than any group of barbaric terrorists," he said. "I want the people of New York to be an example to the rest of the country and the rest of the world that terrorism can't stop us."

The plan Giuliani and his team devised on the fly was brilliant. They determined that the best way to keep the city running was immediately to evacuate and cordon off the bottom three miles of Manhattan. Unless you lived below Houston Street, your living arrangements weren't affected. And the only substantial inconvenience from interrupted subway and transportation service has been for those (like me) who live in southwestern Brooklyn.

God knows it could have been otherwise. During major disruptions at other times in the city's recent history, looting had been commonplace. This time there was none. It is possible to ascribe the social peace to the wonderful élan shown by New Yorkers after this catastrophe, but it is far more likely that the NYPD's powerhouse reputation contributed to quelling any trouble before trouble could start. For instance: Arab neighborhoods in Brooklyn were flooded with police officers to prevent the outbreak of any rioting or casual violence. And, mindful of the calamity of a subway terror attack, there were several cops in every station and at every exit everywhere in the city.

The mere fact that millions of New Yorkers felt safe enough to ride the subways after the events of last week is testimony to the almost unimaginable change in the city's consciousness over the seven years and nine months of the Giuliani mayoralty. His crisis management is of a piece with his overall method of governance. He was elected in 1993 to bring order to a city on the verge of civil collapse, and he hasn't just done it through his brilliant stewardship of the police department. He's done it by making clear that what had seemed like an ungovernable city was in fact governable in all ways—if only by the sheer force of will he could bring to the job.

New Yorkers of all political stripes have rallied to Giuliani. Ed Koch, who supported Rudy and then came to hate him so much he wrote a book about his successor called *Giuliani: Nasty Man*, said: "There's no question many people who disliked him personally but admired his work have been converted." A city councilwoman named Christine Quinn, who won her office by proving to her constituents in Chelsea that she was the most gay candidate in the race, told the *Daily News*: "I feel a level of empathy and connection with the mayor I never felt before. . . . He seemed so much the human being—struggling with his own losses, but most importantly staying strong for this city. I do think this will forever . . . change the way I see this mayor."

As it happens, the primary to choose the Democratic and Republican candidates for the upcoming mayoral election had been underway for almost three hours on September 11 when the first plane crashed into the World Trade Center. The primary was postponed, and for the first time it became perfectly clear to New Yorkers that come January, Rudolph Giuliani really will no longer be the mayor.

Still, Giuliani's mayoralty has been so successful, and his stewardship of the city during the past week so heartening, that it offers hope for the future. For if indeed one man proved he could govern a city that seemed ungovernable before him, then any man can govern the city—provided he believes in the right things and does the right things at the right times.

But it's the rarest of men whose mere image on television offers the powerful consolation of calm leadership in a nightmarish time. In the first few days of this crisis, it was the mayor of the city of New York who offered that consolation.

# THE NEOCONSERVATIVE PERSUASION

## By Irving Kristol

≈

*(Originally published in the August 25, 2003, issue)*

"[President Bush is] an engaging person, but I think for some reason he's been captured by the neoconservatives around him."
Howard Dean, *U.S. News & World Report*,
August 11, 2003

WHAT EXACTLY IS NEOCONSERVATISM? Journalists, and now even presidential candidates, speak with an enviable confidence on who or what is "neoconservative," and seem to assume the meaning is fully revealed in the name. Those of us who are designated as "neocons" are amused, flattered, or dismissive, depending on the context. It is reasonable to wonder: Is there any "there" there?

Even I, frequently referred to as the "godfather" of all those neocons, have had my moments of wonderment. A few years ago I said (and, alas, wrote) that neoconservatism had had its own distinctive qualities in its early years, but by now had been absorbed into the mainstream of American conservatism. I was wrong, and the reason I was wrong is that, ever since its origin among disillusioned liberal intellectuals in the 1970s, what we call neoconservatism has been one of those intellectual undercurrents that surface only intermittently. It is not a "movement," as the conspiratorial critics would have it.

Neoconservatism is what the late historian of Jacksonian America, Marvin Meyers, called a "persuasion," one that manifests itself over time, but erratically, and one whose meaning we clearly glimpse only in retrospect.

Viewed in this way, one can say that the historical task and political purpose of neoconservatism would seem to be this: to convert the Republican party, and American conservatism in general, against their respective wills, into a new kind of conservative politics suitable to governing a modern democracy. That this new conservative politics is distinctly American is beyond doubt. There is nothing like neoconservatism in Europe, and most European conservatives are highly skeptical of its legitimacy. The fact that conservatism in the United States is so much healthier than in Europe, so much more politically effective, surely has something to do with the existence of neoconservatism. But Europeans, who think it absurd to look to the United States for lessons in political innovation, resolutely refuse to consider this possibility.

Neoconservatism is the first variant of American conservatism in the past century that is in the "American grain." It is hopeful, not lugubrious; forward-looking, not nostalgic; and its general tone is cheerful, not grim or dyspeptic. Its 20th-century heroes tend to be TR, FDR, and Ronald Reagan. Such Republican and conservative worthies as Calvin Coolidge, Herbert Hoover, Dwight Eisenhower, and Barry Goldwater are politely overlooked. Of course, those worthies are in no way overlooked by a large, probably the largest, segment of the Republican party, with the result that most Republican politicians know nothing and could not care less about neoconservatism. Nevertheless, they cannot be blind to the fact that neoconservative policies, reaching out beyond the traditional political and financial base, have helped make the very idea of political conservatism more acceptable to a majority of American voters. Nor has it passed official notice that it is the neoconservative public policies, not the traditional Republican ones, that result in popular Republican presidencies.

One of these policies, most visible and controversial, is cutting tax rates in order to stimulate steady economic growth. This policy was not invented by neocons, and it was not the particularities of tax cuts that interested them, but rather the steady focus on economic growth.

Neocons are familiar with intellectual history and aware that it is only in the last two centuries that democracy has become a respectable option among political thinkers. In earlier times, democracy meant an inherently turbulent political regime, with the "have-nots" and the "haves" engaged in a perpetual and utterly destructive class struggle. It was only the prospect of economic growth in which everyone prospered, if not equally or simultaneously, that gave modern democracies their legitimacy and durability.

The cost of this emphasis on economic growth has been an attitude toward public finance that is far less risk averse than is the case among more traditional conservatives. Neocons would prefer not to have large budget deficits, but it is in the nature of democracy—because it seems to be in the nature of human nature—that political demagogy will frequently result in economic recklessness, so that one sometimes must shoulder budgetary deficits as the cost (temporary, one hopes) of pursuing economic growth. It is a basic assumption of neoconservatism that, as a consequence of the spread of affluence among all classes, a property-owning and taxpaying population will, in time, become less vulnerable to egalitarian illusions and demagogic appeals and more sensible about the fundamentals of economic reckoning.

This leads to the issue of the role of the state. Neocons do not like the concentration of services in the welfare state and are happy to study alternative ways of delivering these services. But they are impatient with the Hayekian notion that we are on "the road to serfdom." Neocons do not feel that kind of alarm or anxiety about the growth of the state in the past century, seeing it as natural, indeed inevitable. Because they tend to be more interested in history than economics or sociology, they know that the 19th-century idea, so neatly propounded by Herbert Spencer in his *The Man Versus the State*, was a historical eccentricity. People have always preferred strong government to weak government, although they certainly have no liking for anything that smacks of overly intrusive government. Neocons feel at home in today's America to a degree that more traditional conservatives do not. Though they find much to be critical about, they tend to seek intellectual guidance in the democratic wisdom of Tocqueville, rather than in the Tory nostalgia of, say, Russell Kirk.

But it is only to a degree that neocons are comfortable in modern America. The steady decline in our democratic culture, sinking to new levels of vulgarity, does unite neocons with traditional conservatives—though not with those libertarian conservatives who are conservative in economics but unmindful of the culture. The upshot is a quite unexpected alliance between neocons, who include a fair proportion of secular intellectuals, and religious traditionalists. They are united on issues concerning the quality of education, the relations of church and state, the regulation of pornography, and the like, all of which they regard as proper candidates for the government's attention. And since the Republican party now has a substantial base among the religious, this gives neocons a certain influence and even power. Because religious conservatism is so feeble in Europe, the neoconservative potential there is correspondingly weak.

And then, of course, there is foreign policy, the area of American politics where neoconservatism has recently been the focus of media attention. This is surprising since there is no set of neoconservative beliefs concerning foreign policy, only a set of attitudes derived from historical experience. (The favorite neoconservative text on foreign affairs, thanks to professors Leo Strauss of Chicago and Donald Kagan of Yale, is Thucydides on the Peloponnesian War.) These attitudes can be summarized in the following "theses" (as a Marxist would say): First, patriotism is a natural and healthy sentiment and should be encouraged by both private and public institutions. Precisely because we are a nation of immigrants, this is a powerful American sentiment. Second, world government is a terrible idea since it can lead to world tyranny. International institutions that point to an ultimate world government should be regarded with the deepest suspicion. Third, statesmen should, above all, have the ability to distinguish friends from enemies. This is not as easy as it sounds, as the history of the Cold War revealed. The number of intelligent men who could not count the Soviet Union as an enemy, even though this was its own self-definition, was absolutely astonishing.

Finally, for a great power, the "national interest" is not a geographical term, except for fairly prosaic matters like trade and environmental regulation. A smaller nation might appropriately feel that its national interest begins and ends at its borders, so that its foreign policy is almost always in a defensive mode. A larger nation has more extensive interests. And large nations, whose identity is ideological, like the Soviet Union of yesteryear and the United States of today, inevitably have ideological interests in addition to more material concerns. Barring extraordinary events, the United States will always feel obliged to defend, if possible, a democratic nation under attack from nondemocratic forces, external or internal. That is why it was in our national interest to come to the defense of France and Britain in World War II. That is why we feel it necessary to defend Israel today, when its survival is threatened. No complicated geopolitical calculations of national interest are necessary.

Behind all this is a fact: the incredible military superiority of the United States vis-à-vis the nations of the rest of the world, in any imaginable combination. This superiority was planned by no one, and even today there are many Americans who are in denial. To a large extent, it all happened as a result of our bad luck. During the 50 years after World War II, while Europe was at peace and the Soviet Union largely relied on surrogates to do its fighting, the United States was involved in a whole series of wars: the Korean War, the Vietnam War, the Gulf War, the Kosovo conflict, the Afghan War, and the Iraq War. The result was that our military spending expanded more or less in line with our economic growth, while Europe's democracies cut back their military spending in favor of social welfare programs. The Soviet Union spent profusely but wastefully, so that its military collapsed along with its economy.

Suddenly, after two decades during which "imperial decline" and "imperial overstretch" were the academic and journalistic watchwords, the United States emerged as uniquely powerful. The "magic" of compound interest over half a century had its effect on our military budget, as did the cumulative scientific and technological research of our armed forces. With power come responsibilities, whether sought or not, whether welcome or not. And it is a fact that if you have the kind of power we now have, either you will find opportunities to use it, or the world will discover them for you.

The older, traditional elements in the Republican party have difficulty coming to terms with this new reality in foreign affairs, just as they cannot reconcile economic conservatism with social and cultural conservatism. But by one of those accidents historians ponder, our current president and his administration turn out to be quite at home in this new political environment, although it is clear they did not anticipate this role any more than their party as a whole did. As a result, neoconservatism began enjoying a second life, at a time when its obituaries were still being published.

# BUSH'S GOSPEL

## By Terry Eastland

*(Originally published in the March 1, 2004, issue)*

AMONG THE EVENTS THAT DOOMED Howard Dean's candidacy, one that has been insufficiently parsed took place on January 11 during a question-and-answer session in Oelwein, Iowa. A Bush supporter, Dale Ungerer, got up and condemned the press and the Democratic candidates for over-the-top criticisms of the president. Ungerer invoked the biblical imperative to "love thy neighbor," telling Dean, "Please tone down the garbage. . . . You should help your neighbor and not tear him down." Dean responded, "George Bush is not my neighbor."

Ungerer protested, "Yes, he is," but Dean said, "You sit down. You had your say, and now I'm going to have my say." And he did, identifying ways Bush hadn't been "a good neighbor" to his fellow Americans. Dean added, "Under the guise of supporting your neighbor, we're all expected not to criticize the president because it's unpatriotic. I think it's unpatriotic to do some of the things that this president has done to the country. It is time not to put up [with] any of this 'love thy neighbor' stuff."

Press accounts of the exchange tended to frame it as another instance of Dean's temper flaring, while commentators wondered whether the candidate's treatment of "love thy neighbor" as mere "stuff" wasn't at odds with his recent expressions of respect for religion.

Unnoticed, however, was the fact that Dean had made a frontal attack on the Bush presidency. For if you look closely at the president's speeches and remarks and consider carefully the sweep of his policies, both domestic and foreign, it becomes clear that Bush thinks of his presidency in terms of the commandment invoked in the Oelwein exchange. Indeed, central to George W. Bush's motivation as president is the ethic of "neighbor-love," as it is called in Christian circles.

We're not accustomed to a theological reading of a presidency. Yet it's evident, as Bill Keller of the *New York Times* wrote last year, that Bush's faith is "the animating force of his presidency." What hasn't been recognized is that neighbor-love in particular is what moves Bush and has helped shape his presidency. His faith teaches him to "love thy neighbor as thyself," and he approaches his job with that imperative in mind.

What this means in practice may surprise supporters and critics of the president alike. Bush's neighbor-love presidency envisions not merely a more compassionate citizenry, but a more compassionate government. It sees a larger role for religion in public life. It does not seek to establish any particular religion but is friendly to all faiths and vigilant about protecting the free exercise of religion. The trademarks of this presidency are religious pluralism and religious freedom. Overseas, the neighbor-love presidency is remarkably ambitious. It seeks to ameliorate human suffering, whatever its cause, and it is not reluctant to wage war on behalf of innocent people oppressed by the likes of Saddam Hussein. It stands for the defense and spread of freedom, because it believes that freedom is the God-given right of men and women everywhere.

The neighbor-love presidency is worth elaborating in detail, especially since we haven't seen its likes before, and because its implication for politics and policy is not a simple matter. It represents a modification, even a diminution, of American conservatism. And while its greatest triumphs have been abroad, Democrats believe it is vulnerable on the home front. The fall campaign could become an argument—like the one Dean initiated in Iowa—about what kind of neighbor Bush has been.

George W. Bush grew up in mainline Presbyterian and Episcopal churches, and as an adult became a member of a United Methodist Church in Midland, Texas. But the turning point in his life—actually a turning period, by Bush's account—occurred in the mid-1980s, when, after a conversation with Billy Graham, he renewed his faith. He began weekly Bible studies with a group of men in Midland, and, after an especially wet celebration of his 40th birthday in 1986, he completely quit drinking.

Bush has not embraced the terms "born-again" or "evangelical" to describe his faith, though he has said he wouldn't reject the appellations, either. His faith appears to be what theologically conservative Christians believe, and he expresses his beliefs in a straightforward manner. Bush attends services at the chapel at Camp David and occasionally at St. John's Episcopal Church near the White House. He has a large number of Christian friends, including several pastors, many of whom he sees from time to time, and his closest friend, also a Christian from Midland, is in his cabinet, Commerce Secretary Donald Evans. Bush reads the Bible every morning, and he has said that he reads it through every other year.

Three aspects of Bush's faith stand out. One is his belief that God is in providential control over all that happens, including in his own life. Bush, who describes himself as a "lowly sinner," has told friends and associates that but for God's intervention he would now be in some bar in Texas, not the Oval Office. A second is his belief that, whatever happens in God's providence, he is to accept and carry out each task set before him. Not incidentally, the title of Bush's campaign biography, *A Charge to Keep*, was drawn from "A Charge to Keep I Have," the Charles Wesley hymn, which speaks of doing "my Master's will" and fulfilling "my calling." After the attacks of September 11, Bush believed that the charge of defending freedom had fallen providentially to him, as commander in chief of the United States, and this remains for Bush his highest priority. Yet even this task he sees in terms of a third aspect of his faith: neighbor-love. For Bush, "love your neighbor"—the second great commandment for Christians—is an injunction to be followed in every human task, however big or small it may be. In this understanding, Bush is hardly exceptional, for loving your neighbor is the calling of every Christian.

In his inaugural address, Bush made reference to the parable of the Good Samaritan, which Jesus related in response to a question based on the second great commandment, that question being the obvious one, "Who is my neighbor?" Bush pledged the nation to a goal: "When we see that wounded traveler on the road to Jericho, we will not pass to the other side." In making that pledge, Bush assumed that people can need help for many reasons. Their wounds can be self-inflicted or inflicted by others (as was the case with the traveler helped by the Samaritan). In any case, as he said in the speech, "where there is suffering, there is duty."

In the second week of his presidency, Bush announced the Faith-Based and Community Initiative, which he touted as a new approach to helping Americans who are homeless, fatherless, addicted to drugs or alcohol, or otherwise in desperate need. He has described the initiative as "good public policy based upon the willingness of our citizens to love a neighbor just like you'd like to be loved yourself." Bush sees the initiative as one part of an effort to fight poverty, the other part being welfare reform. And he regards fighting poverty as flowing from an approach to governing that, during the campaign, he dubbed "compassionate conservatism." Bush defines compassionate conservatism this way: "It is compassionate to actively help our fellow citizens in need. It is conservative to insist on responsibility and results." Bush also sees the No Child Left Behind Act, which Congress passed in 2002, as compassionate conservatism. And there are many other policies that he has accorded that label—ones dealing with health care, the environment, home ownership, and Social Security. Nor does compassionate conservatism stop at the water's edge, for it includes government aid to poor countries.

Moreover, not every "compassionate" policy is accompanied by the word "conservative." In 2001, when he signed his first tax-cut bill into law, Bush said that "tax relief is compassionate," explaining that it helps "families struggling to enter the middle class" and "middle-class families squeezed by high energy prices and credit-card debt." Likewise, in December 2003, when he signed the new prescription-drug benefit into law, he said that the reforms in the bill "are the act of a vibrant and compassionate government." He explained: "We show our concern for the dignity of our seniors by giving them quality health care" and "our respect for seniors by giving them more choices and more control over their decision-making." Or consider the Partial

Birth Abortion Ban Act of 2003. In his signing statement, Bush said, the bill "protecting innocent new life from [partial-birth abortion] reflects the compassion and humanity of America." Just recently Bush added to his list of compassionate policies yet another—"defending the sanctity of life," which may entail support for a constitutional amendment declaring that marriage is the union of a man and a woman.

In his 2003 State of the Union address, Bush said, "The qualities of courage and compassion that we strive for in America also determine our conduct abroad. . . . Our founders dedicated this country to the cause of human dignity, the rights of every person, and the possibilities of every life. This conviction leads us into the world to help the afflicted and defend the peace, and confound the designs of evil men." Bush has sought to help the afflicted by stepping up U.S. involvement in the international battles against famine and AIDS. To fight "desperate hunger," as Bush has put it, the United States is now providing more than $1.4 billion a year in global emergency food aid. To fight AIDS, Bush has begun carrying out a relief program designed to prevent the disease from breaking out on a massive scale and to treat millions who already have what he calls "a plague of nature." Bush surprised almost everyone by the magnitude of the $15 billion request he submitted to Congress last year for this program. Asked to explain his decision to insert the United States so deeply into what he has called "a work of mercy," Bush told the *Ladies' Home Journal*, "The Bible talks about love and compassion. That's really a lot behind my passion on AIDS policy."

As for confounding the designs of evil men, Bush has moved Washington into the peace talks in Sudan, where the National Islamic Front government has waged a brutal civil war against a largely Christian and animist population that has claimed the lives of more than two million people. And he has ramped up the government's efforts to curtail human and sex trafficking, which, in his speech last fall to the United Nations General Assembly, he condemned as "a form of slavery." Bush has pledged $50 million to support "the good work of organizations that are rescuing women and children from exploitation and giving them shelter and medical treatment and the hope of a new life."

Of course, the universe of evil men includes terrorists, who have designs upon innocent people beyond the more than 3,000 killed by the attacks of September 11. They have continued to murder innocent people, a point Bush made last year in his speech at Whitehall when he cited the post-9/11 terrorist attacks in Bali, Jakarta, Casablanca, Bombay, Mombassa, Najaf, Jerusalem, Riyadh, Baghdad, and Istanbul. And, by every reckoning, the terrorists intend to kill more innocents. As Bush sees it, both justice, because what the terrorists do is evil, and compassion, because their evil is committed against innocent people, demand a military response.

The universe of evil men also includes oppressive rulers. Discussing the war in Afghanistan, Bush told a Connecticut audience that the United States liberated an innocent people oppressed by a barbaric regime. "We're compassionate," he said. "We care deeply about our fellow citizens in this world." While Bush justified the war in Iraq mainly on grounds involving weapons of mass destruction, he also thought he was saving the Iraqi people from an evil man. Over the decades, Saddam Hussein had killed and maimed millions of Iraqis. During a press conference in December, Bush said, "I believe, firmly believe—and you've heard me say this a lot, and I say it a lot because I truly believe it—that freedom is the almighty God's gift to every person—every man and woman who lives in this world. That's what I believe." He added that "the arrest of Saddam Hussein changed the equation in Iraq. Justice was being delivered to a man who denied that gift from the Almighty to the people of Iraq." Justice was being delivered to Saddam, and, to place Bush's remarks in larger context, compassion was being shown to the Iraqi people. Nor does compassion stop with liberation. For Bush, it includes efforts to establish the kind of institutions in which "the rights of every person" can be protected. It envisions the spread of democracy.

Asked last summer by *Christianity Today* to describe Bush's foreign policy, Don Evans said, "It's love your neighbor like yourself. The neighbor happens to be everyone on the planet."

In a speech last summer to leaders of faith-based organizations, Bush fairly summed up both halves of his neighbor-love presidency: "The mission at home is to help those who hurt, and make the vast potential of America available to every citizen. The mission abroad is to use our good heart and good conscience and not turn our back away when we see suffering."

This, then, is Bush's love-thy-neighbor presidency, and there are aspects of it that deserve closer scrutiny. Consider, to begin with, that in many instances government employees are the ones showing compassion—prosecutors who bring sex-trafficking cases, say, or soldiers who fight terrorists. In still other contexts, compassion lies in a remodeled government—in lowered tax rates, for example. With the faith-based and community initiative, however, the point is to rally private "armies of compassion." As Bush said in his inaugural address, the work of compassion is that of a people, not just a government.

Historically, the work of compassion in America has been mainly that of a religious people acting through private groups. But with the rise of big government and the welfare state, religious charities have played a less prominent role. Bush's initiative contemplates a fuller employment of those organizations. Funding is crucial, and Bush has moved to stimulate charitable giving in the hope that more contributions will flow to religious charities, so as to enlarge their stores of compassion. He also wants to ensure that religious charities can compete for government grants on an equal basis with secular ones.

In 2001 the White House issued a report concluding that federal grant-making procedures often discriminated against religious charities on account of their religion. The administration thus has embraced the principle that where government assistance is generally available, religious groups eligible for it can't be discriminated against on account of religion. Because Congress has refused to legislate this principle, Bush has resorted to executive orders and regulations to establish it throughout the government. He has also adopted rules designed to protect the rights of religious charities to hire the individuals who in their judgment are best able to further their goals.

Bush often talks about "the power of faith to change a life," and he believes that faith-based groups can make a "unique contribution" in ameliorating stubborn social problems. Indeed, he has characterized himself as a "one-man faith-based initiative," the point being that his faith, and perhaps also the community of faith in Midland, helped him pull his life together. But the faith-based initiative isn't designed for Christian charities only and, indeed, given the principles that inform the initiative, it couldn't be. Bush himself has been quite clear on this point. "It doesn't matter what the religion is," he said on one occasion. And, on another, "I don't talk a particular faith." Religious charities of any and all faiths may apply for grants. Nor is there a preference for religious charities over nonreligious ones. "Our plan," he said early in 2001, "will not favor religious institutions over nonreligious institutions." In sum, as he said in a speech to leaders of charities last summer: "All groups should be able to compete on a level playing field, whether faith-based or secular."

Precisely because there is now a level playing field—because the field no longer tilts against religious charities—the likelihood is that they will receive more grants than they have in recent decades. Whether we will go all the way back to a future in which the work of compassion is once again mainly the work of religious charities, though now receiving more government funding, is unclear. Yet the future, as Bush would have it, will unfold guided by principles of pluralism. As he said in a 1999 speech on compassionate conservatism, "We will keep a commitment to pluralism—not discriminating for or against Methodists or Mormons or Muslims, or good people of no faith at all."

There happens to be a compelling theological argument for Bush's "principled pluralism," as the president's chief speechwriter, Michael Gerson, has called it. But the deeper point about Bush's faith-based and community initiative lies in its assumption that neighbor-love is a precept intelligible to all individuals, regardless of what they believe or don't believe. Bush routinely speaks of "the universal call" to love one's neighbor, meaning that it is found in all faiths as well as in secular teachings. (He has not elaborated the point, but he would find the resources to do so in C. S. Lewis's *The Abolition of Man*, which, under "the law of general beneficence," cites a variety of both religious and nonreligious sources.) Bush believes not only that persons of all faiths—or no faith—can respond to that universal call, but also that they can act upon it and do good works. More than a few religious conservatives would doubt whether "unregenerate" man can do much good. On the other hand, a defense of Bush might be that what motivates him as a theological virtue he commends generally as the civic virtue of neighborliness. In any case, Bush's bottom line is practical. Faith without works is dead. "The measure of true compassion," Bush often says, "is results."

The principles that the Bush administration is advancing on behalf of the faith-based and community initiative are being applied in many other contexts. Consider that the Federal Emergency Management Agency has changed the way it awards direct aid to properties damaged by natural disaster. Now it will provide aid under "genuinely neutral criteria" so that religious institutions are not discriminated against. The National Park Service has made a similar policy change; no longer does it refuse to award historic preservation grants to churches or other religious institutions simply because they are religious. And in *Zelman* v. *Harris*, decided in 2002, the Justice Department won from the Supreme Court a vindication of the nondiscrimination principle in a school voucher case. *Zelman* means that voucher programs encompassing church-related schools are constitutional so long as they are "neutral"—i.e., offering a "genuine choice among options, public and private, secular and religious."

Meanwhile, the Justice Department has sought to protect religious liberty by more aggressive enforcement of statutes forbidding religion-based discrimination. In an increasingly pluralistic nation, the beneficiaries of the department's actions have included Christians, Muslims, Jews, and Buddhists.

Bush is also interested in advancing religious liberty overseas. Indeed, it is in the context of foreign policy that the president has spoken most often about religious liberty, no doubt because it is abroad that so many denials of it, many even unto death, routinely occur. In his remarks on this subject, Bush has called religious liberty "the first freedom of the human soul." And when you open up his National Security Strategy of the United States, you find that it calls for "special efforts to promote freedom of religion and conscience and defend it from encroachment by repressive governments." For no other freedom does the strategy call for such "special efforts"—efforts, it is fair to say, that Bush sees as benefiting wounded travelers on many roads.

Of course, Bush will not find in the Scriptures a doctrine of religious liberty, for the Bible advises instead as to what man owes God. Nonetheless, a doctrine of liberty can be developed from the need to ensure that man can in fact exercise "the duty which we owe to our Creator," as the Virginia Declaration of Rights famously put it. Bush stands in a long and distinguished tradition stretching back to that document (and before) when he says, as he did in that December press conference after Saddam's capture, that "freedom is the almighty God's gift to every person." (Compare Thomas Jefferson, who said man's "liberties are the gift of God," and John F. Kennedy, who said "the rights of man come . . . from the hand of God.") For Bush, the freedom God gives includes religious liberty, probably first, but also freedom of speech, assembly, and the rest. It is on behalf of human rights abroad that Bush's love-thy-neighbor presidency shows ardent, if not always consistent, zeal.

And, as the case of Iraq demonstrates, it also wields a sword. Bush believes the war satisfied "just war" principles. The most important just-war theorist was Augustine, who argued that love does not foreclose "a war of mercy," indeed that it is in the nature of love to protect an innocent third party from oppression, by force if necessary. Liberating the people of Iraq from Saddam wasn't Bush's main public argument for the war. But it may have been a powerful motivation. Bush conceives of the United States as "a power," as he put it in his inaugural address, "that [goes] into the world to protect, but not possess, to defend, but not to conquer." In his 2003 State of the Union, he observed that "we sacrifice for the liberty of strangers." The ultimate sacrifice—the Scripture says there is no greater love than this—is to lay down one's life for another.

To say that neighbor-love motivates Bush is not to say that it justifies particular policies or actions he's described as compassionate. Neighbor-love is a principle of high generality. Put a bunch of people around a table, give them the principle, ask them to devise a policy to address Problem X, and you may get as many proposals as you have people. Most of Bush's "compassionate" policies have drawn disagreement, often sharp. Consider, for example, the debate over tax cuts or the No Child Left Behind legislation. Moreover, the faith-based initiative itself has divided religious conservatives in his own party: Fearing the church would be entangled with the state, not a few have objected to the use of direct grants and contended that vouchers and tax credits should be emphasized instead. Or consider the war in Iraq, to which there is outright objection in some evangelical circles. The work of justifying a particular policy is the business of politics, not faith, though faith can suggest, as it has for Bush, areas where policy might be needed.

The wonder is how many policies Bush presents as compassionate. In some cases—home ownership, for example—you wonder why the label is even there, unless it's to impress on voters that the Democrats aren't the only party of concern and care. In any case, Bush's emphasis on compassion seems to have come at some cost to conservatism. Bush hasn't vetoed a single spending bill (or any bill, for that matter), and his compassionate prescription-drug benefit is going to cost not what he first said it would, $400 million over the next 10 years but, according to the latest administration estimate, at least $500 million. Overall government spending has grown substantially, bulking up the deficit.

Conservative arguments about the size and limits of government, including those based on constitutional authority, are seldom heard from this administration. Indeed, the rhetoric sometimes cuts the other way. Bush has gone so far as to proclaim that "there is no question that we can rid this nation of hopelessness and despair." Granted, he says we don't need big government programs to do that—he thinks we the people can achieve it on our own, or perhaps with assistance from faith-based charities. But the goal he has articulated is one Democrats can enthusiastically agree with, even as they outbid him with more federal money and programs.

It is overseas where the Bush presidency is most ambitious and would appear in need of a limiting principle. For if, as Donald Evans says, our neighbor is "everyone on the planet," then the work of compassion is a daunting, even bankrupting, task, more than our military as currently supported can take on, and perhaps more than the American people are willing to support. Even so, it is the work of the Bush presidency abroad that truly distinguishes it. The Framers of the Constitution conceived of the executive as an office suitable for "arduous" and "extensive" undertakings of great public benefit, and there can be no question that the effort not just to defend the nation against further terrorist attack but also to liberate countries from oppressive regimes and plant seeds of democracy in places where terrorists take sanctuary and breed is arduous and extensive. And the efforts to counter AIDS and curtail human and sex trafficking aren't exactly minor.

Suffice it to say, the Democrats aren't intimidated by Bush's neighbor-love presidency, at least not by the domestic side of it. With his "Two Americas" speech—"one favored, the other forgotten"—John Edwards has made compassion his central theme, and now that Edwards is nipping at John Kerry's heels, the Massachusetts senator is copying the North Carolinian. After narrowly winning the primary in Wisconsin, Kerry talked about how "you could just feel the pain" Americans were experiencing. He said "the heart had been ripped out of the heartland," and that, after repealing the tax-rate cuts for those making more than $200,000, he would invest in education, health care, and other programs. The fall campaign is likely to be fought over the very issues on which Bush has taken "compassionate" positions— taxes, poverty, health care, education, and the environment. Indeed, the Democrats may take the next step and declare, as Howard Dean did, that Bush has not been "my neighbor" because his policies have been inadequate or injurious. A party struggling for a way to peel off more moderate, religiously observant voters, who supported Bush in 2000, very well might take that step.

Whether the Democrats can win on economic and domestic issues alone is another matter. As Bush insists, the measure of compassion is results, and even if the Democrats can persuade voters on health care or education or jobs, the results of Bush's national security efforts have benefited the American people. They are concrete and, indeed, of world-historical importance, for they include the liberation of Afghanistan and Iraq and the spread of human rights to places where no one would have thought that possible.

In any case, the big story this year will be either the rejection or the reelection of a Republican president motivated by an ancient yet enduring ethic.

# FEAR AND INTIMIDATION
# AT HARVARD

## By Harvey Mansfield

———

*(Originally published in the March 7, 2005, issue)*

*Cambridge, Massachusetts*

At last week's Harvard faculty meeting, President Larry Summers saved his job, but he took a pummeling from his angry critics. Summers is easily the most outstanding of the major university presidents now on the scene—the most intelligent, the most energetic, as well as the most prominent. So, alarmed at his abilities and intentions, the Harvard faculty decided it would be a good idea to humiliate him.

Summers has supporters, and not all the faculty joined in the game of making him look sick. But the supporters, like Summers himself, were on the defensive, making concessions, and the critics were not. The critics consist of feminist women and their male consorts on the left. But since the left these days looks opportunistically for any promising cause, it is the feminists who are the core opposed to Summers. Together the feminists and the left make up perhaps half the faculty, the other half being moderate liberals who are afraid of the feminists rather than with them.

Summers saved his job by skating backward, listening to his critics without demur and occasionally accepting their harsh words by saying he agreed with them. At no point did he feel able to say yes, *but* . . . in order to introduce a point of his own in response. His accusers were relentless and, as always with feminists, humorless. They complained of being humiliated, but they took no care not to humiliate a proud man. They complained too of being intimidated, but they were doing their best to intimidate Summers—and they succeeded.

At the meeting many said that the issue was not academic freedom vs. political correctness, as portrayed by the media, but Summers's style of governing. The point has a bit of truth. Summers is an economist, and there is almost no such thing as a suave economist. The great Joseph Schumpeter, a Harvard economist of long ago, claimed to be the world's greatest lover as well as the world's greatest economist (it is said), but he was a singular marvel. The reason why economists are blunt is that words of honey seem to them mere diversion from reason and self-interest, which are the only sure guides in life.

More than most people—to say nothing of university presidents—Summers lives by straightforward argument. He doesn't care whether he convinces you or you convince him. He isn't looking for victory in argument. But his forceful intelligence often produces it, in the view of those with whom he reasons. Sometimes the professors he speaks with come out feeling that they are victims of "bullying," as one of his feminist critics stated. As if to reason were to bully.

One faculty colleague said in response to this, "Can anybody on earth have less reason to fear than a tenured Harvard professor?" True enough, a Harvard professor has both the prominence to awe and, if that doesn't work, the security to escape. But feminists do not think like this. They insist on a welcoming atmosphere of encouragement to themselves and to their plans. If they do not get it, they will with a straight face accuse you of intimidating them even as they are intimidating you.

It takes one's breath away to watch feminist women at work. At the same time that they denounce traditional stereotypes they conform to them. If at the back of your sexist mind you think that women are emotional, you listen agape as professor Nancy Hopkins of MIT comes out with the threat that she will be sick if she has to hear too much of what she doesn't agree with. If you think women are suggestible, you hear it said that the mere suggestion of an innate inequality in women will keep them from stirring themselves to excel. While denouncing the feminine mystique, feminists behave as if they were devoted to it. They are women who assert their independence but still depend on men to keep women secure and comfortable while admiring their independence. Even in the gender-neutral society, men are expected by feminists to open doors for women. If men do not, they are intimidating women.

Thus the issue of Summers's supposedly intimidating style of governance is really the issue of the political correctness by which Summers has been intimidated. Political correctness is the leading form of intimidation in all of American education today, and this incident at Harvard is a pure case of it. The phrase has been around since the 1980s, and the media have become bored with it. But the fact of political correctness is before us in the refusal of feminist women professors even to consider the possibility that women might be at any natural disadvantage in mathematics as compared with men. No, more than that: They refuse to allow that possibility to be entertained even in a private meeting. And still more: They are not ashamed to be seen as suppressing any inquiry into such a possibility. For the demand that Summers be more "responsible" in what he says applies to any inquiry that he or anyone else might cite.

Of course, if you make a study of differences between the sexes with a view to the possibility that some of them might be innate, no violence will come to you. You will not be lynched. But you will be disliked, and you will have a hard time getting appointed at a major (or a minor) university. Feminists do not like to argue, and they consider you a case if you do not immediately agree with them. "Raising consciousness" is their way of getting you to fall in with their plans, and "tsk, tsk" is the only signal you should need and will get. Anyone who requires evidence and argument is already an enemy because he is considering a possibility hurtful to women.

Feminist women rest their cause on "social construction" as opposed to nature. The patriarchal society that has been made by humans can be unmade and remade by humans. But how do we know that the reconstruction will be favorable to women and not a new version of patriarchy? To avoid a resurgent patriarchy or other injustice, society, it would seem, needs to be guided by a principle beyond human making, the natural equality of men and women.

Accepting that principle would require, however, thinking about how far it goes and what natural inequalities in the sexes might exist. This might in fact be a benefit if it induced women to think more about what they want and like, and about what is fair to men and good for children. We do need feminism, because women are now in a new situation. But we need a new feminism conceived by women more favorable to liberty and the common good than the "feminists" of today.

# THE GREAT WHITE WASTE OF TIME

## By Matt Labash

◈

*(Originally published, as "Welcome to Canada,"*
*in the March 21, 2005, issue)*

If the national mental illness of the United States is mega-
lomania, that of Canada is paranoid schizophrenia.
—Margaret Atwood, Canadian writer

*Vancouver, British Columbia*

Whenever I think of Canada . . . strike that. I'm an American,
therefore I tend not to think of Canada. On the rare occasion when I
have considered the country that Fleet Streeters call "The Great White
Waste of Time," I've regarded it, as most Americans do, as North
America's attic, a mildewy recess that adds little value to the house,
but serves as an excellent dead space for stashing Nazi war criminals,
drawing-room socialists, and hockey goons.

Henry David Thoreau nicely summed up Americans' indifference
toward our country's little buddy when he wrote, "I fear that I have
not got much to say about Canada. . . . What I got by going to Canada
was a cold." For the most part, Canadians occupy little disk space on
our collective hard drive. Not for nothing did MTV have a game show
that made contestants identify washed-up celebrities under the category "Dead or Canadian?"

If we have bothered forming opinions at all about Canadians, they've tended toward easy-pickings: that they are a docile, Zamboni-driving people who subsist on seal casserole and Molson. Their hobbies include wearing flannel, obsessing over American hegemony, exporting deadly Mad Cow disease and even deadlier Gordon Lightfoot and Nickelback albums. You can tell a lot about a nation's mediocrity index by learning that they invented synchronized swimming. Even more, by the fact that they're proud of it.

But ever since George W. Bush's reelection, news accounts have been rolling in that disillusioned Americans are running for the border in protest. This prompts the thought that it may be time to stop treating Our Canadian Problem with such cavalier disregard. In fact, largely as a result of Bush and his foreign policy, what was once a polite rivalry has become a poisoned well of hurt feelings and recriminations.

These days, Canadian publications are chockablock with surveys showing that Canadians see themselves as something akin to a superior race. The prime ministers of what was once a reliable ally that ponied up in times of war have treated us like traffic-light squeegee-men when we've stopped at their corner, asking for assistance with our latest military adventure. They have spurned our missile-defense shield out of spite, even knowing it would save their Canadian bacon. Their legislators have publicly called us "bastards" and stomped on our president in effigy. Their citizens have booed our children at peewee hockey games.

Being bloodthirsty Americans, we have naturally fired a few warning volleys in lieu of slapping them with a restraining order. A few years ago, my friend Jonah Goldberg from *National Review* wrote a piece elegantly titled "Bomb Canada," encouraging us to smack Soviet Canuckistan, as Pat Buchanan calls it, "out of its shame-spiral" since "that's what big brothers do." Canadians responded as Canadians always will when faced with overt aggression. They wrote inordinate numbers of letters of concern, exercising what Canadian writer Douglas Coupland calls their "almost universal editorial-page need to make disapproving clucks."

Equal outrage was caused when Conan O'Brien showed up to help boost tourism after the SARS crisis. Along for the ride came a Conan staple, Triumph the Insult Comic Dog, who in dog-on-the-street interviews relentlessly mocked French Canadians. When one pudgy Quebecer admitted he was a separatist, Triumph suggested he might want to "separate himself from doughnuts for a while."

Canadians seethed—though polls show they pride themselves on being much funnier than Americans (don't ask me why, when they're responsible for Dan Aykroyd, John Candy, and Alan Thicke). One MP from the socialist New Democratic party called the show "vile and vicious," and said it was tantamount to hatemongering. Historians believe this to be the first time a member of parliament has so categorically denounced a hand puppet.

With the reelection of Bush, however, this poor man's Cold War may be swinging Canada's way. Trend-spotters on both sides of the 49th parallel have taken note of "the Bush refugee," the American progressive who has decided to flee to Canada after growing heartsick at the soul-crushing death knell of liberalism that pundits declared after the president's two-point victory.

A cottage industry was born. Anti-American/pro-Canadian blogs proliferated, as blogs unfortunately do. Websites like canadianalternative.com are open for business, trying to entice emotionally vulnerable Americans to turn their backs on family, friends, and country with boasts that Canada has signed the Kyoto protocol, legalized gay marriage in six provinces, and seen its Senate recommend legalizing marijuana. Vancouver immigration lawyer Rudi Kischer took a whole team, complete with realtors and money-managers, to recruit in American cities, helping potential defectors overcome immigration concerns, such as how to pass Canada's elitist skilled-worker test for entry (*Give us your affluent, your overeducated, your Unitarian masses yearning for socialized medicine*).

Dejected Americans, most of whom already live in progressive enclaves, began sounding off to reporters, vowing to check out of the Red-American wasteland before true misfortune befell them. In footage of a Kischer seminar in San Francisco that I obtained from a Canadian documentary film crew (working title of the piece: "Escaping America"), one attendee who looked like a lost Gabor sister but with more plastic surgery said, "I really can't stand George Bush. I can't stand this culture, which is very selfish, aggressive, and mean, violent I think." After going to Canada for just a half an hour from Buffalo, she concluded, "It was like a completely different country. . . . The people seemed more internationally aware, not so isolated and unilateral. There was less evidence of commercialism and corporations. People were friendly."

It sounded like such an idyllic Rainbowland that I had to see it for myself. So I flew to Vancouver in late January to get a closer look and to meet up with several already-arrived and soon-to-be American expatriates. Taking a day or so to get acclimated, I threw myself into this unspoiled Eden by going to the multinational Virgin megastore to purchase some Joni Mitchell and Leonard Cohen CDs (buy Canadian!). I also looked for Canada's greatest (only?) contribution to world cuisine, Tim Hortons donuts, which is owned by the American fast-food behemoth Wendy's.

With nothing but a Lonely Planet map and a thirst for knowledge, I sought out Vancouver's landmarks—the Gastown clock, which lets out steam and whistle-toots every 15 minutes, Brandi's Exotic Nightclub, where Ben Affleck fraternizes with strippers when in town, and the Amsterdam Café, where potheads openly smoke the potent BC bud, taunting tobacco users, who are confined to a glass cage like common criminals. (The tokers snack on "jones soda" and "chronic candy" and other foodstuffs so cutely named it could make even the most maniacal libertarian cheer for mandatory minimums.)

To see Canadian progressivism in action, though, I trekked down to the East Side, Vancouver's Compton, where the storefront Supervised Injection Site caters to junkies on the government teat. With the surrounding streets hosting an open-air drug market, the Site was conceived as a way to rid the neighborhood of discarded drug paraphernalia and promote "safe" drug-taking practices. In typical Canadian fashion, it's a long way around the barn to get rid of litter.

If the Site has in fact encouraged addicts to do their drugs off the streets, they still buy them right outside. To reach the place, I have to pass through a herd of about 100 junkies over a four-block radius. They offer to sell me all manner of substances my company won't let me expense. When I make it inside the Site, along with several itchy, twitchy customers in search of free cookers and needles and a clean booth to shoot themselves silly, an attendant tells me that unless I'm there to take drugs, I can't stay without a media relations escort. "What we do here is important, so we try to keep a low profile," he says, perhaps oblivious to the hypodermic needle that's embossed on the door.

The staffers aren't rude, however, and retrieve for me a helpful government brochure called "The Safer Fix" that has made me something of an expert on the proper way to tie off. Though it's a bit mind-blowing to a law-and-order American, this is actually pretty small beer, compared with a new Canadian government-funded study called the North American Opiate Medication Initiative. While the Supervised Injection Site is strictly a bring-your-own-smack affair, the new experiment will study the effects of giving half of the drug-addicted research subjects heroin, while the other half get methadone. As a female attendant describes it to me, we agree that it must really suck for the methadoners. But for the other side? "Dude!" she says, stating the obvious, "free drugs for a year!"

Rudi Kischer, the immigration lawyer who went trolling for clients south of the border, has probably done more than any single person besides George Bush to induce Americans to become former Americans. At the top of a high-rise building overlooking Coal Harbor, where seaplanes land in steady succession, Kischer invites me into his office. He is tall, with the bland good looks of a soap-opera extra. By way of an icebreaker, I tell him I flunked the skilled-worker test, and so became a journalist. He says not to worry. Up until a few years ago, lawyers were completely banned from immigrating, the first fact I've heard that recommends his country.

While numbers are hard to come by, it is generally thought that some thousands of Americans are poised to change countries, making them the largest influx Canada has seen since our draft dodgers came this way during Vietnam—much less since Brit-loving Loyalists were shown the door to what was then New France by American revolutionaries. Whether or not this is true, Kischer has plenty of horror stories from interested clients: concerned parents who are moving so their children won't be drafted into Bush's war machine, the rich guy who lives on a yacht and would rather pay exorbitant Canadian taxes than bear the shame of flashing his imperialist American passport when sailing into foreign ports.

I tell Kischer it's a bit much to swallow that so many Americans are being persecuted for disagreeing with the president, since we live in what most regard as a fly-your-freak-flag country. Take me. I wasn't keen on the war in Iraq, and I work in the belly of the neocon beast that gets partial credit for hatching it, yet I've never felt a lick of persecution for offering dissent. Kischer studied briefly at Duke (former basketball great Danny Ferry was in his poli-sci class, he says excitedly), so when I ask him if he ever felt oppressed in America, he laughs as if I've asked a ridiculous question. Of course not, he says, "but it depends on personality types, too. I'm a lawyer, so I've had worse things said to me by better people, right?"

When in America, he blended seamlessly, he says, with everyone else who shops at the same khaki-shorts store. People didn't really suspect he was Canadian, since Canada's not on the radar. "I read one article about Canada in four months," he adds. "It said the socialists are about to take over the government. From the American viewpoint, maybe they already have." Kischer voices a typical concern. Canadians are traditionally so insecure about the lack of attention we pay them that their government has even paid American universities $300,000 to study them. One of the foremost Canadian Studies programs in the country is at Duke. A professor in the program has said, "We're the most important university to make a serious effort to study Canada. That's like being the best hockey team in Zimbabwe."

My first interview with an American comes not in Canada, but in Bellingham, Washington, about 90 minutes from Vancouver. I drive south and clear the Peace Arch border faster than I could a McDonald's drive-thru line (note to Homeland Security), and meet up with Christopher Key in his middle-class rambler with a for-sale sign in the yard. Key is still a patriot, but he hopes to soon be an expatriate. He's descended from "Star Spangled Banner" writer Francis Scott Key, who he admits "wasn't much of a poet."

He has become a minor celebrity of sorts, profiled by everyone from the Canadian Broadcasting Corporation to the *New York Times* (whose reporter flies in the day after me). The silver-haired Key looks like a Chamber of Commerce burgher. He likes to point out he's not some stereotypical longhair, having just left his editor's gig at a failing business magazine. He's had several other career incarnations too: everything from art gallery owner to charterboat skipper.

But Key's weirdest job was in the military, when he served in Vietnam. "They called it 'press liaison,' I think, but I was a news censor," he says. As a wet-behind-the-ears 19-year-old, he was supposed to tell media bigshots like Ed Bradley what they could and could not cover. They all ignored him. "My take," he says, "is that while I had an odious job, I managed to do it very poorly."

Key caught shrapnel on one mission, and later was ambushed in the central highlands. While running to his truck, he felt a stitch in his side. The wound took out a good bit of his right kidney, and served as his ticket home.

Though Key wasn't bullish on the war when he was drafted, he never thought of fleeing to Canada to beat it. He comes from a long military line, and running isn't what his family does. But since Bush was elected in 2000, he says he's watched the country go into a tailspin, becoming less tolerant, more mean-spirited, more judgmental. In the past, Key waited out Nixon and Reagan. "I voted for Dukakis," he says. "I'm used to losing." But the war in Iraq pushed him closer to the edge, and at about 3 a.m. the day after the election, he made his decision to eject. "All the voices of moderation—Colin Powell—were going to be replaced by yes-people like Condoleezza Rice. It's going to get worse."

He says that when satellite trucks first started showing up in his driveway, the neighbors were atwitter. He loves his neighbors, a healthy mix of Republicans and Democrats. They regularly get together for barbecues, and come see him perform in community theater. As a Universal Life Church minister—he secured his ordination certificate off the Internet for 25 bucks—he's performed their weddings and funerals. But they couldn't talk him into staying, even though his adult daughter lives next door with her family, and her former twin sister, now her twin brother, lives in Seattle. How could he stay in a place that would frown on his performing the wedding of his own daughter/son?

"Come again?" I ask.

"This gets confusing," he apologizes. His second daughter, Bonnie, it seems, "who became my son, was a lesbian before he went transgender," making him heterosexual. The twins are actually scheduled to go on *Oprah* to discuss this. I think I understand, but ask for a flow chart to make sure. "Listen, it won't help. It looks like an explosion in a spaghetti factory," he says. "I can't keep up—how the hell can you?"

While Key puts a premium on Canadian tolerance, he's spent long enough in the country to understand it's not Canaan. A part-time blogger, he's even written pieces with titles like "The Canadian Identity Crisis," in which he tweaks his future compatriots for being America-fixated ninnies, and for coasting on their reputation for politeness. While Canadians don't exhibit road rage, he says, they are carpool-lane cheaters and worse: "Victoria dumps its untreated sewage into the waters off Vancouver Island. How impolite can you get?"

Still, Key is leaving his homeland, and he's sick of hearing from talk-show types who say good riddance on the one hand and he should stay and fight on the other. "Shouldn't you?" I ask, picking up the latter sentiment. After all, he gets along beautifully even with his Republican neighbors, and nobody except a few journalists has questioned his patriotism. So how bad, really, is the alleged cauldron of intolerance known as America? Isn't he boxing with Sean Hannity's shadow, responding not to the America he actually knows, but to the polarized version of it that lives in his cable box?

Besides, I suggest in a windy disquisition (I've had wine with lunch) after hearing at length how he once marched for civil rights and against Vietnam, even if this ugly America is as pervasive as he says, isn't it our duty as Americans to get in on the debate, to jump into the sandbox and hit somebody on the head with a shovel while no one's looking? It's what made our country great. Our forefathers may have quit their home countries once upon a time, but they came here to build a better one.

He isn't buying. "I'm f—ing tired," he says, "and I don't need to rebuild the country. There's a perfectly good one 30 miles away."

Just how perfectly good a country Canada is, is a matter of dispute. The expats I eventually meet buy into Canadian self-mythologizing without so much as giving the tires a kick. Yet even some Canadians gag on the constant stream of virtue-proclaiming advertorials that are, for lack of a better word, a crock. This is self-evident in the pathological Canadian claims of modesty and politeness.

Will Ferguson is a cockeyed nationalist and brilliant satirist who calls his country "a nation of associate professors." In his book *Why I Hate Canadians*, he writes that his countrymen even boast about their Great Canadian Inferiority Complex. While it's difficult to go five minutes without hearing how collectively nice Canadians are, Ferguson says, "what we fail to realize is that self-conscious niceness is not niceness at all; it is a form of smugness. Is there anything more insufferable than someone saying, 'Gosh, I sure am a sweet person, don'tcha think?'"

This strain of nails-on-the-blackboard nationalism is most evident in the recent bestseller *Fire and Ice*, an Americans-are-from-Mars, Canadians-are-from-Venus study of the two countries' values by Canadian sociologist Michael Adams. Based on three head-to-head values surveys done over a decade, it shows Americans coming up short on matters from militarism to materialism. This is hardly news. But Adams pushes his luck, giving conventional wisdom a twirl by advancing that it is the Americans who are actually the slavish followers of an established order, while Canadians are rugged individualists and autonomous free thinkers.

Give Adams points for cheek. His is, after all, a country that didn't bother to draft its own constitution until 1982, that kept "God Save the Queen" as its national anthem until 1980, and that still enshrines its former master's monarch as its head of state. Her Canadian title is "Elizabeth the Second, by the Grace of God, of the United Kingdom, Canada and Her other Realms and Territories Queen [breath], Head of the Commonwealth, Defender of the Faith." Maybe they should change their national anthem again, to Britney Spears's "I'm A Slave 4 U."

After suffering through Adams's book, I decided two can practice snake-oil sociology. So I spent three days on Nexis kicking up every comparison-survey and statistic I could find on American/Canadian values. I became so gripped with the subject I could have been mistaken for a Canadian.

This unscientific research quickly confirmed that Canadians are bizarrely obsessed with us, binge-eating out of our cultural trough, then pretending it tastes bad. Plainly the two things Canada needs most are a mirror and a good psychiatrist.

Though they don't know who they are, they know they're not us (roughly 9 out of 10 comparison surveys are done by Canadians), so they bang that drum until their hands bleed. Still, it seems there is almost nothing Canadian that isn't informed in some way by America. When the late Canadian radio host Peter Gzowski had a competition to come up with a phrase comparable to "American as apple pie," the winner was "As Canadian as possible, under the circumstances." In 1996, when Canadians were asked to name both the greatest living and the all-time greatest Canadian, 76 percent said "no one comes to mind." Another survey showed them to believe that the most famous Canadian was Pamela Anderson, star of America's *Baywatch*. When Canadians were asked to name their favorite song, they settled on one by a good Canadian band, The Guess Who. The song: "American Woman."

Several years ago, Molson beer aired a commercial featuring Joe Canadian, a regular beerdrinking Joe who went on a rant aboot what Canadians are and aren't (not fur traders or dog sledders; they pronounce it "about," not "aboot"). He became a media darling and a national mascot. Then the actor who played Joe moved to Hollywood to find work. When he returned, tail tucked between legs, even he admitted, "I think, yeah, it is a little sad that Canadians draw their identity not so much from 'I am Canadian' as 'I am not American.'"

While Canadians pride themselves on knowing more about us than we do about them (undoubtedly true), the problem—captured in a survey done for Canada Day in 2000—is that even historically challenged Americans know more about ourselves than Canadians do about themselves. In parallel 10-question quizzes on everything from our first president/prime minister to the words of our respective national anthems, 63 percent of Americans scored five or more right answers. Only 39 percent of Canadians did. One Canadian television critic expressed disbelief, writing, "Average Americans appear to be in worse shape—judging by the evidence on TV, anyway." She would know, since at the time of her comment, 92 percent of the comedies and 85 percent of the dramas on Canadian television were made elsewhere, mainly in America.

Where Canada fails is no big secret. Most of us know that its universal health care is a great thing, if you don't mind waiting, say, nine months for an MRI on your spinal cord injury. We all know Canadians are overregulated, to the point that Canadian rocker Bryan Adams was denied "Canadian content status" for cowriting an album with a British producer, limiting the play his songs could receive on the radio (a policy that's supposed to encourage Canadian talent, but that in Adams's words "encourage[s] mediocrity. People don't have to compete in the real world. . . . F—ing absurd").

We all know the Canadian military has become a shadow of itself. Things have gotten so dire that a Queen's University study (titled "Canada Without Armed Forces?") predicted the imminent extinction of the air force. This unpreparedness has become such a joke that Ferguson says their military ranks just above Tonga's, which consists of nothing more than "a tape-recorded message yelling 'I surrender!' in thirty-two languages."

What many don't consider is how much Canada has oversold itself in the areas where it purportedly does succeed. While it's true that the government has been much friendlier than ours to gay marriage, only 39 percent of Canadians decidedly support it. While Canada is supposedly more environment-friendly, it has been cited for producing more waste per person than any other country. While Canada is supposedly safer, a 1996 study showed its banks had the highest stick-up rate of any industrialized nation (one in every six was robbed). And while a great deal is made of Americans' passion for firearms, the *Edmonton Sun*, citing *Statistics Canada*, reported that Canada has a higher crime rate than we do.

Canadians are supposedly less greedy than Americans, yet they lead the world in telemarketing fraud, and most of their victims are Americans. Are they more generous? Not by a long shot. The Vancouver-based Fraser Institute publishes a Generosity Index, which shows that more Americans give to charity, and give more when they do.

Is the Canadian "mosaic" more successful than the American

"melting pot," a distinction they constantly make? You be the judge. Imagine every decade or so America's Spanish-speaking southwesterners holding a referendum over whether to secede. It's happened twice since 1980 among the Francophones of Quebec, and some say it's going to happen again. While America has figurative language police on its college campuses, Quebec has literal ones—"tongue troopers," the locals call them—who ruthlessly enforce absurd language laws requiring, for example, that restaurant trash cans feature the word "push" on their lids in French instead of English.

Apart from the Anglo/Franco teeter-totter that Canada can't ever seem to get off, are Canadians less racist, as many of them claim? Well, like America, they saw both slavery and segregation. If Canadians today are less racist, someone ought to tell their aboriginal peoples, who've spent centuries getting their land annexed and being generally mistreated (as of 2000 in Nova Scotia, there was still a law on the books offering hunters a bounty for Indian scalps).

Recent polling shows 35 percent of Canada's "visible minorities" (such as blacks and Asians) have experienced discrimination in the last five years. Another poll showed 54 percent of Canadians believe anti-Semitism is a serious problem in Canadian society today. It certainly was yesterday. Around World War II, a few Jews did manage to squeak in—despite the policy summed up by Canada's director of immigration as "None is too many." Will Ferguson points out that more Nazi war criminals are thought to have found sanctuary in Canada than refugees fleeing the Holocaust.

But even when Canada succeeds, it carries the whiff of failure. For nearly a decade, the country sat atop the United Nations quality-of-life index, a fact that Canadian schoolchildren could parrot in their sleep. When Canada dropped to eighth, just behind the United States, its collective psyche took a beating. The next year, Canada shot past us again, but not back to the top. The headline in Ontario's *Windsor Star* tells you all you need to know about Canadian triumphalism: "Cheers to us, we're No. 4."

In a sense, Canada is the perfect place for American quitters, as it evidences self-loathing masquerading as self-congratulation. This I learn over dinner in Vancouver. A delightful realtor named Elizabeth

McQueen has enticed me with a promise any American boy likes to hear—that we'd be dining with "two very attractive lesbians." She didn't lie. One of them could make a killing as a Courteney Cox celebrity impersonator. Besides, they're psychotherapists from San Francisco. They ask me to change their names to Cocoa and Satchi since their patients don't yet know they're leaving America.

They've come to Vancouver to look for real estate, having gotten married on an earlier trip to Canada. They were politically active back home. They wrote letters to the editor for every cause: "Save the whales, save the trees, save the lesbians," says Cocoa. They hate the war and the Patriot Act and the results of the gay-marriage resolutions. They hate the conservative agenda and fundamentalist crackers and all the other usual suspects. They hate it that Karl Rove, in Cocoa's words, helped to elect "an alcoholic butthead who can't put two sentences together, cocaine addict, married to a frigid drunk-driver-murderer-Martha-Stewart wannabe."

But beneath all her gracious sentiments is something else: a loss of faith. When describing how she feels traveling abroad, Cocoa sounds like the old joke about how Canadians apologize when you step on their shoes: "I felt ashamed as I was going everywhere with my American passport. It was just like 'I'm so sorry.' . . . After the last election, I kind of lost faith in what we Americans are doing in our country."

Even many Canadians recognize that theirs is a faithless country compared with America. Not just in terms of religious belief—though they are much less fervent. As *National Post* columnist Andrew Coyne recently wrote in a piece chiding his countrymen for regarding American patriotism as cheap sentiment, "You see, in Canada we gave up believing years ago: in religion, in ideals, in much of anything, really. Secure as we were under the American defense umbrella, we were infantilized; having no need to defend ourselves, we could not understand why anyone else would have more. Or perhaps it was this: Having renounced even the wish to defend ourselves, having absorbed the notion that the country could be destroyed at any moment by a vote of half the population of one province [Quebec], what was left to believe?"

There are some American expats, however, who are of more robust stock. I journey out to a hippie leftover, New Agey enclave on British Columbia's far western shore, Quadra Island, where I actually smell spliffy smoke on the ferry ride over. At the island's edge lies the Heriot Bay Inn, owned by American Lorraine Wright. She bought it last year after moving north a while ago, partly for business opportunities, partly because of the political climate back home.

On this island, natural beauty surrounds us. Nick, her boat captain (she also owns a whale and grizzly-watching adventure tours business) takes me out through choppy coastal sounds to deliver explosives to a remote construction site, since he doubles as a water-taxi in the off-season. We look for sea lions and seals, which the locals call "rock sausages." They often serve as finger-food for transient killer whales.

One night in Wright's wood-paneled, maritime-themed bar, I meet her cast of regulars. There's the bar curmudgeon Bruce, who shakes hands with one that has lost a finger to a circular saw. He seems to like being on my tab, but spends most of the night whispering anti-Americanisms in my ear. There's oyster farmer Brian, whose border collie lolls between the tables. He offers to call up a Vietnam draft-dodger friend who might make a good interview, though the friend's not home, just as he wasn't when his country called.

Lorraine, who has a large personality and a barbed wit, blows in like a northerly in an orange Arc'teryx jacket. We tuck into a discreet corner where the barmaid keeps finding us with a steady supply of my whiskeys (bourbon, not Canadian—drink American!) and her cosmopolitans. For the next four hours, I and this former surfer girl from California, born to a Republican family before she became a bleeding heart, go at it like two drunks in a bar fight, which come to think of it, we half resemble.

She calls herself a "compassionate capitalist" and clowns on my old Clinton-bashing pieces, which she's pulled off the Internet. I try my level best to make her feel like Benedict Arnold, who lost the fight when we invaded Quebec during the Revolution, before he slunk off to England. Instant friends, with similar sensibilities, we throw flurries of rabbit and kidney punches. But just when I think my roundhouse is going to drop her like a sack of potatoes—after I posit that real

Americans, whatever their political persuasion, are fighters, not run-
ners—Wright clocks me with this:

"America is built on people leaving places. We're a country of people who've left. Constitutionally, the pursuit of happiness is something
we not only honor, but something we legally protect. This ain't
Russia. I don't have to stay. This ain't Cuba. I can leave.

"In fact, find me one American who would make me stay and
fight. They'd say no, go, do what's right for you. I found happiness
here. I'll be in BC the rest of my life. I pray to God that I don't die
somewhere else, that I'm not vacationing somewhere when I die,
because that would bum me out. . . .

"Pursue your happiness. We were the first country to do it. And
we live for that, the fact that people have personal rights. Go where
you want. Do what you want. The fact that I chose Canada is almost a
bigger embodiment of the American dream. . . . I still love America."

"So you're saying being unpatriotic is an act of patriotism?" I
counter, though my heart is no longer in it. "I've had too many cocktails for that one," Wright says.

I settle the tab, and the next morning I'm off, promising that
someday I'll come back to visit with my family. By then, with any luck,
she'll have had a chance to explain America to her new countrymen.

# HOW LIBERALISM
# FAILED TERRI SCHIAVO

## By Eric Cohen

*(Originally published in the April 4, 2005, issue)*

THE STORY OF TERRI SCHIAVO is both peculiar in its details and paradigmatic in its meaning. The legal twists, political turns, and central characters are so odd that one hesitates to draw any broader conclusions. But the Schiavo case is also a tragic example of the moral and legal confusions that govern how we care for those who cannot speak for themselves, especially those whose lives might seem less than fully human. And so we have a responsibility to confront what has happened and why—especially if we are to understand our moral obligation as caregivers for incapacitated persons, and our civic obligation to protect those who lack the capacity to express their will but are still human, still living, and still deserving of equal protection under the law.

In February 1990, a sudden loss of oxygen to the brain left Theresa Marie Schiavo in a coma and eventually in a profoundly incapacitated state. Terri's husband, Michael Schiavo, took care of her, working alongside Terri's parents. He took her to numerous doctors; he pursued experimental treatments; he sought at least some modest restoration of her self-awareness. In November 1992, he testified at a malpractice hearing that he would care for Terri for the rest of her life, that he "wouldn't trade her for the world," that he was going to nursing school to become a better caregiver. He explicitly reaffirmed his marriage vow, "through sickness, in health."

But the lonely husband eventually began seeing other women. His frustration with his wife's lack of improvement seemed to grow. When Terri suffered a urinary tract infection in the summer of 1993, he decided to cease all treatment, believing that her time to die had come, that this was what Terri would have wanted. But Terri's caregivers refused to let her die, and Michael Schiavo relented—for the time being. Not all Terri's doctors, however, saw their medical obligation in the same way; one physician declared that Terri had basically been dead for years, and told Michael that he should remove her feeding tube. Michael responded that he "couldn't do that to Terri," that he could never leave his wife to die of dehydration. But at some point, his heart changed. He decided that it was time for her final exit and his new beginning. He decided that his own wishes—for children, for a new family, for new love unclouded by old obligations—were also her wishes. He decided that she had a right to die and that he had a right to let her die.

Terri's parents, Robert and Mary Schindler, objected. They claimed that their son-in-law was no longer a fit guardian; that he was motivated by the money he would inherit at Terri's death; that Terri could improve with more love and better care. And so a long legal drama ensued, making its way through the Florida court system, centered on two sets of questions: First, what would Terri Schiavo have wanted? Would she want to die rather than live in a profoundly incapacitated condition? Was Michael Schiavo's decision to remove her feeding tube an act of fidelity to his wife's prior wishes or an act of betrayal of the woman entrusted to his care? Second, what was Terri Schiavo's precise medical condition? Did she have any hope of recovery or improvement? If her condition was unalterable—the persistence of sleeping and waking, the inscrutable moans, the uncontrolled movement of her bladder, the apparent absence of any self-awareness—was her life still meaningful?

The first question—what would Terri Schiavo have wanted?—is the central question of modern liberalism when it comes to caring for those who cannot speak for themselves. It is the autonomy question, the self-determination question, the right to privacy question. At its best, the liberal autonomy regime protects the disabled from having other people's wishes wrongly imposed on them—whether in the form

of overtreatment or undertreatment. And it affirms the "liberty interest" of those who no longer possess the capacity to act freely, by allowing the past self to speak for the present one. In legal terms, this is
called the "substituted judgment" standard: We must do what the
incompetent patient would have wanted; we must pretend that she
could pass judgment on the worth of the person she is now, according
to the interests and values of the person she once was.

The right to have medical treatment withheld on one's behalf was
codified in a string of legal cases over the last few decades. Ideally, the
individual's wishes would be laid out in an advance directive or living
will, describing in detail what kind of care a person would want under
various conditions. This is procedural liberalism's ideal of autonomy in
action: The caregiver simply executes the dependent person's prior
orders, like a lawyer representing his client. But even persons without
living wills still have a legal right to have their wishes respected, so
long as those wishes can be discovered. Each state establishes specific
criteria and procedures for adjudicating the incompetent individual's
wishes in cases where these wishes are not clear, especially when there
is a dispute between family members (as in the Schiavo case) or
between the family and the doctors.

Under the law of Florida, where the Schiavo case was adjudicated,
the patient's prior wishes must be demonstrated with "clear and convincing evidence"—the highest standard of legal certainty in civil
cases. In cases where this standard of proof is not met, the court must
"err on the side of life," on the assumption that most people, even
those who are profoundly disabled, would choose life rather than
death. In other words, the state is not supposed to judge the compara-
tive worth of different human beings, but to protect the right of individuals to decide for themselves when their lives would still have
meaning. And in cases where the individual's wishes are uncertain, the
state of Florida is charged to remain neutral by not imposing death.
This is the aim of procedural liberalism—and this is where things
went terribly wrong in the Schiavo case.

With scant evidence, a Florida district court concluded that Terri
Schiavo would clearly choose death over life in a profoundly incapacitated state. There was no living will, no advance directive, no formal

instructions left by Terri Schiavo about what to do for her under such circumstances. Instead, the court relied entirely on Michael Schiavo's recollection of a few casual conversations, on a train and watching television, in which Terri supposedly said that she wouldn't want to live "if I ever have to be a burden to anybody" or be kept alive "on anything artificial." This was evidence of her possible wishes, to be sure. But in light of Michael Schiavo's own earlier statements and behavior—including his pledge to care for Terri for the rest of her days, his unwillingness to remove her feeding tube when the idea was first suggested, his shifting sense of moral obligation as he realized that Terri's condition was probably permanent, and his romantic involvement with multiple other women—these recollections hardly constituted "clear and convincing evidence" of Terri's wishes. In this case, the court had a legal obligation to "err on the side of life." Instead, it chose to allow Michael Schiavo to choose death.

Part of the problem was simply judicial incompetence—especially the court's decision, in direct violation of Florida law, to act as Terri Schiavo's guardian at key moments of the case rather than appoint an independent guardian to represent her interests, separate from the interests of her husband and her parents. But the problem went deeper than incompetence: It also had to do with ideology—with a set of assumptions about what makes life worth living and thus worth protecting. *Procedural liberalism* (discerning and respecting the prior wishes of the incompetent person; preserving life when such wishes are not clear) gave way to *ideological liberalism* (treating incompetence itself as reasonable grounds for assuming that life is not worth living). When the district court's decision to allow Michael Schiavo to remove the feeding tube was challenged, a Florida appeals court framed the question before it as follows:

> [W]hether Theresa Marie Schindler Schiavo, not after a few weeks in a coma, but after ten years in a persistent vegetative state that has robbed her of most of her cerebrum and all but the most instinctive of neurological functions, with no hope of a medical cure but with sufficient money and strength of body to live indefinitely, would choose to continue the constant nursing care and the supporting tubes in hopes

*that a miracle would somehow recreate her missing brain tissue, or*
*whether she would wish to permit a natural death process to take its*
*course and for her family members and loved ones <u>to be free to continue</u>*
*<u>their lives</u>.* [Emphasis added.]

Now, one could surely read this as an effort to get inside Terri's
once competent mind. But more likely, it expresses the court's own
view of Terri's now incompetent and incapacitated existence as a meaningless burden, a barrier to her husband's freedom. The court's obligation to discern objectively what Terri's wishes were and whether they
were clear—a question of fact—morphed into an inquiry as to whether
she could ever get better, with the subjective assumption that life in
her present condition was not meaningful life. The question became:
Was she in a persistent vegetative state (PVS), and if so, can't we
assume that Terri believed death to be preferable to life in such a state?

In response, both sides brought out their best medical experts:
Michael Schiavo's doctors to quiet our consciences and assure us that
Terri was already long gone, a mere ghost of her former self; the
Schindlers' doctors to tell us that she was still responsive to her environment and still might get better, even after years of not improving.
Clearly, for many years, Terri's treatment was subpar, and to this day
many tests that could clarify her diagnosis have not been done. At the
same time, a conservative estimate of her prospects for recovery suggests that her chances were slim, and that she would remain in her
profoundly incapacitated state till the end of her days. The court
finally ruled that she was indeed in a PVS, and that her feeding tube
should be removed—which it was on October 15, 2003.

By then, of course, the Schiavo case had become a public drama,
and the outcry at the prospect of leaving Terri to die was overwhelming. The Florida legislature sprang into action, and on October 21,
2003, it passed "Terri's Law," giving the governor authority to stay the
court's judgment, order the feeding tube back in, and order a review of
the case by an independent guardian charged to report on Terri's
behalf. So began the next round of court fights and political battles.
The ACLU joined Michael Schiavo in challenging the constitutional-
ity of Terri's Law. Terri's court-appointed guardian issued a largely

unhelpful report. And eventually, the Florida court overturned Terri's Law, rejected the Schindlers' appeals, and ordered that the feeding tube once again be removed—which it was the other day, on March 18, 2005. And despite Congress's dramatic effort to restart the case in federal court and Gov. Jeb Bush's continued encouragement to the Florida legislature to act again on her behalf, the most likely outcome at this writing is death by dehydration—the final triumph of Michael Schiavo's will, and supposedly what Terri Schiavo herself would have wanted.

For all the attention we have paid to the Schiavo case, we have asked many of the wrong questions, living as we do on the playing field of modern liberalism. We have asked whether she is really in a persistent vegetative state, instead of reflecting on what we owe people in a persistent vegetative state. We have asked what she would have wanted as a competent person imagining herself in such a condition, instead of asking what we owe the person who is now with us, a person who can no longer speak for herself, a person entrusted to the care of her family and the protection of her society.

Imagine, for example, that the Schindlers had agreed with Michael Schiavo that Terri's time had come, that she would never have wanted to live like this, that the feeding tube keeping her alive needed to come out. Chances are, there would have been no federal case, no national story, no political controversy. Terri Schiavo would have been buried long ago, mourned by the family that decided on her behalf that death was preferable to life in her incapacitated state. Under law, such an outcome would have been unproblematic and uneventful, so long as no one had claimed that Terri Schiavo's previous wishes were being violated. But morally, the deepest problem would remain: What do we owe those who are not dead or dying but profoundly disabled and permanently dependent? And even if such individuals made their desires clearly known while they were still competent, is it always right to follow their instructions—to be the executors of their living wills—even if it means being their willing executioners?

For some, it is an article of faith that individuals should decide for themselves how to be cared for in such cases. And no doubt one response to the Schiavo case will be a renewed call for living wills and advance directives—as if the tragedy here were that Michael Schiavo did not have written proof of Terri's desires. But the real lesson of the Schiavo case is not that we all need living wills; it is that our dignity does not reside in our will alone, and that it is foolish to believe that the competent person I am now can establish, in advance, how I should be cared for if I become incapacitated and incompetent. The real lesson is that we are not mere creatures of the will: We still possess dignity and rights even when our capacity to make free choices is gone; and we do not possess the right to demand that others treat us as less worthy of care than we really are.

A true adherence to procedural liberalism—respecting a person's clear wishes when they can be discovered, erring on the side of life when they cannot—would have led to a much better outcome in this case. It would have led the court to preserve Terri Schiavo's life and deny Michael Schiavo's request to let her die. But as we have learned, the descent from procedural liberalism's respect for a person's wishes to ideological liberalism's lack of respect for incapacitated persons is relatively swift. Treating autonomy as an absolute makes a person's dignity turn entirely on his or her capacity to act autonomously. It leads to the view that only those with the ability to express their will possess any dignity at all—everyone else is "life unworthy of life."

This is what ideological liberalism now seems to believe—whether in regard to early human embryos, or late-stage dementia patients, or fetuses with Down syndrome. And in the end, the Schiavo case is just one more act in modern liberalism's betrayal of the vulnerable people it once claimed to speak for. Instead of sympathizing with Terri Schiavo—a disabled woman, abandoned by her husband, seen by many as a burden on society—modern liberalism now sympathizes with Michael Schiavo, a healthy man seeking freedom from the burden of his disabled wife and self-fulfillment in the arms of another. And while one would think that divorce was the obvious solution, this was more than Michael Schiavo apparently could bear, since it would require a definitive act of betrayal instead of a supposed demonstration of loyalty to Terri's wishes.

Perhaps we can fashion better laws or better procedures to ensure that vulnerable persons get the care they deserve. But even truly loving caregivers will face hard decisions—decisions best left in their hands, not turned over to the state. And in reality, most decisions will be made at the bedside, where the reach of the law will always be limited, and usually should be. Moreover, the autonomy regime, at its best, prevents the worst abuses—like involuntary euthanasia, where doctors or public officials decide whose life is worth living. But the autonomy regime, even at its best, is deeply inadequate. It is based on a failure to recognize that the human condition involves both giving and needing care, and not always being morally free to decide our own fate.

In the end, the only alternative is a renewed understanding of both the family and human equality—two things ideological liberalism has now abandoned and modern conservatism now defends. Living in a family means accepting the burdens of caring for those bound to us in ties of fidelity—whether parent for child, child for parent, or spouse for spouse. The human answer to our dependency is not living wills but loving surrogates. And for those who believe in human equality, this means treating even the profoundly disabled—people like Terri Schiavo, who are not dead and are not dying—as deserving of at least basic care, so long as the care itself is not the cause of additional suffering. Of course, this does not mean that keeping our loved ones alive is our only goal. But neither can we treat a person's life as a disease in need of a cure, or aim at death as a means of ending suffering—even if a loved one asks us to do so.

Perhaps we should not be surprised at the immovable desire of Terri's parents to keep her alive and the willingness of Terri's husband to let her go. Parental love and spousal love take shape in fundamentally different ways. Parents first know their children as helpless beings, totally dependent on their care. Husbands first know their wives as attractive, autonomous beings who both give and receive love, and who enter into marriage as willing partners. But to marry means pledging one's fidelity despite the uncertainties of fortune. The beautiful wife may become disfigured, the wished-for mother may prove to be infertile, the young woman teeming with life may be plunged into a persistent vegetative state. Marriage often demands heroism, and we can hardly condemn those who fall short of it. But we can surely fault those, like Michael Schiavo, who claim to speak in the name of loved ones they have abandoned, and insist that letting them die is what they desire or deserve.

To question whether Michael Schiavo has his wife's best interests at heart is not to make this case ethically or humanly easy. The decision to continue feeding a person in a profoundly incapacitated state is always wrenching. We must at least wonder whether ensuring years or decades with a feeding tube, with no self-control, and with virtually no possibility of improvement is not love but torture, not respect for life but forced degradation. We, too, must tremble when we demand that people like Terri be fed. But in the end, the obligation to feed should win out, because the living humanity of the disabled person is undeniably real.

On March 18, 2005, the day her feeding tube was removed, Terri Schiavo was not dead or dying. She was a profoundly disabled person in need of constant care. And despite the hopes of her parents, it was unlikely that her medical condition would improve, even with the best possible care administered by those with her best interests at heart. But even in her incapacitated state, Terri Schiavo was still a human being, a member of the Schindler family and the human family. As such, she was still worthy of protection and care, even if some of those closest to her wished to deny it.

# III

PEACE AND WAR AT
HOME AND ABROAD

# BOSNIA: THE REPUBLICAN CHALLENGE

## By David Tell

*(Originally published in the December 11, 1995, issue)*

PRESIDENT CLINTON HAS DECIDED to deploy U.S. troops in Bosnia. By doing so, he tests Republicans on a yea-or-nay question concerning America's continued engagement with the rest of the world. At this point, all too many of them are flunking that test.

One would prefer a situation in which, long before a final troop commitment was imminent, the general thrust of a president's international judgments appeared clearly correct, and the other party *said* so, working cooperatively to rally the country behind him. But we have never been blessed with such luck on the question of Bosnia, a horror of Byzantine complexity to which the United States has responded with a dizzying series of false starts and hard swerves. There has been legitimate Republican-led opposition to the administration's Balkan policy. And it was therefore almost inevitable, if and when the call for American ground forces arose, that some measure of partisan conflict would arise with it.

That conflict, *per se*, is nothing to celebrate. Domestic politics invites, even requires, all manner of bare-knuckle fighting between Congress and the White House. But except in the rarest of circumstances, only a president's *personal* stature is invested in those battles, and the "worst" possible outcome is that his party loses its next national campaign. Where international security determinations are concerned, however, a president's individual authority is significantly inseparable from the authority of the presidency generally. Any major rejection of the man also, unavoidably, impeaches his office—the institution against which foreign governments judge American resolve. If the president must make his way overseas against furious opposition, or fails to make his way at all, then U.S. international credibility and influence are damaged, at least in the short run.

That will be one sad, undeniable result if, in the next few weeks, Congress fails to support the Bosnian peace plan—and U.S. troop commitment—initialed in Dayton, Ohio. But it is rapidly becoming apparent that something even more momentous than a temporary wound to American prestige is at issue here. The entire structure and purpose of post-1945 American foreign relations, our posture of energetic international engagement, is implicated in Bosnia. And with distressingly few exceptions, *Republicans*, who have worked hardest to maintain that posture these past 20 years, are behaving as though they may no longer care.

If the United States has no "vital national interest" in leading a coherent, effective NATO, we have no vital interest in anything beyond our shores. Bosnia is the victim of brutal aggression across internationally recognized borders on a European continent over which NATO necessarily claims protective dominion. The war has been exacerbated by past NATO actions; enforcement of the arms embargo has worked to Serbian advantage. The current cease-fire is the product of NATO will. The prospective peace is entirely dependent on NATO force. And without a U.S. deployment, that force will not exist. Our British and French allies, already on the ground in Bosnia, insist on it. Bosnia's president refused to come to Dayton unless it was guaranteed. No U.S. troops, no peace. And because Bosnia has, like it or not, become a NATO responsibility, any failure of peace threatens NATO's effective survival.

President Clinton, unfortunately, is an imperfect guide for such a dark forest. Hansel-like in his national television address last week, he dropped dozens of pebbles by which he might find his way back to domestic political safety in next year's reelection campaign. No imaginable Republican attack line was left unanticipated. His Bosnia mission will not be about war, but about peace, and "especially children."

It will be "under the command of an American general." It will have "clear, limited, and achievable" goals, and "should and will take about one year," at which point Bosnia can once again become "a shining symbol of multiethnic tolerance." *But*—over and over again he conceded—"that doesn't mean we can solve every problem." America "cannot and must not be the world's policeman."

It was a timid speech. The president sang his score with few technical mistakes. But in its weird combination of soaring, excessive promise and painfully obvious pleading, its required theme—that we either go to Bosnia or signal the beginning of an American military and diplomatic retreat from the entire world—was barely audible.

Republicans might be expected to amplify that internationalist chorus; until recently, they had it memorized. Bob Dole still does. He will encourage other Republicans to "support the president," words that just a handful of them are now prepared to use. Most of Dole's presidential primary opponents excoriate him for offering even the *hope* of eventual Republican agreement on Bosnia. Phil Gramm promises that Bosnia will "define this race." Other leading Republicans claim still not to see the American security interest in the Balkans and pose ultimately unanswerable questions about "exit strategy." Further down the leadership ladder, undisciplined by their seniors, the vocal mass of Republicans take daily aim at the president, and make grotesquely casual references to "body bags" and "Vietnam."

What's got into them? If it is public opinion on Bosnia that Republicans fear, they are fearful too quickly. Most Americans do not want to send troops. They almost never do. But a plurality of CBS survey respondents say they at least *understand* why we might go. A plurality of respondents to the CNN/*USA Today*/Gallup poll already *favor* deployment, and a majority of them feel an American moral obligation to help keep a Bosnian peace. Widespread, visceral opposition to an American Bosnia mission seems still restricted to what might be called the populist "conservative street." The newsletter *Talk Daily* reports 85 percent opposition to the Dayton accord among call-in radio listeners, most of whom appear to believe that President Clinton's Bosnia policy was invented by Democratic campaign consultants.

Liberal critics to the contrary notwithstanding, Republicans did not take control of Congress last fall by pandering to populism's least sophisticated, most crudely nativist impulses. A gestural anti-Clinton politics on Bosnia is something Republicans do not need; the president has given them all the domestic policy opportunities they could ever ask for in next year's election. And at its current volume, such a pandering, populist politics is bad for the country. When the "conservative street" is wrong, it should be corrected—or ignored.

The current Washington consensus is that American troops *will* go to Bosnia, one way or the other, as they must. Most expect the Senate to endorse the deployment. But the House of Representatives remains very much in doubt, and in private conversation administration officials admit they would, if faced with explicit rejection, prefer to see no House vote at all. An unpleasant prospect, that: the Congress essentially holding its nose in grudging acquiescence as American soldiers march into harm's way. If the president is to lead us overseas, the maintenance of American international standing requires that we succeed. Boxing Clinton in and carping at him won't help achieve that result. Instead, Republicans can and should improve America's position on Bosnia—in two particular respects.

The worst conceivable disaster that might befall America in Bosnia is a chaotic military pullout, under fire, in a breakdown of the peace. And such hostilities are made more likely, not less, if potential combatants are convinced that U.S. domestic politics will force us to withdraw at the first hint of trouble, or the end of Dayton diplomacy's one-year limit, whichever comes first. The president himself knows his time limit must be elastic; he told Senate Democrats as much behind closed doors last Tuesday, reminding them that his speech had promised a homecoming in "about" one year, not strictly on the 365th day. In his mouth, that sounds like a "didn't inhale" equivocation. But it's true, just the same. And Republicans should give him cover, loudly announcing that once our troops go in, they will not be pushed out by any hostile force—or any arbitrary deadline that becomes inconvenient.

Diplomatic niceties aside, the bulk of U.S. forces will leave Bosnia only when they can do so with a reasonable expectation that interethnic carnage will not instantly resume. *That* is the much sought-for "exit strategy"; there is no other realistic one. And diplomatic niceties aside once more, it is not "mutual trust" based on "NATO neutrality" that will allow such an exit. The Serbs do not put down their guns because they trust America will treat them fairly. They do so because they know we sympathize with Bosnia, and they trust only that we will kick their skulls in if they break the peace. In the absence of NATO force, an equal deterrent function can only be served by a rearmed Bosnian Federation. Here, too, Republicans should give the president cover, justifying and strengthening his determination to pursue an American-led rearmament effort.

This is asking most congressional Republicans to change the spirit of their Bosnia rhetoric rather dramatically, to be sure. It will be awkward for many of them. But that's a small price to pay given the stakes involved. The alternative, a body blow against the perceived American commitment to international leadership, would be a grave shame. More than a small bit of which would justly attach to the GOP.

# SADDAM MUST GO

## By William Kristol and Robert Kagan

*(Originally published in the November 17, 1997, issue)*

SOME NATIONS CAN AFFORD to suffer more humiliation than others. When you're the United States, even a little humiliation exacts too high a price. This isn't just a matter of national pride. When the world's strongest power abases itself, allies begin to worry, adversaries start whetting their appetites, and pretty soon America's international credibility—a big and important component of national power—starts taking a dive.

This past week, Iraq's Saddam Hussein humiliated the United States: First he ordered the expulsion of American officials from a United Nations team charged with ensuring that Iraq is not producing weapons of mass destruction. Then he demanded an end to all flights by American U-2 surveillance aircraft over Iraq and threatened to shoot them down. Then he moved some equipment that could be used to manufacture weapons out of the range of video cameras that had been installed by the U.N. inspection team to keep watch over them.

A few observers, including some administration officials, have described Saddam's actions as foolish. Some fool. Saddam's actions are well calibrated to achieve three important aims: to embarrass and thereby weaken the United States; to exploit divisions in the international coalition that defeated him in the Gulf War but has been fraying ever since; and last but certainly not least, to build as rapidly as possible the weapons of mass destruction that can put him back in the driver's seat in the Middle East—a scant six years after his armies were decimated in Operation Desert Storm.

Despite the Clinton administration's denials, Saddam appears to be succeeding on all three fronts. The last is particularly alarming. According to a report in the *New York Times*, U.N. inspectors believe that Iraq now possesses "the elements of a deadly germ warfare arsenal and perhaps poison gases, as well as the rudiments of a missile system" that can launch the warheads. Thanks to Saddam's recent actions, the U.N. inspection team "can no longer verify that Iraq is not making weapons of mass destruction" and specifically cannot monitor "equipment that could grow seed stocks of biological agents in a matter of hours."

The Clinton administration's response to Saddam so far has compounded the humiliation, and the danger. On the one hand, officials trying to sound ominous in warning Saddam against a wrong step have succeeded only in sounding ridiculous—as when President Clinton declared it would be a "big mistake" for Saddam to shoot down an American U-2. On the other hand, the administration has agreed—or worse still, has been forced to agree—to a number of concessions to Saddam's bullying. Rather than simply telling Saddam to shove it and preparing the first wave of air and missile strikes, the United Nations dispatched a team last week to "talk" with Saddam about the importance of complying with U.N. resolutions. The Clinton administration insisted that these talks were not "negotiations," but that pretense was all but exploded when the U.N. and the United States agreed to suspend the U-2 flights Saddam had complained about. This appalling concession, intended to improve the atmosphere for these nonnegotiations, was the worst of the administration's missteps so far.

All these concessions were evidence, moreover, that the old Gulf War coalition is indeed collapsing. Apparently, the United States has been having a devil of a time convincing other Security Council members to approve any kind of military action against Saddam, no matter how long he defies the international community. At the end of last week, administration officials started talking about trying to persuade them at least to impose new sanctions on Iraq. Even that action, however, pitiful as it is, would be difficult given the clear determination of the French and Russians to remove sanctions altogether.

But here's the really bad news. Even if the United States summoned the courage, alone or with U.N. approval, to launch a missile strike against Iraq this week or next, such an attack would gain only a brief pause in the downward slide of U.S. policy in the Gulf. Saddam has already calculated that he can survive another cruise-missile strike, as he survived the last, and may even come out of it in a stronger position. Once the assault has ended, the situation will return to the status quo ante: The international coalition will continue to collapse, Saddam will continue to probe for weaknesses, and U.S. credibility will continue to erode. Indeed, a U.S. attack that leaves Saddam in charge of Iraq, no matter how much damage it does to his country, might serve only to expose the futility of American power.

So there is really only one alternative now. It has become increasingly clear ever since the Gulf War ended that the Gulf War ended badly. The decision to leave Saddam in control of Iraq, and to hope vainly that he would be overthrown or assassinated by his own people, was a mistake—an understandable mistake, perhaps, but a mistake nevertheless. We were sorry to see former President Bush last week denounce those who are now coming to this conclusion. The fact that he erred in letting Saddam remain in power does not detract from his magnificent accomplishment in fighting the Gulf War and liberating Kuwait. It would be a real service to the nation if Bush could acknowledge his error. Because what we most need now is to take the difficult but inescapable next step of finishing the job Bush started.

American policy toward Iraq should aim at removing Saddam from power. We are under no illusions about what will be required to accomplish this goal. There will be no coup against Saddam and no assassination at the hands of his own lieutenants. Nor, unfortunately, will an air and missile strike do the job. In a sustained air campaign, we might get lucky and hit Saddam by accident, but if we didn't get him during the weeks-long barrage of air and missile attacks in Desert Storm, we're unlikely to succeed in a shorter and smaller attack today.

We would certainly support a serious and sustained air attack on Iraq, and the sooner the better. But the only sure way to take Saddam out is on the ground. We know it seems unthinkable to propose another ground attack to take Baghdad. But it's time to start thinking the unthinkable. The fact is, it would take fewer than the half-million troops deployed in Desert Storm to roll into Baghdad today, especially after an air campaign scattered or destroyed whatever resistance Saddam might be able to throw up. Who knows how many Iraqi soldiers would even fight in a Desert Storm II? Their last experience against American forces and weapons was not such as to encourage exceptional valor.

If you don't like this option, we've got another one for you: continue along the present course and get ready for the day when Saddam has biological and chemical weapons at the tips of missiles aimed at Israel and at American forces in the Gulf. That day may not be far off.

# AT LAST, ZION

## By Charles Krauthammer

*(Originally published in the May 11, 1998, issue)*

## I. A Small Nation

Milan Kundera once defined a small nation as "one whose very existence may be put in question at any moment; a small nation can disappear, and it knows it."

The United States is not a small nation. Neither is Japan. Or France. These nations may suffer defeats. They may even be occupied. But they cannot disappear. Kundera's Czechoslovakia could—and once did. Prewar Czechoslovakia is the paradigmatic small nation: a liberal democracy created in the ashes of war by a world determined to let little nations live free; threatened by the covetousness and sheer mass of a rising neighbor; compromised fatally by a West grown weary "of a quarrel in a faraway country between people of whom we know nothing"; left truncated and defenseless, succumbing finally to conquest. When Hitler entered Prague in March 1939, he declared, "Czechoslovakia has ceased to exist."

Israel too is a small country. This is not to say that extinction is its fate. Only that it can be.

Moreover, in its vulnerability to extinction, Israel is not just any small country. It is the only small country—the only country, period—whose neighbors publicly declare its very existence an affront to law, morality, and religion, and make its extinction an explicit, paramount national goal. Nor is the goal merely declarative. Iran, Libya, and Iraq conduct foreign policies designed for the killing of Israelis and the destruction of their state. They choose their allies (Hamas, Hezbollah) and develop their weapons (suicide bombs, poison gas, anthrax, nuclear missiles) accordingly. Countries as far away as Malaysia will not allow a representative of Israel on their soil nor even permit the showing of *Schindler's List* lest it engender sympathy for Zion.

Others are more circumspect in their declarations. No longer is the destruction of Israel the unanimous goal of the Arab League, as it was for the thirty years before Camp David. Syria, for example, no longer explicitly enunciates it. Yet Syria would destroy Israel tomorrow if it had the power. (Its current reticence on the subject is largely due to its post-Cold War need for the American connection.)

Even Egypt, first to make peace with Israel and the presumed model for peacemaking, has built a vast U.S.-equipped army that conducts military exercises obviously designed for fighting Israel. Its huge "Badr '96" exercises, for example, Egypt's largest since the 1973 war, featured simulated crossings of the Suez Canal.

And even the PLO, which was forced into ostensible recognition of Israel in the Oslo Agreements of 1993, is still ruled by a national charter that calls in at least fourteen places for Israel's eradication. The fact that after five years and four specific promises to amend the charter it remains unamended is a sign of how deeply engraved the dream of eradicating Israel remains in the Arab consciousness.

## II. The Stakes

The contemplation of Israel's disappearance is very difficult for this generation. For fifty years, Israel has been a fixture. Most people cannot remember living in a world without Israel.

Nonetheless, this feeling of permanence has more than once been rudely interrupted—during the first few days of the Yom Kippur War when it seemed as if Israel might be overrun, or those few weeks in May and early June 1967 when Nasser blockaded the Straits of Tiran and marched 100,000 troops into Sinai to drive the Jews into the sea.

Yet Israel's stunning victory in 1967, its superiority in conventional weaponry, its success in every war in which its existence was at stake, has bred complacency. Some ridicule the very idea of Israel's impermanence. Israel, wrote one Diaspora intellectual, "is fundamentally indestructible. Yitzhak Rabin knew this. The Arab leaders on

Mount Herzl [at Rabin's funeral] knew this. Only the land-grabbing, trigger-happy saints of the right do not know this. They are animated by the imagination of catastrophe, by the thrill of attending the end."

Thrill was not exactly the feeling Israelis had when during the Gulf War they entered sealed rooms and donned gas masks to protect themselves from mass death—in a war in which Israel was not even engaged. The feeling was fear, dread, helplessness—old existential Jewish feelings that post-Zionist fashion today deems anachronistic, if not reactionary. But wish does not overthrow reality. The Gulf War reminded even the most wishful that in an age of nerve gas, missiles, and nukes, an age in which no country is completely safe from weapons of mass destruction, Israel with its compact population and tiny area is particularly vulnerable to extinction.

Israel is not on the edge. It is not on the brink. This is not '48 or '67 or '73. But Israel is a small country. It can disappear. And it knows it.

It may seem odd to begin an examination of the meaning of Israel and the future of the Jews by contemplating the end. But it does concentrate the mind. And it underscores the stakes. The stakes could not be higher. It is my contention that on Israel—on its existence and survival—hangs the very existence and survival of the Jewish people. Or, to put the thesis in the negative, that the end of Israel means the end of the Jewish people. They survived destruction and exile at the hands of Babylon in 586 BC. They survived destruction and exile at the hands of Rome in AD 70, and finally in AD 132. They cannot survive another destruction and exile. The Third Commonwealth—modern Israel, born just 50 years ago—is the last.

The return to Zion is now the principal drama of Jewish history. What began as an experiment has become the very heart of the Jewish people—its cultural, spiritual, and psychological center, soon to become its demographic center as well. Israel is the hinge. Upon it rest the hopes—the only hope—for Jewish continuity and survival.

## III. The Dying Diaspora

In 1950, there were 5 million Jews in the United States. In 1990, the number was a slightly higher 5.5 million. In the intervening decades, overall U.S. population rose 65 percent. The Jews essentially tread water. In fact, in the last half-century Jews have shrunk from 3 percent to 2 percent of the American population. And now they are headed for not just relative but absolute decline. What sustained the Jewish population at its current level was, first, the postwar baby boom, then the influx of 400,000 Jews, mostly from the Soviet Union.

Well, the baby boom is over. And Russian immigration is drying up. There are only so many Jews where they came from. Take away these historical anomalies, and the American Jewish population would be smaller today than in 1950. It will certainly be smaller tomorrow than today. In fact, it is now headed for catastrophic decline. Steven Bayme, director of Jewish Communal Affairs at the American Jewish Committee, flatly predicts that in twenty years the Jewish population will be down to four million, a loss of nearly 30 percent. In twenty years! Projecting just a few decades further yields an even more chilling future.

How does a community decimate itself in the benign conditions of the United States? Easy: low fertility and endemic intermarriage.

The fertility rate among American Jews is 1.6 children per woman. The replacement rate (the rate required for the population to remain constant) is 2.1. The current rate is thus 20 percent below what is needed for zero growth. Thus fertility rates alone would cause a 20 percent decline in every generation. In three generations, the population would be cut in half.

The low birth rate does not stem from some peculiar aversion of Jewish women to children. It is merely a striking case of the well-known and universal phenomenon of birth rates declining with rising education and socioeconomic class. Educated, successful working women tend to marry late and have fewer babies.

Add now a second factor, intermarriage. In the United States today more Jews marry Christians than marry Jews. The intermarriage rate is 52 percent. (A more conservative calculation yields 47 percent; the demographic effect is basically the same.) In 1970, the rate was 8 percent.

Most important for Jewish continuity, however, is the ultimate identity of the children born to these marriages. Only about one in four is raised Jewish. Thus two-thirds of Jewish marriages are producing children three-quarters of whom are lost to the Jewish people. Intermarriage rates alone would cause a 25 percent decline in population in every generation. (Math available upon request.) In two generations, half the Jews would disappear.

Now combine the effects of fertility and intermarriage and make the overly optimistic assumption that every child raised Jewish will grow up to retain his Jewish identity (i.e., a zero dropout rate). You start out with 100 American Jews; you end up with 60. In one generation, more than a third have disappeared. In just two generations, two out of every three will vanish.

One can reach this same conclusion by a different route (bypassing the intermarriage rates entirely). A *Los Angeles Times* poll of American Jews conducted in March 1998 asked a simple question: Are you raising your children as Jews? Only 70 percent said yes. A population in which the biological replacement rate is 80 percent and the cultural replacement rate is 70 percent is headed for extinction. By this calculation, every 100 Jews are raising 56 Jewish children. In just two generations, 7 out of every 10 Jews will vanish.

The demographic trends in the rest of the Diaspora are equally unencouraging. In Western Europe, fertility and intermarriage rates mirror those of the United States. Take Britain. Over the last generation, British Jewry has acted as a kind of controlled experiment: a Diaspora community living in an open society, but, unlike that in the United States, not artificially sustained by immigration. What happened? Over the last quarter-century, the number of British Jews declined by over 25 percent.

Over the same interval, France's Jewish population declined only slightly. The reason for this relative stability, however, is a one-time factor: the influx of North African Jewry. That influx is over. In France today only a minority of Jews between the ages of twenty and forty-four live in a conventional family with two Jewish parents. France, too, will go the way of the rest.

"The dissolution of European Jewry," observes Bernard Wasserstein in *Vanishing Diaspora: The Jews in Europe since 1945*, "is not situated at some point in a hypothetical future. The process is taking place before our eyes and is already far advanced." Under present trends, "the number of Jews in Europe by the year 2000 would then be not much more than one million—the lowest figure since the late Middle Ages."

In 1900, there were eight million.

The story elsewhere is even more dispiriting. The rest of what was once the Diaspora is now either a museum or a graveyard. Eastern Europe has been effectively emptied of its Jews. In 1939, Poland had 3.2 million Jews. Today it is home to 3,500. The story is much the same in the other capitals of Eastern Europe.

The Islamic world, cradle to the great Sephardic Jewish tradition and home to one-third of world Jewry three centuries ago, is now practically *Judenrein*. Not a single country in the Islamic world is home to more than 20,000 Jews. After Turkey with 19,000 and Iran with 14,000, the country with the largest Jewish community in the entire Islamic world is Morocco with 6,100. There are more Jews in Omaha, Nebraska.

These communities do not figure in projections. There is nothing to project. They are fit subjects not for counting but for remembering. Their very sound has vanished. Yiddish and Ladino, the distinctive languages of the European and Sephardic Diasporas, like the communities that invented them, are nearly extinct.

## IV. The Dynamics of Assimilation

Is it not risky to assume that current trends will continue? No. Nothing will revive the Jewish communities of Eastern Europe and the Islamic world. And nothing will stop the rapid decline by assimilation of Western Jewry. On the contrary. Projecting current trends—assuming, as I have done, that rates remain constant—is rather conservative: It is risky to assume that assimilation will not accelerate. There is nothing on the horizon to reverse the integration of Jews into Western culture. The attraction of Jews *to* the larger culture and the level of acceptance of Jews *by* the larger culture are historically unprecedented. If anything, the trends augur an intensification of assimilation.

It stands to reason. As each generation becomes progressively more assimilated, the ties to tradition grow weaker (as measured, for example, by synagogue attendance and number of children receiving some kind of Jewish education). This dilution of identity, in turn, leads to a greater tendency to intermarriage and assimilation. Why not? What, after all, are they giving up? The circle is complete and self-reinforcing.

Consider two cultural artifacts. With the birth of television a half-century ago, Jewish life in America was represented by *The Goldbergs*: urban Jews, decidedly ethnic, heavily accented, socially distinct. Forty years later *The Goldbergs* begat *Seinfeld*, the most popular entertainment in America today. The Seinfeld character is nominally Jewish. He might cite his Jewish identity on occasion without apology or self-consciousness—but, even more important, without consequence: It has not the slightest influence on any aspect of his life.

Assimilation of this sort is not entirely unprecedented. In some ways, it parallels the pattern in Western Europe after the emancipation of the Jews in the late 18th and 19th centuries. The French Revolution marks the turning point in the granting of civil rights to Jews. As they began to emerge from the ghetto, at first they found resistance to their integration and advancement. They were still excluded from the professions, higher education, and much of society. But as these barriers began gradually to erode and Jews advanced socially, Jews began a remarkable embrace of European culture and, for many, Christianity. In *A History of Zionism*, Walter Laqueur notes the view of Gabriel Riesser, an eloquent and courageous mid-19th-century advocate of emancipation, that a Jew who preferred the nonexistent state and nation of Israel to Germany should be put under police protection not because he was dangerous but because he was obviously insane.

Moses Mendelssohn (1729–1786) was a harbinger. Cultured, cosmopolitan, though firmly Jewish, he was the quintessence of early emancipation. Yet his story became emblematic of the rapid historical progression from emancipation to assimilation: Four of his six children and eight of his nine grandchildren were baptized.

In that more religious, more Christian age, assimilation took the form of baptism, what Heinrich Heine called the admission ticket to European society. In the far more secular late-20th century, assimilation merely means giving up the quaint name, the rituals, and the other accouterments and identifiers of one's Jewish past. Assimilation today is totally passive. Indeed, apart from the trip to the county courthouse to transform, say, (*shmattes* by) Ralph Lifshitz into (Polo by) Ralph Lauren, it is marked by an absence of action rather than the active embrace of some other faith. Unlike Mendelssohn's children, Seinfeld required no baptism.

We now know, of course, that in Europe, emancipation through assimilation proved a cruel hoax. The rise of anti-Semitism, particularly late-19th-century racial anti-Semitism culminating in Nazism, disabused Jews of the notion that assimilation provided escape from the liabilities and dangers of being Jewish. The saga of the family of Madeleine Albright is emblematic. Of her four Jewish grandparents— highly assimilated, with children some of whom actually converted and erased their Jewish past—three went to their deaths in Nazi concentration camps *as Jews.*

Nonetheless, the American context is different. There is no American history of anti-Semitism remotely resembling Europe's. The American tradition of tolerance goes back 200 years to the very founding of the country. Washington's letter to the synagogue in Newport pledges not tolerance—tolerance bespeaks nonpersecution bestowed as a favor by the dominant upon the deviant—but equality. It finds no parallel in the history of Europe. In such a country, assimilation seems a reasonable solution to one's Jewish problem. One could do worse than merge one's destiny with that of a great and humane nation dedicated to the proposition of human dignity and equality.

Nonetheless, while assimilation may be a solution for individual Jews, it clearly is a disaster for Jews as a collective with a memory, a language, a tradition, a liturgy, a history, a faith, a patrimony that will all perish as a result.

Whatever value one might assign to assimilation, one cannot deny its reality. The trends, demographic and cultural, are stark. Not just in the long-lost outlands of the Diaspora, not just in its erstwhile European center, but even in its new American heartland, the future will be one of diminution, decline, and virtual disappearance. This will not occur overnight. But it will occur soon—in but two or three generations, a time not much further removed from ours today than the founding of Israel fifty years ago.

## V. Israeli Exceptionalism

Israel is different. In Israel the great temptation of modernity— assimilation—simply does not exist. Israel is the very embodiment of Jewish continuity: It is the only nation on earth that inhabits the same land, bears the same name, speaks the same language, and worships the same God that it did 3,000 years ago. You dig the soil and you find pottery from Davidic times, coins from Bar Kokhba, and 2,000-year-old scrolls written in a script remarkably like the one that today advertises ice cream at the corner candy store.

Because most Israelis are secular, however, some ultra-religious Jews dispute Israel's claim to carry on an authentically Jewish history. So do some secular Jews. A French critic (sociologist Georges Friedmann) once called Israelis "Hebrew-speaking gentiles." In fact, there was once a fashion among a group of militantly secular Israeli intellectuals to call themselves "Canaanites," i.e., people rooted in the land but entirely denying the religious tradition from which they came.

Well then, call these people what you will. "Jews," after all, is a relatively recent name for this people. They started out as Hebrews, then became Israelites. "Jew" (derived from the Kingdom of Judah, one of the two successor states to the Davidic and Solomonic Kingdom of Israel) is the post-exilic term for Israelite. It is a latecomer to history.

What to call the Israeli who does not observe the dietary laws, has no use for the synagogue, and regards the Sabbath as the day for a drive to the beach—a fair description, by the way, of most of the prime ministers of Israel? It does not matter. Plant a Jewish people in a country that comes to a standstill on Yom Kippur; speaks the language of the Bible; moves to the rhythms of the Hebrew (lunar) calendar; builds cities with the stones of its ancestors; produces Hebrew poetry and literature, Jewish scholarship and learning unmatched anywhere in the world—and you have continuity.

Israelis could use a new name. Perhaps we will one day relegate the word Jew to the 2,000-year exilic experience and once again call these people Hebrews. The term has a nice historical echo, being the name by which Joseph and Jonah answered the question: "Who are you?"

In the cultural milieu of modern Israel, assimilation is hardly the problem. Of course Israelis eat McDonald's and watch *Dallas* reruns. But so do Russians and Chinese and Danes. To say that there are heavy Western (read: American) influences on Israeli culture is to say nothing more than that Israel is as subject to the pressures of globalization as any other country. But that hardly denies its cultural distinctiveness, a fact testified to by the great difficulty immigrants have in adapting to Israel.

In the Israeli context, assimilation means the (re)attachment of Russian and Romanian, Uzbeki and Iraqi, Algerian and Argentinian Jews to a distinctively Hebraic culture. It means the exact opposite of what it means in the Diaspora: It means *giving up* alien languages, customs, and traditions. It means giving up Christmas and Easter for Hanukkah and Passover. It means giving up ancestral memories of the steppes and the pampas and the savannas of the world for Galilean hills and Jerusalem stone and Dead Sea desolation. That is what these new Israelis learn. That is what is transmitted to their children. That is why their survival as Jews is secure. Does anyone doubt that the near-million Soviet immigrants to Israel would have been largely lost to the Jewish people had they remained in Russia—and that now they will not be lost?

Some object to the idea of Israel as carrier of Jewish continuity because of the myriad splits and fractures among Israelis: Orthodox versus secular, Ashkenazi versus Sephardi, Russian versus sabra, and so on. Israel is now engaged in bitter debates over the legitimacy of Conservative and Reform Judaism and the encroachment of Orthodoxy upon the civic and social life of the country.

So what's new? Israel is simply recapitulating the Jewish norm. There are equally serious divisions in the Diaspora, as there were within the last Jewish Commonwealth: "Before the ascendancy of the Pharisees and the emergence of Rabbinic orthodoxy after the fall of the Second Temple," writes Harvard Near East scholar Frank Cross, "Judaism was more complex and variegated than we had supposed." The Dead Sea Scrolls, explains Hershel Shanks, "emphasize a hitherto unappreciated variety in Judaism of the late Second Temple period, so much so that scholars often speak not simply of Judaism, but of Judaisms."

The Second Commonwealth was a riot of Jewish sectarianism: Pharisees, Sadducees, Essenes, apocalyptics of every stripe, sects now lost to history, to say nothing of the early Christians. Those concerned about the secular-religious tensions in Israel might contemplate the centuries-long struggle between Hellenizers and traditionalists during the Second Commonwealth. The Maccabean revolt of 167–164 BC, now celebrated as Hanukkah, was, among other things, a religious civil war among Jews.

Yes, it is unlikely that Israel will produce a single Jewish identity. But that is unnecessary. The relative monolith of Rabbinic Judaism in the Middle Ages is the exception. Fracture and division is a fact of life during the modern era, as during the First and Second Commonwealths. Indeed, during the period of the First Temple, the people of Israel were actually split into two often warring states. The current divisions within Israel pale in comparison.

Whatever identity or identities are ultimately adopted by Israelis, the fact remains that for them the central problem of Diaspora Jewry—suicide by assimilation—simply does not exist. Blessed with this security of identity, Israel is growing. As a result, Israel is not just the cultural center of the Jewish world, it is rapidly becoming its demographic center as well. The relatively high birth rate yields a natural increase in population. Add a steady net rate of immigration (nearly a million since the late 1980s), and Israel's numbers rise inexorably even as the Diaspora declines.

Within a decade Israel will pass the United States as the most populous Jewish community on the globe. Within our lifetime a majority of the world's Jews will be living in Israel. That has not happened since well before Christ.

A century ago, Europe was the center of Jewish life. More than 80 percent of world Jewry lived there. The Second World War destroyed European Jewry and dispersed the survivors to the New World (mainly the United States) and to Israel. Today, 80 percent of world Jewry lives either in the United States or in Israel. Today we have a bipolar Jewish universe with two centers of gravity of approximately equal size. It is a transitional stage, however. One star is gradually dimming, the other brightening.

Soon and inevitably the cosmology of the Jewish people will have been transformed again, turned into a single-star system with a dwindling Diaspora orbiting around. It will be a return to the ancient norm: The Jewish people will be centered—not just spiritually but physically—in their ancient homeland.

## VI. The End of Dispersion

The consequences of this transformation are enormous. Israel's centrality is more than just a question of demography. It represents a bold and dangerous new strategy for Jewish survival.

For two millennia, the Jewish people survived by means of dispersion and isolation. Following the first exile in 586 BC and the second exile in AD 70 and AD 132, Jews spread first throughout Mesopotamia and the Mediterranean basin, then to northern and eastern Europe and eventually west to the New World, with communities in practically every corner of the earth, even unto India and China.

Throughout this time, the Jewish people survived the immense pressures of persecution, massacre, and forced conversion not just by faith and courage, but by geographic dispersion. Decimated here, they would survive there. The thousands of Jewish villages and towns spread across the face of Europe, the Islamic world, and the New World provided a kind of demographic insurance. However many Jews were massacred in the First Crusade along the Rhine, however many villages were destroyed in the 1648–1649 pogroms in Ukraine, there were always thousands of others spread around the globe to carry on.

This dispersion made for weakness and vulnerability for individual Jewish communities. Paradoxically, however, it made for endurance and strength for the Jewish people as a whole. No tyrant could amass enough power to threaten Jewish survival everywhere.

Until Hitler. The Nazis managed to destroy most everything Jewish from the Pyrenees to the gates of Stalingrad, an entire civilization a thousand years old. There were nine million Jews in Europe when Hitler came to power. He killed two-thirds of them. Fifty years later, the Jews have yet to recover. There were sixteen million Jews in the world in 1939. Today, there are thirteen million.

The effect of the Holocaust was not just demographic, however. It was psychological, indeed ideological, as well. It demonstrated once and for all the catastrophic danger of powerlessness. The solution was self-defense, and that meant a demographic reconcentration in a place endowed with sovereignty, statehood, and arms.

Before World War II there was great debate in the Jewish world over Zionism. Reform Judaism, for example, was for decades anti-Zionist. The Holocaust resolved that debate. Except for those at the extremes—the ultra-Orthodox right and far left—Zionism became the accepted solution to Jewish powerlessness and vulnerability. Amid the ruins, Jews made a collective decision that their future lay in self-defense and territoriality, in the ingathering of the exiles to a place where they could finally acquire the means to defend themselves.

It was the right decision, the only possible decision. But oh so perilous. What a choice of place to make one's final stand: a dot on the map, a tiny patch of near-desert, a thin ribbon of Jewish habitation behind the flimsiest of natural barriers (which the world demands that Israel relinquish). One determined tank thrust can tear it in half. One small battery of nuclear-tipped Scuds can obliterate it entirely.

To destroy the Jewish people, Hitler needed to conquer the world. All that is needed today is to conquer a territory smaller than Vermont. The terrible irony is that in solving the problem of powerlessness, the Jews have necessarily put all their eggs in one basket, a small basket hard by the waters of the Mediterranean. And on its fate hinges everything Jewish.

## VII. Thinking the Unthinkable

What if the Third Jewish Commonwealth meets the fate of the first two? The scenario is not that far-fetched: A Palestinian state is born, arms itself, concludes alliances with, say, Iraq and Syria. War breaks out between Palestine and Israel (over borders or water or terrorism). Syria and Iraq attack from without. Egypt and Saudi Arabia join the battle. The home front comes under guerrilla attack from

Palestine. Chemical and biological weapons rain down from Syria, Iraq, and Iran. Israel is overrun.

Why is this the end? Can the Jewish people not survive as they did when their homeland was destroyed and their political independence extinguished twice before? Why not a new exile, a new Diaspora, a new cycle of Jewish history?

First, because the cultural conditions of exile would be vastly different. The first exiles occurred at a time when identity was nearly coterminous with religion. An expulsion two millennia later into a secularized world affords no footing for a reestablished Jewish identity.

But more important: Why retain such an identity? Beyond the dislocation would be the sheer demoralization. Such an event would simply break the spirit. No people could survive it. Not even the Jews. This is a people that miraculously survived two previous destructions and two millennia of persecution in the hope of ultimate return and restoration. Israel is that hope. To see it destroyed, to have Isaiahs and Jeremiahs lamenting the widows of Zion once again amid the ruins of Jerusalem, is more than one people could bear.

Particularly coming after the Holocaust, the worst calamity in Jewish history. To have survived it is miracle enough. Then to survive the destruction of that which arose to redeem it—the new Jewish state—is to attribute to Jewish nationhood and survival supernatural power.

Some Jews and some scattered communities would, of course, survive. The most devout, already a minority, would carry on—as an exotic tribe, a picturesque Amish-like anachronism, a dispersed and pitied remnant of a remnant. But the Jews as a people would have retired from history.

We assume that Jewish history is cyclical: Babylonian exile in 586 BC, followed by return in 538 BC, Roman exile in AD 135, followed by return, somewhat delayed, in 1948. We forget a linear part of Jewish history: There was one other destruction, a century and a half before the fall of the First Temple. It went unrepaired. In 722 BC, the Assyrians conquered the other, larger Jewish state, the northern kingdom of Israel. (Judah, from which modern Jews are descended, was the southern kingdom.) This is the Israel of the Ten Tribes, exiled and lost forever.

So enduring is their mystery that when Lewis and Clark set off on their expedition, one of the many questions prepared for them by Dr. Benjamin Rush at Jefferson's behest was this: "What Affinity between their [the Indians'] religious Ceremonies & those of the Jews?" "Jefferson and Lewis had talked at length about these tribes," explains Stephen Ambrose. "They speculated that the lost tribes of Israel could be out there on the Plains."

Alas, not. The Ten Tribes had melted away into history. As such, they represent the historical norm. Every other people so conquered and exiled has in time disappeared. Only the Jews defied the norm. Twice. But never, I fear, again.

# A TALE OF THE NEW CHINA

## By Ethan Gutmann

*(Originally published in the May 24, 1999, issue)*

*Beijing, China—Saturday, May 8, 1999*

I really wasn't in the mood to go to the "happening," a modern art show, but my wife insisted. So we began pedaling to a gallery in a hidden courtyard just west of the Forbidden City. As we rode side by side, I asked her, unironically, "Are we having fun yet?" And her eyes smiled and she said yes, and the spring air seemed to fill with the barely held-in satisfaction of two foreigners making it in a strange culture.

We weren't living the most glamorous life, it was true. I came here to do my own TV documentary and ended up creating feel-good talk shows for Chinese-style wages—the only white man at an independent but exclusively Chinese television production company. And my wife—let's call her Betsy—was pursuing her scholarly career in the Asian way: poring through the moldering lists of the Qing emperors, hobnobbing with the poor academics who had made it through the Cultural Revolution to emerge as slightly less poor academics in the New China.

But I had been named executive producer of a new television show—a Chinese attempt to place themselves in the American market—and a top state-run TV network had just signed on. That meant, down the road, 100 million plus viewers! 150 million! More if Shanghai picked it up! True, I could only do shows about divorce, and pollution, and other "nonsensitive" topics. But still, I was building the

New China, working with enlightened Chinese producers. And Betsy was methodically building her *guanxi*, her connections. Evidence of her success was clear: the occasional lavish banquet at Deng Xiaoping's favorite Sichuan restaurant, the growing trust between her masters and her, the cultural exchanges that seemed more liquid every day. What's more, art, Betsy's kind of high art, was becoming a kind of cornerstone of legitimacy for the New China, as it drugged the populace with larger and larger doses of nationalism and nostalgia for imperial China's tokens of power, culture, and authority.

The contemporary Chinese art scene was really a sideshow for Betsy, but the invites kept coming. Politeness had become warmth, had become something close to actual friendship. And I had basked in that reflected warmth! We were turning down invitations, we were a happening couple . . . these were my heady thoughts as we arrived at the art show near the Forbidden City.

We were greeted by the kind of eager, young, lithe beauty that you come to expect at these kinds of events. Always wearing a black bodysuit, the universal symbol for sophisticated, international culture. Always: "Please sign the book, please!" Always the shy and expectant smile.

We grabbed wine off a makeshift table and quickly toured the exhibits: huge carefully posed photographs of a thin naked Chinese man and a white girl with dark roots wearing little see-through rain-coats, battery-operated dildos undulating in raw chunks of meat, an imperial robe constructed entirely of lime-green plastic, and plastic models of various state buildings filled with birds, goldfish, and—gosh, how egg roll—crickets. The guests were piling in: half Chinese artist types with bohemian hair configs and half expats, white girls with more black bodysuits and short haircuts, white men with tex-tured, checked shirts and skinny Chinese girlfriends, everyone talking excitedly, pleased to be at the Very Center of the New China on a bril-liant Saturday afternoon.

A new acquaintance, a jocund American art broker, hailed me from across the garden and I joined him under a finely painted Chinese canopy—"How are you doing?" He beamed at me. I beamed back and, after a short interlude of small talk—I couldn't wait!—I mentioned my new show. He began nodding knowingly, very good, very good, national TV, eh? You hear the news? No? I just heard about it on the way over—a bomb hit the Chinese embassy in Belgrade last night. Just a little damage, I guess, but they say 18 people were hurt. There's bound to be some trouble.

I was surprised—God, another stupid accident—but I was relieved as well. No deaths, that's the main thing. I flashed back to the previous weeks: lunching with my Chinese coworkers as they occasionally tried to rake me over the coals for airstrikes in Kosovo. I always asked the same things: "Do you know how long this war has been going on? Do you know what it's about? The Opium war had an economic element; do you think that America is trying to make money in Kosovo?" Followed by my statement that while I understood the moral impetus for the NATO intervention, I personally thought the war was a mistake because we couldn't win it; of course, the silver lining was that a Republican would probably win the election. All this was poorly translated at best, and this last comment in particular tended to elicit rather chilly expressions from some. Once a girl from accounting with Trotsky glasses closed the lunch by shouting something translated roughly as: "America is the worst country!" But in general, my coworkers were a thoughtful and likable bunch—they tended to give my position, as well as they understood it, a kind of compartmentalized respect. It was a door that would be opened and quickly shut again on the occasional rainy day lunch just to clear the air.

No deaths, that's the main thing. Could have been caused by a leaky gas stove. The Chinese tend to crowd around the kitchen and . . . anyway, no deaths, I repeated as a small, mostly naked Chinese man locked himself into a plastic bubble, painted himself green, and began to fill the bubble with water as the cameras rolled—the "happening," however, seemed to be slowing down, freezing up. Rumors began spreading around the party. First, it was one dead, and 23 injured. Then two dead. Finally a Yugoslav journalist informed me rather definitively (he saw CNN): three dead, hit by three missiles, and one was a journalist from the Xinhua news agency, top of the state heap.

It dawned on me that we should do something, and I clamored for Betsy to kiss cheeks and exchange cards. We rode east with the

Yugoslav, thinking about heading straight for the American embassy. We waved to a Chinese friend who was on her way to a massage. In what now seems like a particularly surreal moment, we almost chose to go to a book fair instead, except that the Yugoslav's cell phone tinkled, and he was informed that, indeed, something was going on near the embassy.

On Jianguomen Avenue, we passed our first rapidly marching squadron of police. Turning the corner at the Citic building, the police began to rapidly increase in density: one every ten yards, then one in five, then one in three. As we turned the corner down embassy row, we heard a strange sound in the distance, the roar, the sound that calls a man as surely as bagpipes.

My heart raced now, and we pedaled fast down the block, suddenly running into a police barricade, with about 50 Chinese onlookers trying to peer down the block toward the embassy. Betsy and I simply walked past the police cordon, making a big show of chaining our bikes up. The police would not stop people like us from entering, because foreigners could still do what they wanted in Beijing, because they weren't Chinese.

Immediately the first battalion of young student protesters from some obscure university advanced down the block. You've seen the pictures, I imagine. It was textbook: long red banners with Chinese characters splashed on in black paint. Waving little fists mechanically. Freshmen and sophomores mostly. The hipper Chinese students had torn scarves wrapped around the head. The majority were clean-cut specimens. We ran past them toward the embassy, actually finding ourselves a cramped space just in front of the gate.

Beyond the gate, all was normal. Spreading trees turned the area into a kind of large grotto, gentle sunlight, although the wind was picking up. The U.S. embassy was well kept, as always, its facade showing no signs of life—why would it on a golfer's Saturday? Surrounding it were the cameras, mainly foreign, although a few from Chinese state television, CCTV. About 15 policemen in their green uniforms lined the gate, relaxed, almost in a holiday mood. As Betsy pointed out, any break in the monotony of life in an authoritarian state is so lovely! They were finally being given something to do!

About 500 student demonstrators stood in front, with long red and gold banners and hand-painted signs: USA go to hell, F— USA, US Killer, F— NATO, NATO=Nazi. The chants were translated for me: PLEAD GUILTY! U.S. KILLER! GO HOME! MURDERERS! U.S. PIGS GO HOME!—COME OUT COME OUT OR WE'LL GO IN! AMBASSADOR COME OUT! PEOPLE'S REPUBLIC OF CHINA BANZAI! Very loud and high-pitched, over and over, led by an individual with a megaphone. The screaming would change pitch and pace every minute, and when it did, the faces would relax—were they having fun yet? Their eyes smiled yes. Then the megaphone organizer would start pumping his arms, and the teeth would retract, and the mouths would start, and the hands would start spasmodic thrusting, and the testosterone would pump, and the eyes would start rolling. All in precision drill and extremely responsive to orders, as if they were being given mild electric shocks. Still, the holiday atmosphere prevailed— after all, this is the pure, angry, righteousness-defining moment that college students the world over dream of! But in China, for ten long years nothing—and now this!

No one touched us, no one shoved, and yet, behind the police, behind the fence, inside the courtyard was a flag—mine—and a plot of land, safe land. Yet I felt heady and faint just for being here: the capital of next century's Superpower, the center of the world for a day, its youth, Borg-like in their unified loathing of our flag and our little plot.

After a while, when the chanting lost its steam, the megaphone leader would strike up a short sing-along of the national anthem. This was the signal to leave, to shuffle along and give the next university its chance to demonstrate.

The cycle continued, fresh waves of students, monotony. Several British journalists discussed the numbers: They felt it was low, about 3,000; in a kind of Chinese scarf trick, the same student groups kept reappearing after an hour or so. The students, when isolated and interviewed, were naively forthcoming; the university authorities had told them to come, told them to make banners, arranged the buses. The whole demonstration was canned, and yet . . .

Fresh chanters had started from Beijing University. As the major

instigators of Tiananmen, they had a legacy to uphold. Their demonstration went through the cycles, the patriotic song drifted off, time to
leave, but suddenly someone sat. Immediately, 50 more sat, and then
the rest, with the organizer yelling impotently. From the moment of
arrival in Beijing, I had always sensed the weird political static electricity that seemed to surround Chinese crowds—a split-second deteri-
oration of the rules, a tendency for aggression, unpredictability. As one
China hand had put it to me, "If left to their own devices, would the
Chinese people have Li Peng hanging from a lamp post within ten
minutes?"

The Beijing U. students sat down, and we wondered, were we present at the birth of a new Tiananmen? Just as quickly, it became clear
that they were the wrong cast: too young, too well scrubbed, and too
neophyte. Pleased, excited by their own petty audacity, they stayed put
for a minute, and then the Tiananmen wannabes were herded off. The
cops doubled in front of the embassy and locked arms. I told Betsy to
conserve our film.

Next up, Qinghai U., "the MIT of China," was back. The chanting
reached a fever pitch, and then a lull . . . something flew out of the
crowd and crashed against the embassy. Whoops of joy. Then another,
then another, sounding like bugs crashing on a windshield. Now we
could see the hands releasing the chunks of concrete. The lamps top-
ping the fence were quickly destroyed . . . the cops impassive.

And the day was really over, and a night of destruction of the
embassy and sport was beginning. I groggily realized that my new TV
show was probably gone, maybe my job, too . . . that I was reduced,
but also less compromised. After ten years the State was showing its
fist to the world again, not just to a few China watchers and China
hands.

The rest, you know. But how could I have known what would follow? The nonstop xenophobic and racist exhortations on TV, with
weeping relatives of the dead in Belgrade holding bloody clothes to
their chests. The total blackout of American statements of explanation, apology, regret. The cancellation of all American movies and
music. The burning of the Chengdu consulate. The anti-American war
films in the afternoon. The beating of journalists. The sanctioned

racism on the streets. The condescending "tolerance" at work. My slowly awakening comprehension of the leadership struggle that manufactured many of these events. Most of all, the feeling that something had shifted under my feet.

China was discarding the foreign devil, like a used shell dropping off a cicada's back. Fun was fun, but all around us, the wings were suddenly beating way too fast.

# A COWERING SUPERPOWER

## By Reuel Marc Gerecht

*(Originally published in the July 30, 2001, issue)*

IN DECEMBER 1999, THE Clinton administration issued a worldwide terrorist alert to Americans overseas advising them to avoid crowded millennial celebrations. Bomb-toting Islamic militants under the banner of the Saudi terrorist Osama bin Laden had declared war, so Americans were to stay discreetly indoors while other Westerners partied. In Israel and Jordan, American Christians were strongly advised to avoid any public manifestation of their faith. Vexed by the growing number, geographical range, and fearfulness of Washington's warnings, one senior Foreign Service officer declared the millennial alarm "the chicken-little PR finale of America's cover-your-ass foreign policy."

Unfortunately, this hard-nosed diplomat was wrong. The policy he deplored was not about to end. The Bush administration has continued and actually surpassed its predecessor's display of timidity in the Middle East. The possibility of terrorist attacks recently prompted the Pentagon to withdraw U.S. Marines from military exercises in Jordan and hastily move ships anchored in Bahrain, the home base of the U.S. Navy in the Persian Gulf. Likewise, pistol-packing FBI officials investigating the October 2000 attack on the USS *Cole* in Aden, Yemen, decided to scoot—against the counsel of the State Department and the U.S. embassy in San'a—when they thought a terrorist attack might be imminent.

Which prompts the question: Are we a great power or not? If we are, then what in the world are we doing running from men whose mission in life it is to make us flee? If Marines and men-of-war cannot hold their own against the specter of a Saudi terrorist, how will our friends, let alone our enemies, in the macho Middle East measure us against real heavyweights like Saddam Hussein or the clerics of Iran?

Osama bin Laden and his terrorist organization, al Qaeda, scored an impressive victory by nearly sinking the *Cole*, yet Washington still has not responded. Our fear is pure oxygen to Islamic militants. Every alert, particularly when it panics U.S. military and diplomatic personnel, sends an adrenaline rush into the central nervous system of men truly convinced that with God's help and the right explosives they can crack the will of the infidels who are, in their eyes, destroying the one true faith.

Secretary of defense Donald Rumsfeld's decision to yank the Marines out of Jordan is, when viewed from the mud-brick and cinder-block ghettos of the Middle East, an extraordinary triumph, further proof that the martyrs of the *Cole* attack died gloriously. America's military leaders may think that they're being prudent with our soldiers; the average man in the streets of Amman certainly knows better. Terrorism is war by unconventional means. Its ultimate objective is the psychological debilitation of the enemy through fear. In the fight against terrorism, the U.S. military's ever-more exclusive focus on "force protection" diminishes the awe in which America is held abroad, the ultimate guarantor of the safety of U.S. civilians and soldiers, especially in lands where hostility to the West rests near the surface.

Martyrdom has a long and complex history in the Muslim world. It began with God's promise of paradise to the seventh-century warriors who died expanding the first Islamic state. Over the centuries, rules and understandings evolved about the pivotal difference between combatants and civilians, but these have evaporated in the fundamentalists' radical modernity, which divides the world cleanly and brutally between good and evil. If we want to play hardball with Islamic militants—and the Bush administration isn't spending billions of dollars on counterterrorism to be nice—we need to pay more attention to the history and metaphysics of Islamic extremism. In other words, we need to take bin Laden's men apart psychologically. Cutting off the flow of oxygen to the Muslim world's anti-American radicals isn't an impossible task, so long as we patiently hold our ground.

Osama bin Laden and his men are, or at least aspire to be, contemporary "Assassins," the medieval founding fathers of modern political terrorism, who from their mountain redoubts in Iran and Syria first showed the possibilities of purposeful, disciplined terrorism. For a time, great sultanates and kingdoms lived in profound fear of men who gladly sacrificed themselves to kill their enemies. The word "assassin" entered Western languages because the originality and shock of the Assassins' assaults were sufficient to embed the word permanently into the consciousness of the region's Muslims and Christians. The allure of the Assassins' propaganda, which depicted acts of violence as acts of divine love and anger, tapped into strong currents within Islam that see God's justice continuously betrayed by the *ulu al-amr*, "the men who hold the reins." Bin Laden might not like being paired religiously with the "Old Man of the Mountain," the mysterious Shi'ite overlord of the Assassins, but the Sunni Arab militant wouldn't mind at all the geopolitical comparison, which, given his own mountain hideaway and his faithful kamikazes, has no doubt already occurred to him.

Though there have been times when large numbers of young Muslim men felt the thrill of a charismatic calling—the early years of Iran's Islamic revolution is the most recent case—the contemporary Sunni Arab world, where bin Laden draws most of his strength, hasn't experienced a similarly infectious wave. One can find many angry young men in Yemen, Egypt, Saudi Arabia, Algeria, or Gaza; few want to vent their emotions against the age-old Western enemy by vaporizing themselves in a truck or skiff. Jihad, the moral and spiritual obligation of a Muslim to wage war to protect (and, in Islam's ascendant days, expand) the faith, is no longer understood by most Muslims as denoting anything more than an individual's duty to survey his soul.

Drawing in good new recruits to al Qaeda's cause thus isn't, as many Westerners might assume, an easy task. In Afghanistan, a broken, barren country far from the crossroads of the Muslim world, it probably seems daunting, which is one reason why so many of bin Laden's foot soldiers are hapless, ill-educated misfits who get themselves arrested when they stray too far from their native stomping grounds in the disorganized, listless Third World.

Islamic militants, like everybody else, must have hope. They, like everybody else, believe in winning. Israel's most determined enemies—Lebanon's Hezbollah, Iran's mullahs, the Palestinian fundamentalists in Hamas and Islamic Jihad, and Yasser Arafat's protégés in his security and intelligence services—constantly underscore Israel's decision to withdraw unilaterally from southern Lebanon in their clarion calls for more martyrs. This Israeli action, widely applauded in the West as strategically astute and morally estimable, was seen (correctly) in the Middle East as an astonishing retreat by a once seemingly unbeatable Western power. Israeli weakness, not Israeli "intransigence," is what heats the militant's death-wish dreams red-hot.

We need to remember that al Qaeda, like its allied fundamentalist organizations, has to survive on little regular positive feedback. For Hamas, killing Israelis is easy since the Arab and Jewish communities are geographically and economically intertwined. The body count on the nightly news keeps the spirits up. For anti-American holy warriors based in Afghanistan and the northwest frontier of Pakistan, daily life in comparison is tough. Radios and satellite phones are the only constant links with the outside world. Time passes very slowly. The two years between the bombings of the U.S. embassies in Africa in August 1998 and the *Cole* attack could seem like an eternity to young men who burn to die. When failures supervene—for example, the botched suicide attack on the USS *The Sullivans* in Aden in January 2000—it becomes that much harder to sustain spirit and momentum.

America is, as Muslim militants quite frankly admit, an awesome foe. The allure and mystique of America in the Middle East are nearly impossible to overstate: It's Goliath, Thomas Jefferson, Wall Street, and Madonna rolled together in a cacophony of sound and color that relentlessly fascinates and repels. In the eyes of Islamic fundamentalists, we are worse than the Mongols, who laid low the Muslim heartland and nearly annihilated the faith. As fundamentalists regularly complain, most Muslims are easygoing backsliders, willing victims of Western ways. Even the Saudi royal family—perhaps the folks bin Laden detests the most—who are supposed to maintain the rigorous, funless, Hanbali school of Sunni Islam, have become woefully dependent on the West, in particular the United States.

The Afghan civil war also probably complicates bin Laden's life. His disparate collection of holy warriors fight alongside the fundamentalist Taliban against Ahmad Shah Massoud's Northern Alliance. This gives al Qaeda's guerrillas some combat experience and esprit de corps. Though the main tie between bin Laden and Taliban leader Mullah Omar is spiritual, the war allows the Saudi militant to further secure his exile home by contributing men and materiel to the Taliban campaigns and the Pakistani-approved camps where Kashmiri separatists are sometimes trained.

However, this war *really* isn't fun: ambushes, minefields, artillery barrages, and trench warfare through mountainous countryside increasingly define Afghanistan's strife. The offer of "terrorist training" in Afghanistan, a country where "good" Muslim peasants are fighting "bad" Muslim peasants, isn't a recruitment pitch with lasting appeal for young Arab men who really just want to kill Americans.

For bin Laden's "sleepers"—agents already outside of Afghanistan awaiting the right moment to strike an American target—the situation is probably little better. While terrorists who've implanted themselves into the local environment can obviously be lethally effective (both the embassy attacks in Africa and the operations against *The Sullivans* and the *Cole* in Yemen relied on such people), few men in bin Laden's network are likely to have the fortitude, talent, and discretion to hold themselves in position long, their death-wish intact. Like isolated foreign espionage agents in dangerous areas, they probably need regular spiritual reinforcement and monitoring, perhaps more than their more numerous brethren in Afghanistan, who can counter isolation and ennui through open fraternity.

If these are the terrorists we're up against, what would a successful American counterterrorist policy look like? Obviously, it should play up our strengths and relentlessly play upon our enemy's anxieties and fears.

Bloodied, the crew of the wounded *Cole* did better. When their ship limped out of Aden's mountain-ringed harbor, the sailors played over the loudspeakers hard rock graphically describing what they wanted to do to the terrorists, if not the denizens of Aden. Would that the Clinton White House and the Navy's senior brass had matched the crew's insight into the Middle East's power politics and immediately dispatched other warships to Yemen to demonstrate symbolically the indefatigability of American power.

Once upon a time, the U.S. Navy reacted more astutely to tragedy. After the kamikaze truck-bomb assault on the U.S. Marine barracks in Beirut in 1983, the Navy's planners correctly anticipated kamikaze boat-bombers. The Navy experimented with weaponry and discovered that a .50-caliber round fired into the engine block of a small boat will stop forward momentum quite quickly. Such weaponry on the *Cole*, combined with the shoot-to-kill orders that are standard operating procedure for U.S. diplomatic security officers who determine that a lethal threat exists, would have likely saved the ship.

It's hard to believe that the Navy, which enjoys a relatively isolated and protected preserve for its vessels in Bahrain, couldn't have adopted similar tactics to protect its ships and men. We ought not make our enemies larger than they are: Bin Laden's holy warriors aren't remotely in a class with our SEAL teams, the elite commando strike units of the U.S. Navy which tirelessly train to disable warships in protected harbors. If bin Laden wants to triumph over us again, we should at least make his men do something more stressful than converse menacingly over intercepted telephones—which apparently was enough to provoke the Pentagon's flight from Jordan and Bahrain.

Most American diplomats and intelligence officers unquestionably know there is no efficacy in a bull-horned terrorist warning: It's quiet, bare-knuckled, local police work, not worldwide bulletins on CNN, that saves lives. Yet as another senior Foreign Service officer remarked, "There is no percentage in standing against the tide." Informing American citizens discreetly that a specific and credible threat exists in a certain time and place may have some value (informing terrorists that we are privy to their plans may well incline them to switch targets). But advising Americans that a country the size of Turkey, which always seems to be in some state of alert, may have an anti-American terrorist plotting within its borders is just silly. In this risk-averse quagmire, America's martial virtues and pride inevitably get lost.

Going in the opposite direction, other foreign-affairs circles pooh-pooh the terrorist threat from the Middle East, pointing out that more

Americans kill themselves each year flying kites than die at the hands of holy warriors. Compared with those of the 1970s and '80s—the halcyon days of the Palestine Liberation Organization, Hezbollah, and the intelligence ministries of Syria, Libya, and clerical Iran—today's death tolls and sense of siege really aren't so bad. The issue of terrorism has been hijacked, so these circles often assert, by the 24-hour media maw and intelligence and security bureaucracies eager to encourage Congress's multibillion-dollar counterterrorist budgets.

This critique is statistically correct and bureaucratically astute, but otherwise wrong. Today's radical Islamic terrorism matters because it helps define the way the United States is perceived in the Middle East and beyond. Only 17 sailors died on the *Cole*, but symbolically it was a stunning achievement for a jihadist fraternity that proved it could strike a warship, the historic instrument of Western power. Anyone who has been in the coffeehouses and bazaars of the Middle East since the *Cole* attack knows how ordinary Muslims, who generally don't countenance bin Laden's killing, nevertheless are in awe of him. A good tactician when it comes to Muslim emotions, bin Laden has played well the clash of civilizations.

These are bad days for America in the Middle East. Ali Khamenei, Iran's clerical overlord, isn't alone in seeing the United States on the defensive throughout the region. American policy toward the Israeli-Arab confrontation—keep trading Israeli-held land for the promise of Arab peace—is naive. Yet the Israeli Left adopted this policy and kicked it into overdrive, and now the inevitable dénouement is at hand: a real war between the Israelis and Palestinians. Seemingly endless Israeli concessions, always applauded by the Clinton administration, have undermined America's standing in the Middle East.

The Bush administration, led by an obviously and understandably exasperated Colin Powell, has compounded the problem by endorsing the Mitchell Report, which puts forth the odd, very secular notion that Israeli settlements in the West Bank and Gaza, comprising less than 2 percent of the land, have provoked Palestinian young men to blow themselves to bits. The White House and Foggy Bottom are desperate to "stop the cycle of violence." But only violence—Israeli violence, if prime minister Ariel Sharon still has the stamina and insight at last to unleash it—may recoup the damage that the Labor party, Bill Clinton, and the Near East Bureau of the State Department have done to America's standing in the region.

Farther east, the situation is even worse. From the spring of 1996, the Clinton administration's Iraq policy was in meltdown; under the Bush administration, it has completely liquefied. The administration's retargeted "smart sanctions" are clearly a huge retreat, which the Russians, we can only pray, have turned into a permanent defeat with their threatened veto in the Security Council. All we need is to have two of our principal allies in the region, Turkey and Jordan, further enmeshed in an America-ordained, U.N.-"enforced" sanctions regime that pivots, when all the diplomatic varnish is off, on bribery. Face to face in the Middle East, *rishwa* is often the only expeditious route for virtue to triumph over villainy. But bribery mediated by the United Nations would be a strategic cross-cultural mess. With "smart sanctions" in place, not only would Saddam continue his "illegal" cross-border weapons-related commerce—the allure of Iraqi oil money is just too great—but we would have Turkey and Jordan adamantly seeking financial redress for their efforts to staunch the unstoppable trade. We would again be asking others—in the case of Jordan, a weak kingdom always inclined to appease Saddam Hussein—to bear the burden and responsibility for our failure to confront directly the Iraqi dictator.

Does anyone in the Bush administration remember Madeleine Albright, Sandy Berger, and their minions spinning themselves dizzy trying to deny that Saddam Hussein had outwaited and outplayed Washington? It would be better to see the administration start explaining how we will live with Saddam and his nuclear weapons than to see senior Bush officials, in the manner of the Clintonites, fib to themselves and the public. In any case, in Middle Eastern eyes, the Butcher of Baghdad has checked, if not checkmated, the United States.

Only against this backdrop can we properly assess the threat bin Laden poses. The Saudi militant is unquestionably going to come at us again. If he can find a weak spot, which he probably can, he will target us most likely in the Third World, where his men can maneuver. Then the Bush administration will have to make a defining decision. Will President Bush continue the Clinton administration's preference for putting terrorist strikes into the FBI's investigative hands and, forensic evidence willing, into the courts, thereby avoiding the diplomatically messy question of retaliation? Will the administration forcefully complement the above with another barrage of cruise missiles aimed at rock huts on the thin hope of catching bin Laden and his lieutenants unawares?

Deputy secretary of state Richard Armitage recently warned that the United States would hold the Taliban responsible for future attacks by al Qaeda. We can only hope that this doesn't mean filing some future court case in New York City or bouncing the rubble in makeshift camps in Afghanistan. The Taliban chieftain Mullah Omar ought to discover that dead Americans mean cruise missiles coming through his bedroom window and cluster bombs all over his frontline troops.

The Pentagon's alarms in the Middle East and the fecklessness of the administration's policy toward Saddam Hussein and Yasser Arafat, however, suggest a different chain of events. Odds are, America's position in the Middle East is going to get much worse. In the not too distant future, bin Laden may well rightfully proclaim that he, as much as Saddam Hussein, exposed America's writ and most terrifying principles—liberal, secular democracy—as finished in the Arab world. This would be an amazing accomplishment for a Saudi holy warrior, considering the forces arrayed against him. The Assassins achieved far less and were immortalized by friend and foe alike.

# THE UNITED STATE
# OF AMERICA

## By David Tell

⇒⬥

*(Originally published, as "The End of Illusions,"
in the September 24, 2001, issue)*

EVEN AS THE SKY WAS FALLING Tuesday morning, September 11, visitors to the *Nation* magazine's website could find a freshly posted essay by Edward Said on the intellectual's role in the modern world. A true intellectual, Said declared, now makes it his mission to publicize those injustices that are "occurring in reality"—like the Israeli "occupation" of Palestine. And this work is hard, for the "dominant discourse" has all but smothered reality in a blanket of "counterfeit universals" designed to "create consent and tacit approval." Indeed, Said noted, so thoroughly has the propaganda of "unseen power" penetrated popular consciousness that even some intelligent people have come to believe in its empty "confections": "the West" and "democracy," on the one hand, and "rogue states" and Middle Eastern "terrorism" on the other.

Now that two "confections," each carrying 20,000 gallons of liquid high-explosive, have flown across New York Harbor at 400 miles an hour and slaughtered 3,000 office workers (and counting) in the blink of an eye . . . well, how easy it becomes to forget that anyone in America once took seriously such arguments as these. That until recently, our custodians of respectable opinion, surveying the "dispute" between Arab suicide bombers and the Israeli schoolchildren whose blood they spilled, carefully constrained themselves to revile the sin but not entirely the sinner. That it was the policy of the United

States government, in fact, to divvy up the equities just a bit—to insist that its friends in Jerusalem acknowledge, in the "sources" of antiquity, some claim to justice by the enemies at their gates.

No more, all gone, goodbye—vanished in the fireballs and, later that same day, in the television image of ululating hags and Palestinian Authority policemen in Nablus, dancing with joy on lower Manhattan's grave. It seems there is, after all, something very properly called Middle Eastern terrorism and something else, starkly different, very properly called the West. It seems the former means indiscriminately to kill anything associated with the latter—means to kill *us*. No, *has* killed us, and will no doubt eagerly kill us *again*, huge numbers of us at a time, given the slightest opportunity. What more bracing piece of information could there be than that? And who among us, having now absorbed it, is any longer susceptible to the imbecile morality of Edward Said's black-is-white dialectics—which would explicitly extenuate such obvious evil? No one, really. No one at all.

This unanimity itself bears inspection. Pearl Harbor is widely invoked as a historical parallel to last week's World Trade Center and Pentagon disasters. As a practical matter, the analogy is well short of perfect. But it is true, all the same, that one must reach back as far as December 1941—and reaches in vain any further—to find any other instance in which Americans, as with a single mind and heart, were led to think and feel the same things, with such clarity and conviction, about something so important. And in the space of an hour.

We fret a great deal these days that the nation isn't indivisible any more, that our "mosaic" is badly frayed, that we no longer know who we are. It turns out that is nonsense. At the moment, America fairly vibrates with an almost tribal sense of identity, a fraternal concern that can barely be contained. We know exactly who we are. And we love ourselves as we should and must. Had they been asked for direct and personal assistance in the New York rescue effort last week, millions of Californians would have set out across the continent immediately, even if they'd had to walk.

And so with our politics. They are "fractured," we tell ourselves, over and over. A giant crevice of ideology and partisanship has opened beneath our feet, so broad and deep that critical decision-making, by voters or their government, has become all but impossible. Has it, though? Honestly, now: Isn't it true that, from the moment we started lashing ourselves about this alleged incapacity, no genuinely critical decision has actually confronted the United States—until now? And isn't it true, too, that we have just managed to defy our own well-practiced pessimism not once or twice but three times in a matter of days?

Thus, the United States has announced that it considers terrorist attacks on its citizens and property an act of war, not a crime. The United States has announced that it will prosecute this war unilaterally if need be. The United States has announced that its targets are not simply the men directly responsible for mass murders in New York City and Arlington, Virginia—but any group or government that has supported or sustained them, *and any group or government inclined to support or sustain others like them in the future.* Each of these announcements represents a striking and hugely consequential departure from past policy and practice. And the entire exercise has been effected without ordinary public debate. Not because debate was suppressed or obscured by the emotion of the moment. But because debate was unnecessary. We are all thinking the same things, and reaching the same conclusions, and all by ourselves, individually, at lightning speed. Imagine: American politics is operating with supranormal efficiency and effectiveness at the moment.

But not without self-doubt. The educated man mistrusts his passions, however appropriate, and the democratic man mistrusts consensus, however perfectly reasoned. No course of politics is ever entirely right, we figure, so we have already begun looking for errors at the margin, even prospective ones so improbable as to approach the ludicrous. It won't do indiscriminately to bomb Arabs overseas, or to lynch those of Arab descent who are our fellow citizens here at home, or to purchase the heightened security we need by mortgaging the First and Fourth Amendments—so we murmur in the op-ed pages of our leading newspapers. Odd that we should insult ourselves this way, cautioning against misdeeds that none of us has even contemplated. In the movie version, perhaps, martial law would be declared and nukes would drop. But this is not a movie, and the United States has no reason whatsoever to suspect that the real-life hero's role it now assumes might wobble into ambiguity.

Nor should President Bush, as he embarks on the Good War of a new century, have serious cause to fear that he will fail the standards established in the last. They are the wrong standards, in any case: This is nothing close to Pearl Harbor when you get right down to it. We are not the unarmed and inexperienced America of 1941; we are a global colossus. We are stunned and mourning, but we are not for a moment afraid of defeat. We do not need to hear, as White House counselor Karen Hughes bizarrely reassured us last Tuesday, that "your government continues to function"—for who among us worried that the government was broken? We do not need to hear from Bush himself that "my resolve is steady and strong"; we expect and demand as much. We really don't need to hear anything but the news, in fact. Rooseveltian eloquence would be almost frivolous at this point. Where America might be led, she has already arrived on her own.

Behind us, in a preformed coalition, stands the civilized world—the West. Before us are our enemies, and they are . . . what? In the first hours after the towers fell in Manhattan, television analysts narrowed the list of likely suspects by repeatedly dubbing their operation "sophisticated." The usage was spectacularly misapplied. A group of men truly intent on the massacre of innocents can usually pull it off. What distinguishes them is not their "sophistication," but instead—and simply—their willingness to do it. It is an impulse unique to barbarians. And in the present circumstance, we know with all necessary precision who the barbarians are, who their friends are, and where they live. Most of all, we know that they will conduct some future massacre—or massacres—unless we destroy them first.

So we will destroy them. And we have only to worry that we will slow ourselves down by worrying overmuch about how.

# VALOR AND VICTIMHOOD AFTER SEPTEMBER 11

## By Tod Lindberg

*(Originally published, as "Rebirth of a Nation,"*
*in the March 4, 2002, issue)*

THERE ARE NO MORE YELLOW RIBBONS. For more than 20 years, in times of travail, the yellow ribbons have come out. The Iranian hostage crisis of 1979–80 called forth a nationwide flowering of yellow ribbons. And at one time or another since then—can this really all have been wrought by Tony Orlando and Dawn singing "Tie a Yellow Ribbon Round the Ole Oak Tree"?—the yellow ribbon has been pressed into service as a symbol of hope amid adversity, an expression of longing for the return of those who are not home. In accordance with past practice, the aftermath of the attack on the Twin Towers could surely have been an occasion for yellow ribbons: thousands lost and feared dead, the uncertainty of the families of the missing, the conclusion growing inevitable that even the bodies might never be recovered. And in fact, in the first day or two, one did see a few yellow ribbons, usually in a collage with a photograph of someone missing, held desperately by a loved one still in shock. But then, without comment, the yellow ribbons were gone. All the ribbons now are red, white, and blue.

The difference between a country full of yellow ribbons and a country full of red, white, and blue ribbons—and buttons, bumper stickers, lapel pins, scarves, neckties, billboards, and flags of all size and description—neatly captures the passing of one era and the birth of another, as well as the character of each. The yellow ribbon is the symbol of the victim—of the aggrieved individual, someone powerless at the hands of the powerful. The victim's opposite number is the self-satisfied individual, master of his own life and times. The United States of September 10 was a place peopled amply with both types. The private concerns of people, whether satisfied or unsatisfied, were at the forefront of daily life.

The red, white, and blue ribbons are the symbol of something different: a nation. Which is to say, Americans with a sense of themselves as a people, countrymen, united by something that is precisely not private. The red, white, and blue were a product of a sudden sense of solidarity, the felt need to express the view that an attack on one is an attack on all. It wasn't that nearly 3,000 individuals died in the Twin Towers. It was that they died in an attack on the United States. American solidarity wasn't born that day; it was revealed. After a long absence, Americans returned to the public square they had left for their private gardens, and to make sure everyone knew, they draped it in red, white, and blue.

The last two decades of the twentieth century saw the apparent triumph of classical liberalism. The old collectivist aspirations of communism and socialism, as well as the political control over society they entailed, gave way to a new respect for the individual and for the system in which individuals seem to prosper collectively, namely, democratic capitalism. These decades saw a tremendous flourishing of human freedom, and while the gains were neither universal nor uniform across the globe, they were unmistakable where they occurred.

Now, this was all to the good. But it did give rise to certain distortions of perspective, themselves the product of the focal point of all the accomplishment, the individual.

In 1992, my Hoover Institution colleague Charles J. Sykes published an incisive book called *A Nation of Victims: The Decay of the American Character*. In it, he argued that "perhaps the most extraordinary phenomenon of our time has been the eagerness with which more and more groups and individuals—members of the middle class, auto company executives and pampered academics included—have defined themselves as victims of one sort or another." He noted, "In a culture of soundbites and slogans substituted for rational argument, the claim that one is a victim has become one of the few universally recognized currencies of intellectual exchange."

The victim is the supreme authority on his own grievance. Others with something to say on the subject of the grievance in question must defer to the victim, whose unique experience as victim lends him an unimpeachable righteousness, which he does not hesitate to assert. Thus, one may be told that until one has gone through what the victim has gone through, one cannot really know what it's like to be victimized in this way—until one is black, a rape victim, gay, disabled, a war veteran, a cancer survivor, a family member of someone who died in a massacre, and so on, one cannot know what it is like, and so one ought to defer, if not shut up. It's worth noting that the appeal of victimhood transcends political divisions in the United States. Republicans on Capitol Hill have been tireless champions of "victim's rights" where the victimization is due to crime. And I have been at more than one black-tie Washington event in which a roomful of people including current and former senior government officials have stood up to cheer a conservative who has just finished describing victimization at the hands of the left.

Victims come in groups, and typically these groups are minorities as against a majority that is responsible in some collective sense for the victims' unjust treatment. But this is not necessarily so. Women outnumber men, but it is no rebuttal to the claim that women are the victims of men to cite the greater number of women. The essential element is an imbalance of power, as perceived by the victim.

The paradox of victimization is that claiming this status is actually an assertion of superiority. Whatever handicap one suffers as a result of victimization, because the handicap is unjust, it cannot be said to diminish one. When a victim claims a right to fair treatment, those who have already been treated fairly (or better)—those with greater power—are called upon not only to treat the victim fairly but also to acknowledge the victim's status as someone who has been treated unfairly. This status is permanent, quite apart from the remedy of fair treatment, even if the latter is forthcoming. This status thus confers a permanent claim to speak with righteousness on the subject in question. And it will be up to the victims themselves to decide whether and when and to what degree the underlying power relations that gave rise to their victimization have really changed. This is what Bill Clinton means when he says African Americans are the conscience of the United States.

Yet the group character of expression of a sense of victimization is misleading. Regarding oneself as a victim is a fundamental expression of self. This is not to say that one has a choice; a woman who has been raped is a rape victim and is going to regard herself as such. But it's the rape of the individual that makes the victim, not the relationship between this rape and other rapes. Obviously, the fact that suffering is individual, even if more than one person suffers, is all the more important to cases in which a sense of oneself as a victim is, so to speak, more optional. Although I have ridden bicycles, I have never felt myself to be a victim of "motorism," as one professor quoted by Sykes claimed to be. That someone else should feel himself to have suffered this injustice does not enable him to arouse, against my wishes, such a feeling in me (although he is welcome to try; the activity goes by the name of consciousness-raising).

The danger in making generalizations about victimhood is insensitivity. The conclusion that victims' claims ultimately amount to an assertion of superior status tells us nothing about whether they should be taken seriously as victims or in what way. Sykes, in *A Nation of Victims*, has an answer to this, juxtaposing the newer "victimism" with an older "American character" whose ethic is personal responsibility; he finds the latter to be in a state of "decay," the result being a profusion of bogus "rights" claims.

But it's here, I think, that we meet another character of recent times, inversely related to but less recognized than the victim. He is the "unvictim." At his extreme, he sees himself as "a Master of the Universe," in Tom Wolfe's unforgettable phrase from *The Bonfire of the Vanities*. He is personal responsibility in flesh and blood, in that he believes himself to be something very close to the sole agent of his own achievement. The unvictim sees his success as the product of his hard work, his persistence (especially in adversity), and his determination.

Above all, he has never allowed himself to think of himself as a victim, as powerless before powerful forces. When faced with a set-back, the unvictim dusts himself off and looks forward, not backward. He thinks others should do the same. If others fail to "take responsibil-ity," they make the least of their situation, however bad it may be, when they could make more. In the United States, some measure of success is available to everyone who just buckles down, in the unvic-tim's credo. As for those whom the unvictim deems failures, he does indeed hold them personally responsible for failure. The harshness of this judgment is mitigated in the unvictim's mind by the possibility, available to all, of reversing ill fortune immediately by taking personal responsibility, and also perhaps by the unvictim's memories of his own personal failures as things he was able to overcome. Rush Limbaugh is arguably America's leading unvictim.

Some would say that the unvictim is a member of an oppressor class. Collectively, this class is acting to maintain its power over others and its position of privilege. "Personal responsibility," in this view, is just code for perpetuating the system from which the unvictims gain advantage. They are the authors of their success only collectively, and only at the expense of those whom they collectively oppress.

This argument, based once again on an assumption of false con-sciousness on the part of those it describes, similarly attempts to derive a group identity from what are distinctly individual feelings. Unvictims may have certain political views in common, but at bot-tom, these are convictions in favor of a politics of individuality. Once again, this group of people is not a collective, conscious or uncon-scious, in the sense of acting collectively. What is wrongly described as class-based political behavior—in this case, oppression—is actually the sum of thoroughgoingly individualist concerns.

There is obviously a certain correspondence between "victimism" and the principal concerns of the Democratic party, as well as between "unvictimism" and the concerns of the Republican party. But neither tendency is wholly confined within either party, nor is it the case that the two cannot commingle, or at least exist as contradictory impulses, within individuals.

Notwithstanding our sometime preoccupation with our partisan-ship and division, it is time to quit looking at the differences between the two tendencies and start looking at what they have in common. If we have been a nation of victims, we have also been a nation of unvictims. In each case it is the apotheosis of the individual, of private concern. Domestic political dispute then comes down to a quarrel over whether government is an instrument of self-actualization or an obstacle to self-actualization.

In the pre-September 11 world in which both the victim and the unvictim flourished, so did the yellow ribbon as a symbol of victimization, *as well as* the urge to tear down yellow ribbons as proof of unvictimization. What was really missing all along, as became clear with the appearance of red, white, and blue ribbons, was the nation.

The high-water mark of victimism may have been the solicitude Attorney General John Ashcroft displayed last summer toward the survivors and the families of the dead with regard to the execution of Oklahoma City bomber Timothy McVeigh. Ashcroft repeatedly maintained that he personally drew great strength from his encounters with the victims. He even went so far as to arrange for closed-circuit television transmission of McVeigh's demise by lethal injection. The viewing room in the death house of the Indiana federal prison was, of course, too small to accommodate more than a handful of victims, whom government officials had determined to number in the thousands. Yet viewing McVeigh's end was something Ashcroft actually called a "right" the victims enjoyed (though, of course, not an obligation). In the end, a couple hundred availed themselves of Ashcroft's generosity, gathering at a site in Oklahoma City to watch.

In viewing victims' rights as the primary end of the justice system, Ashcroft (certainly without reflection) placed the U.S. Constitution and laws at the service of the private wishes of victims. Society was set aside; it was as if McVeigh killed 168 people without doing the United States itself an injury. Yet clearly he did. His stated purpose was to do so. But it is not even necessary that he fancied himself a revolutionary in order to see that murder is a crime against society and its laws, not merely against a particular individual.

As for a high-water mark for unvictimism, perhaps the selection of Jeff Bezos, founder and CEO of Amazon.com, as *Time*'s 1999 "person of the year" will do. This was arguably both the culmination of the giddy enthusiasm about the "new economy" and the apotheosis of the entre-preneur. Twenty years earlier, George Gilder claimed in *Wealth and Poverty* to have found a self-sustaining moral basis for capitalism that would ensure its perpetuation despite its supposed contradictions. By the middle of the 1980s, in *The Spirit of Enterprise*, he was describing entrepreneurs in heroic terms, single-minded in their pursuit of visions that would enrich not only themselves, but all the rest of us. The future of capitalism was one of "infinite possibilities" and "ines-timable treasure."

At the time, these ideas met with substantial resistance, to put it mildly—from the criminal prosecution of junk-bond impresario Michael Milken to the widespread use of the sobriquet "Decade of Greed." What changed by 2000 was that consequential resistance had simply ceased. Gilder described billionaires as heroes because of the wealth they created for others. In Jeff Bezos, *Time* was celebrating a billionaire entrepreneur whose company hadn't yet made any money. The pursuit of private gain, once thought to have socially useful conse-quences, was now a subject of celebration in itself, never mind the con-sequences.

These radically private views did not survive September 11. How could they? What on earth could John Ashcroft do to accommodate the "rights" of the victims, i.e., the survivors and the family members of the dead, in the event an al Qaeda member receives a death sentence for complicity in the attack? Will he invite them to the Meadowlands to watch the execution on the JumboTron? Meanwhile, the lesson for the devotees of unvictimism is that no one is ever sole author of his success, that life is never merely private, that self-sufficiency is an illu-sion made possible only by forgetting that individuals live in societies that shape them—and that the security they enjoy as a result of the nation they live in is not a given, but subject to challenge by enemies.

To judge by opinion polls and overwhelming anecdotal evidence, virtually everyone at once understood the importance of the American nation. The terrorist attack was not simply an act affecting individual victims, but an attack on the United States. The dead were not just victims, they were casualties, to be mourned not just in their particu-larity (though the *New York Times* has done a remarkable job with its vignettes) but also abstractly, as any given American in a war in which the enemy's target is all Americans. "There but for the grace of God go I": This is a sentiment centered not on the satisfactions of being oneself but on the plight of others. It was almost universally felt.

It is also striking that Americans seemed to be as one on the question of what to do, and this from the very moment they knew what was hitting them. As others have noted, the counterattack in the terror war began not over Afghanistan on October 7, but on September 11 in the skies over Pennsylvania, when a 32-year-old businessman named Todd Beamer entered the history books with the battle cry, "Let's roll!" The passengers, on cell phones, had figured out what the hijackers intended, and they were determined to prevent it. They lost their lives, but the mission they conceived and executed by themselves, without instruction, was a success.

Was Todd Beamer a victim? Of course—but not in any sense that can be said to diminish the fact that he was a hero. Surely he would rather have been somewhere else that day, and I can't imagine anyone rationally wishing to trade places with him. But I think that when most people think about him, they find not someone to pity but someone to admire, someone who has set an example that, in the event they ever find themselves in a similar position, they would aspire to emulate. The stories since of passengers who have more or less spontaneously risen up and subdued the likes of would-be "shoe bomber" Richard Reid and other persons behaving bizarrely on airplanes suggest that this is something to which many people have now given considerable thought.

These are cases in which people have taken action on their own. But that does not make the actions wholly private. Self-defense (if, indeed, that was the motive of their actions) is not the same as the self-actualization of victims and unvictims. The private focus of the latter presupposes self-preservation, without acknowledging the presupposition. When Beamer and the others rose up, they were responding not only to a harm they themselves were suffering, but also to an act of war, which is to say, an injury to the society of which they were members. They were acting not just to save their own lives, but the lives of other Americans. Self-defense is always socially sanctioned because what gives rise to it is always a wrong against society in addition to a wrong against a particular person.

There are at least some signs that this spirit of self-defense has new resonance well beyond the scope of the war on terrorism. When a disgruntled ex-student returned to the Appalachian School of Law in January and opened fire, killing a student, a professor, and the school's dean, other students there, some armed, quickly tackled him and subdued him. This is not especially unusual in the United States. But it does stand in marked contrast, for example, to the hypercautious response, including from SWAT team members, at Columbine High School in response to a shooting spree by two students there. And if airline passengers were once passive during hijackings on the grounds that this course afforded them the best chance of surviving the ordeal, those days are over. Indeed, it's probably fair to say that any would-be hijackers in the future had better have martyrdom in mind, since passengers will assume they do and will do everything they can to overpower them, if necessary crashing the plane in the belief that they are protecting the White House, the Eiffel Tower, or Westminster. Nonsuicidal hijackers have little credibility with passengers and crew these days.

If people now feel they know what to do to fight back in a way that they didn't before September 11, it's also true that at the national level, there was no doubt about the response. There was no question of shrugging and moving on, or of acting only in a symbolic way. Talk of the "root causes" of terrorism, never popular, receded into the mists once people saw the videotape of Osama bin Laden having a good laugh over the whole thing. It was instantly clear to Americans that the time had come to fight, and the war could be a long one. That sense has not let up.

Poll evidence suggests that most Americans think what we are doing is obviously necessary. We have gathered a substantial amount of international support for our actions, but among our allies, sentiment that we are doing the right thing understandably trails our own. One of the questions is framed roughly as follows: Do you think supporting the United States makes your country more vulnerable to attack? The worry is out there. Interestingly, no one is saying that supporting the United States will make your country *less* vulnerable to attack. This may be the product of a realistic sense among our allies as to what they have to contribute to defeating terrorism, namely, not much beyond domestic law enforcement and intelligence. But it is also indicative of a frame of mind that simply no longer exists in this country, namely, that perhaps a better response would be to do nothing, or to negotiate, or to appease, or to try to address those "root causes"— that these approaches, in the long run, would save more lives. Some of our allies have grown very accustomed to the security we provide them. In any case, Americans never seriously considered anything of the kind. Once again, this is a product of a sense of danger not just to Americans as individuals, but to the nation as a whole.

There has been, of course, some American dissent, from pockets of the far left and far right, but it is noteworthy mainly for its marginality— and for the speed with which other voices with substantial credibility in the political spheres in question have stepped up to dissent from the dissent. Writing in the January/February *Mother Jones*, Todd Gitlin, the New York University professor and veteran leftist activist and social critic, decried "a kind of left-wing fundamentalism, a negative faith in America the ugly." Addressing those who profess it, he asked, "What's offensive about affirming that you belong to a people, that your fate is bound up with theirs? Should it be surprising that suffering close-up is felt more urgently, more deeply, than suffering at a distance? After disaster comes a desire to reassemble the shards of a broken community, withstand the loss, strike back at the enemy. The attack stirs, in other words, patriotism—love of one's people, pride in their endurance, and a desire to keep them from being hurt anymore. . . . [I]t should not be hard to understand that the American flag sprouted in the days after September 11, for many of us, as a badge of belonging. . . ." On the right, one could cite similar denunciations of Jerry Falwell and Pat Robertson, who had described September 11 as divine retribution visited on a sinful country. The element common on left and right was the emphatic rejection of those unwilling to see Americans as a people, a nation, whose divisions on politics and other matters take place only within a broader unity.

There was a time when an American flag on a hard hat—think of the 1970 Peter Boyle movie *Joe*—was a cultural symbol of racism and oppression, the benighted exclusion of others from a common identity as a people. Those days are over. The flag on a hard hat today, or a police officer's or firefighter's uniform, is more likely to be taken as a mark of heroism.

Heroism is famously problematic in democratic societies, where egalitarian impulses as well as the bourgeois fear of violent death drastically circumscribe the desire to, for example, pursue glorious victory on the battlefield and conquer the world. In general, a hero is someone who has proved by his deeds his superiority to others, and this is obviously problematic for us. The usual solution is to define heroism down: hence, the figure of the heroic entrepreneur.

But I do think the vision of a genuinely democratic sort of hero became clear September 11 and after. This kind of heroism has been with us since the nation's beginnings, but it is perhaps easier to see given the volume of it to which we have lately been exposed.

Heroism, in a society such as ours, is risking your life to save a stranger's. This is the ultimate expression of egalitarianism, the conviction that human beings who may be very different in their particularity (rich or poor, black or white, native or foreign-born) share a common humanity—in the American context, as a people, a nation. The heroic aspect of the act is the assertion by deed that one does not value one's own life more highly than that of any other. These are the police and firefighters rushing to the scene, the hard hats searching for signs of life in treacherous debris, the passengers bringing the plane down now rather than clinging in hope and terror to a few more minutes of life at the cost of the lives of others.

And, of course, this quality stands as a corrective to the perspective of both victim and unvictim. Because this quality, finally, is the purest expression of the ties that bind this nation and its people—the sphere of a common public enterprise, not just an agglomeration of individual interests in pursuit of self-actualization. We may be fighting for, among other things, the right to pursue happiness. But while happiness is private and individual, right is a public matter. Sometimes a nation has to come together and fight. And while we are not all cut out for heroism, we know it when we see it, and nowadays, it is wearing red, white, and blue.

# LIBERTÉ, EGALITÉ, JUDÉOPHOBIE

## By Christopher Caldwell

*(Originally published in the May 6, 2002, issue)*

*Strasbourg, France*

The atmosphere of the first round of France's presidential election was captured by candidate François Bayrou's visit to Strasbourg on April 9. Bayrou, who represents Valéry Giscard d'Estaing's center-right Union of French Democracy (UDF), was scheduled to visit a new mayoral suboffice on Strasbourg's outskirts with the city's elegant, Berkeley-educated UDF mayor, Fabienne Keller. Bayrou got hung up campaigning in another city. While Keller waited for him, she was surrounded by a mob of *jeunes des banlieues*—or "suburban youth." This is the euphemism the French use for residents of the crime-infested ring of high-rise housing projects (HLMs) that were built on the outskirts of all French cities in the 1960s and '70s.

The "youth," all of them *beurs*, or Muslims of North African descent, were staging an orchestrated protest against Bayrou, who as education minister in the mid-1990s had opposed letting Muslim girls wear the *hijab*, the Muslim headscarf, to public schools. But Keller was a convenient stand-in. They shouted insults and obscenities at her, one of them threatening (according to an account I was too embarrassed to ask the mayor to confirm specifically when I interviewed her days later) to take a razor to her private parts. When Bayrou arrived, the two went inside for meetings, and the crowd began to pelt the new building with stones, and howl what was really on their minds. First, "Why did you ban the headscarf!" And second, "F— off! We don't want to live anymore in a country that has Jews in it!"

Bayrou emerged from the building while the stones were still flying and told the mob, "Talk about Jews that way today, and you may find people talking about young Muslims the same way tomorrow." At some point during Bayrou's visit, an 11-year-old boy jostled up against him and tried to pick his pocket. Bayrou, heedless that the cameras were running, slapped the kid in the face.

Politicians of the left tried to make hay of the incident, using it to paint Bayrou as some kind of fogey, and themselves as hip to the country's new and "vibrant" youth culture. "Heck, I live in the suburbs, and no one's ever tried to pick *my* pockets," said Communist party presidential candidate Robert Hue. "Me neither," added Socialist prime minister Lionel Jospin, also running for president. The French public didn't see it that way. The more the Bayrou slap played on national television, the higher Bayrou's poll numbers rose—as he was seen as willing to support an assertion of authority against the country's lawless youths. He emerged from deep in the pack of 16 presidential candidates to finish a respectable fourth place, just behind Lionel Jospin. To the extent that he mentioned crime at all (and he never did, preferring the euphemism *insécurité*), Jospin evinced a la-di-da attitude that dropped him to third place and ended his political career.

As French students by the hundreds of thousands stage protest marches across the country, pretty much the entire world knows the result of the first round of the French election. Jacques Chirac, the conservative sitting president, goes into a runoff on May 5 against not Jospin but Jean-Marie Le Pen, leader of the country's fascistic National Front. Le Pen has built his career mimicking the oratory of the rightists who collaborated with Nazi Germany in World War II. He has been a consistent foe of immigration and a practitioner of nudge-nudge, wink-wink cracks against Jews. In the past decade he has added rage against America and the global economy to his oratorical repertory. He is a goon and a gangster, but he had little need to raise divisive issues in the first round. France now has 4,244 crimes per 100,000 residents annually, according to European Union statistics, making it a higher-crime society than even the long-belittled United States. During a week when the top story on tabloid TV was the bloody beating of an 80-year-old man in sleepy Orléans by a gang of *beurs* who had invaded his house, Le Pen focused, as did Chirac, on the dramatic upsurge in violence over the past decade.

But while crime was what brought voters to the polls, France has an even more ominous problem: a wave of attacks and threats against the country's 700,000 Jews that is unprecedented in the last half century of European history. It includes what Rabbi Abraham Cooper of the Simon Wiesenthal Center describes as "the largest onslaught against European synagogues and Jewish schools since Kristallnacht" in 1938. What is surprising and confusing in all of this is that the "new anti-Semitism" in France is a phenomenon of the left. It has practically nothing to do with Le Pen. In fact, its most dangerous practitioners are to be found among the very crowds thronging the streets to protest him.

## "In Paris as in Gaza—Intifada!"

The outbreak began in September 2000, in the days after Palestinians launched the "second intifada" against Israel. The first attacks included firebombings of synagogues in Paris, Villepinte, Creil, Lyons, Ulis (badly damaged), and Trappes (burned to the ground), and other Jewish buildings (high schools, kosher restaurants) throughout France; desecrations of synagogues and cemeteries; widespread stonings of Jews leaving Sabbath worship, death threats, bomb threats, and Nazi and Islamist graffiti of every description: swastikas, "Hitler was right," "F— Your Mother, Jews" (*Nique ta mère les juifs*—a slogan so commonplace that it now appears more usually as *NTM les juifs*), "Death to the Jews," and "In Paris as in Gaza—Intifada!"

Such slogans, particularly the last, now get chanted routinely at pro-Palestinian rallies in Paris and elsewhere. (As do hymns to Osama bin Laden, according to reports of last October's pro-Palestinian march in Paris.) Anti-Jewish violence has indeed tracked the progress of the intifada, rising during violent periods in the Middle East and falling during truces. There was also a spike after September 11; on the following Sabbath alone, worshippers were stoned at synagogues in Clichy, Garges-lès-Gonesse, and Massy; gangs sought to storm a synagogue in Villepinte; and shots were fired outside a Jewish association in Paris. But if it has slowed at times, the cascade of such incidents has never stopped, even for a week, in the last 19 months. At the turn of this year, the League of French Jewish Students and the watchdog agency SOS Racism compiled a list of 406 such incidents.

After Israel's attack on terrorist camps in Jenin and elsewhere, the violence exploded to unheard-of proportions. Over Passover weekend last month, a bomb was found in a cemetery in Schiltigheim, outside Strasbourg, and three synagogues were burned. The authorities seemed to be waking up. While it took 12 days for any national official to even comment on the October 2000 attacks, this time the Ministry of the Interior issued a report showing 395 anti-Jewish incidents in the first half of April alone. Almost two-thirds of these involved graffiti, but the others were more serious, including 16 physical assaults and 14 more firebombings. The Wiesenthal Center circulated an advisory urging Jewish travelers to France to exercise "extreme caution."

What has been most shocking to the Jews of France is that the political class of their country, which has an antiracism establishment to rival any in the world, has been largely silent about their plight. When a Jewish cemetery was defiled at Carpentras in May 1990, and right-wing extremists were (wrongly) suspected of the misdeed, there was a mass demonstration in Paris. Between a quarter and a half million people marched, and President François Mitterrand marched with them—the only demonstration he attended in his presidency.

Yet Jacques Chirac recently announced in front of Israeli foreign minister Shimon Peres that "There is no anti-Semitism and no anti-Semites in France." Every French politician interviewed for this article said pretty much the same. Strasbourg mayor Fabienne Keller says: "There is no significant anti-Semitism." Her deputy mayor Robert Grossmann says: "There is no active anti-Semitism." How can they say this with a straight face?

## Not Your Father's Nazism

One innocent explanation would be that French society has suited up to do battle with the anti-Semitism of 70 years ago, and simply doesn't recognize any other kind. The new anti-Semites are not German-speaking militarists—who were conquered. They are not Catholic traditionalists—whose anti-Semitism rested on doctrines no longer asserted by Catholicism, which, in any case, is a religion the French no longer practice. As such, the French lack the imagination to see that the new anti-Semites—who are primarily radical Muslims—are anti-Semites at all. "Your father's Nazism is dead," says the political scientist Alexandre Del Valle. "It exists in the heads of three or four alcoholic skinheads." In other words, the new anti-Semitism is not coming from the right.

"The danger that looms over the Jewish community is not the danger that threatened us before," says Gilles William Goldnadel, author of an acute study of recent anti-Semitism, *The New Breviary of Hatred*. Goldnadel told a crowd at a B'nai B'rith Center in Paris's sixth arrondissement a few nights before the election, "Worry about the right has turned out to be a decoy—in the military sense—to distract us from the real danger. French antiracists have been parsing the tiniest dictum of Le Pen, while Jewish blood has been spilled by the left in Athens, Istanbul, Rome, Vienna, and Paris." (Particularly by Palestinian terrorists.) There are indications that the government, too, is looking at the wrong target. By the turn of this year, 60 people had been questioned for the hundreds of acts of intimidation. "Only 5 were subject to legal proceedings, being far Right," according to a report prepared by Shimon Samuels of the Wiesenthal Center. "As if the others were not really anti-Semitic and their exactions not just as serious."

There's another way that French politicians can deny that what they are dealing with is an outbreak of anti-Semitism. That is, in the philosopher Pierre-André Taguïeff's memorable phrase, to "dissolve the anti-Jewish acts in a rising tide of delinquency." French foreign minister Hubert Védrine told the Wiesenthal Center last June that the anti-Jewish acts were a matter of "suburban hooliganism." (He continues to hold that view.) The French ambassador to Israel, Jacques

Huntzinger, called them "only part of the general violence manifested by marginal youth in France."

Since France's foreign policy has for the past half decade been built around its role as a force that would "tame" or "bridle" Anglo-Saxon capitalism, it was clearly an embarrassment that the country was unable to bridle anti-Semitic violence in its own backyard. Ignoring anti-Semitism has the advantage of allowing French politicians to proceed as if nothing has happened. In the first weeks of April, while the worst acts of aggression were occurring, Socialist culture minister Catherine Tasca led a march against the "threat" emanating from Italy's conservative prime minister Silvio Berlusconi, and Jospin warned that Berlusconi could serve as a model for his rival Chirac. (Jospin's suggestions for stopping the actual anti-Semitism, meanwhile, went no further than a generalized crime initiative, the highlight of which was a proposal to reduce the number of shotguns a hunter could legally own from 12 to 6.) In the course of the campaign, only 3 of the 16 candidates—Bayrou, the free-marketer Alain Madelin, and the centrist Corinne LePage—condemned the acts unconditionally.

And this unwillingness to call a spade a spade trickled down. The three boys who burned the synagogue at Montpellier—identified as "Morad," "Jamel," and "Hakim"—denied being anti-Semites, and so did those around them. Everyone interviewed about them in the news was content to call them "classic delinquents." The prosecutor described them as "like a lot of petty delinquents, animated by a spirit of revenge, who try to ennoble their excesses by using a political discourse." This seems to apply to all synagogue-burners, if we're to believe the representative from the local office of the mutual-aid society Cimade, who said, "In Montpellier—as in [the synagogue-burning at] Nîmes—more and more kids from the projects are identifying the victimization of the Palestinians with their own. It's a simplistic thing, it's not really an ideology."

This would seem to be immunity on grounds of animality—or at least on grounds of ignorance. Such an understanding appalls Goldnadel. "Delinquents?" he asks. "All anti-Semitic thugs are delinquents. Who do they think was burning down Jews' houses on the

Russian steppes a hundred years ago? Disgruntled architects?" And with immunity comes impunity. In January, the young men who had vandalized a synagogue in Créteil, outside Paris, were convicted of "general violence" and given a sentence of three months—suspended.

## Benladenisation des banlieues

The Jewish attacks—it should be plain by now—are the work of the Muslim minority in France. Let no one doubt the delinquency, though. These neighborhoods are becoming single-race areas, inhabited by North African immigrants and their second- and third-generation descendants. They are zones of drug-dealing, political apathy, unemployment (which stands over 35 percent in such places), and violence. Hence law-enforcement agents, mayors, and politicians refer to the most violent among them as *zones de non-droit* ("lawless areas"), where even the police won't go, except maybe in daylight hours to remove a body. Public powers are resisted with force, and not just the police, who have been targeted for killing by organized "anticop brigades." Even firemen, long a beloved class of public servants in France, have been assaulted in housing projects surrounding Paris.

Law enforcement is underequipped to handle such a challenge. France is supposed to be "the most policed country in Europe," with 130,000 officers—but most of those, thanks largely to a strong union, are employed in administrative or nonbeat tasks, with only 10,000 or so available for duty at any given time. According to an exposé in *Le Point* this spring, units from chilly Normandy are even detailed to the Côte d'Azur to help "reinforce" the beaches there. When police do succeed in making arrests, liberal judges often set criminals free, and 37 percent of sentences passed are not even carried out, according to André-Michel Ventre, secretary-general of the police chiefs' union.

In fact, it would be accurate to describe "suburban" as the French equivalent of the American adjective "inner-city," except for one difference. France's HLMs and other "sensitive neighborhoods" have become missionary fields for professional *re-islamisateurs*—proselytizers, usually financed by Saudi Arabia (which occasionally uses

Algerian foundations as a pass-through for its funding) or Iran, and sometimes by fundamentalist groups in London. These seek to woo young people of Islamic background to a radical political understanding of Islam.

It is such proselytizing that has led to what French people call *la benladenisation des banlieues*, the most famous alumnus of which is Zacarias Moussaoui. But he's not alone. The "Arab" suicide bomber who—to protest Arab countries' "preventing their people from launching jihad against the Jews"—blew up a truck full of explosives in front of a synagogue in Tunisia on April 11, killing a dozen German tourists and six others, was a Franco-Tunisian named Nizar Nawar. His family lives in Lyons, where his uncle, too, was arrested in connection with the attack. One of the four terrorists on trial for trying to blow up Strasbourg's synagogue last year has long lived in France. September 11 saw West Bank–style rejoicing incidents in some Arab neighborhoods. There was also a spectacular terrorist incident a week before. On September 2 in the town of Béziers, along the Belgian border, a hoodlum named Safir Bghouia attacked a group of police with a shoulder-held rocket launcher, phoned in death threats to local officials, machine-gunned the local police constabulary, and executed the town's deputy mayor, before he himself was shot dead the next day, dressed in white and howling that he was a "son of Allah."

With London its only rival, Paris is the media and intellectual capital of the Arab world, much as Miami is capital of the Hispanic world. As a result, beyond terrorism, the weight of fundamentalist Islam—and the anti-Semitism that goes along with it—is making itself felt in ordinary French life. According to the literary scholar Eric Marty, one professor of literature at the University of Paris was unable to teach the works of Primo Levi (including the Auschwitz memoir *If This Is a Man*), because his Arab students booed him out of the classroom. "Kenza," a young *beurette* who was on the French reality-TV show *Loft Story* (a sort of NC-17-rated equivalent of *Survivor*), complains that she got kicked off the show last season because "television is controlled by the Jews." A friend of mine was working out at his gym near Strasbourg and got to talking with a friendly *beur* about British prime minister Tony Blair. "Don't believe anything Blair says," the man told my friend. "Don't you know his real name is actually Bloch?" (Bloch is a common Alsatian Jewish surname.)

That is not the whole story of Arabo-Muslim France, of course. Claude Imbert, editor of *Le Point*, admits that French immigration was badly handled under the Socialist presidency of François Mitterrand, but remains "a resolute partisan of immigration." He notes that *beurs* are among France's most dynamic entrepreneurs. "They're the only ones who have the American-style careers we need," he says. "Taking a pizza delivery with one car and turning it into a big company—that sort of thing." There are others who have courageously stood up against the Islamist wave, like Rachid Kaci, a mayoral aide in Sannois who appeared at a Jewish colloquium east of Paris in mid-April to say, "You have my total solidarity to fight by your side against this new fascism." Kaci urges an Islam cut off from foreign influences, following somewhat the message of Tunisian novelist Abdelwahab Meddeb's cri de coeur, *The Sickness of Islam*.

And others are seeking to make Islam more open to all Muslims, and more transparent in its sources of funding. That includes Strasbourg's mayor Fabienne Keller, who has put on hold a hard-line, Saudi-sponsored mosque project that was approved by the outgoing Socialist mayor (and Jospin's culture minister) Catherine Trautmann, despite the involvement of foundations that now appear on the U.S. government's terrorist blacklist. In general, France is seeking to create an Islam that is in harmony with the country's secular traditions, which is wholly admirable. Unfortunately, that kind of Islam is going to have to be invented, since it has never existed in 1,300 years. It may, indeed, be a logical contradiction. And it is certainly something that the more radical among France's 6 million to 9 million Muslims—who make up close to half the population of the young in the country's cities, and have a birthrate that outstrips that of non-Muslims by 3-to-1—have no reason to work for.

Which brings us to the real reason the French don't think they have a problem with anti-Semitism, and the reason they're wrong.

## *Judéophobie*

Pierre-André Taguïeff, director of research at the Center for the Study of French Political Life (Cevipof), has just published a book called *The New Judeophobia* (*La Nouvelle judéophobie*), which lays it out. The ideology on which the new anti-Semitism rests is largely imported. It has its roots in the anti-Western paranoia that all Americans will recognize (without being able to explain) from the banners carried in the Iranian revolution. It is a hybrid of apocalyptic Islam and pre-Nazi Western anti-Semitism of the *Protocols of the Elders of Zion* type. Taguïeff resists the term "anti-Semitism." First, because, as Bernard Lewis has shown, "Semitic" is a linguistic and not a racial term, allowing people to play inane word games with what is happening in France. ("The *jeunes*/Hamas/Hezbollah can't be anti-Semitic," one reads almost daily in the French press. "They're Semites themselves!") Second, because anti-Semitism is a racial ideology, and today's Jew-hatred is not really a racial ideology. That is why, Taguïeff argues, it is so often found in tandem with anti-Americanism.

Taguïeff's book is brilliant, and extraordinarily well sourced, and will convince any reader who is not already dug in on Middle Eastern questions. It has also infuriated the French intellectuals at whom it is aimed, because Taguïeff's claim is that the two pillars of the new anti-Semitism are anti-Zionism and Holocaust denial. He's right, but this requires some explaining.

The first infuriates the French because they are largely anti-Zionist, to the extent that the word can be used to mean antipathetic to Israel's interests and sympathetic to those of its enemies. Whereas Americans sympathize with Israel in the Middle East conflict by a margin of 41–13, according to a recent *Economist* poll, the French sympathize with the Palestinians over Israel by the widest margin in Europe, 36–19. What's more, the Middle East conflict has become an absolute obsession among the left-wing intelligentsia, of the sort you'd have to sit in a Socialist party hangout in Strasbourg on a Friday night to believe.

Doesn't the citizen of a free country have a right to back whatever side he wants in a foreign war? Of course he does. "Even among Jews," as William Goldnadel says, "You don't have to be a self-hating Jew to view the destiny of the Jews as living in the Diaspora." That's not Taguïeff's target. What he is talking about is "mythic anti-Zionism," which treats Zionism as *absolute evil*, against which only absolute warfare can be raised. In this understanding, Zionism constitutes not just racism but the *ne plus ultra* of racism.

This is a vision that the French—particularly given the French left's obsession with race, and their history of romantic attachments to Third World guerrillas—are in danger of embracing. The philosopher Alain Finkielkraut notes that, in France, "support for the Palestinian cause is not shaken but reinforced by the indiscriminate violence of Palestinians." In particular danger of embracing this Manichaean view of the Arab-Israel conflict are those who support Third-Worldism, neo-communism, and neo-leftism, whom Taguïeff lumps together as the "anti-globalization movement." The Chomskyites . . . the people who think *Empire* is a good book. If you ask them why, of all the dozen conflicts the Muslim world is waging against the civilizations it borders on, *this* one obsesses them (why not Chechnya? why not Sudan? why not Nigeria?), they can give you an answer that stops just this side of anti-Semitism. Israel-Palestine is the one where the "capitalist" world of the West (and, by implication, the Jews who run it) meets the underprivileged victim peoples of the South. Jews thus get to pay the price for the West's depredations since the Middle Ages, most of which they were on the receiving end of.

That, of course, is the great obstacle to this discourse of Jews-as-victimizers: The Jews have been through rather a lot. And that is why denial, or at least minimization, of the Holocaust is an indispensable part of the ideology. Abbé Pierre, a popular priest who became a national hero, lamented in 1991 that "Jews, the victims, have become the executioners." He even embraced the Stalinist-turned-Muslim-radical Roger Garaudy when he was accused of Holocaust-denial. At a pro-Palestinian demonstration at Les Halles in late March, marchers carried a Star of David with a swastika over it, shouting *Jihad, Jihad, Jihad.* If you walk across the pont des Invalides, you can see, in yellow print on black background, a poster that urges that Ariel Sharon be sent to the Hague to be tried on war crimes:

## SHARON À LA HAYE

Invasion - Torture - Humiliation - Epuration éthnique

Colonialisme - Crimes de guerre - Crimes d'état

Violations du droit international - Assassinats

Racisme - Profanations de lieux de culte - Massacres

Extermination - Carnages - Sionisme - Injustice

Crimes contre l'humanité - Déportation - Persécution

### ISRAEL

### COUPABLE

Comité Francais pour la Liberté des Peuples

It's hard to say which is the strangest imposture in the poster: to see "Zionism" ranked next to "Extermination" among crimes, or to see Israel accused of doing in the West Bank what the Nazis did in France. ("Deportation"—whatever that may mean in the context of an anti-terrorist operation in the West Bank—is a word that maintains a terrible resonance for the Jews of France.)

At times the superimposition of Nazi German motifs on Israel takes on aspects of a religious vision. Claude Kei-flin, a political reporter for the *Dernières Nouvelles d'Alsace* who covers Middle Eastern matters for the paper, asked me during an interview, "How could the Jewish people, after having undergone the Holocaust, be putting numbers on the arms of their Palestinian prisoners?"

"What? . . . You mean tattoos?"

"Yes."

"What?"

"Okay, not engraved in the skin, but, still . . ."

France has laws against Holocaust denial. The current climate shows them to be bad laws, not just because they make free-speech heroes of those who are basically mentally ill, but because they can be violated in spirit with impunity. Such a violation was committed by José Bové in the first days of April, when he was expelled from Israel following a visit to Yasser Arafat's compound in Ramallah. Bové, who rose to fame for vandalizing a McDonald's in southern France as a protest against American influence, is not merely the informal leader of the younger French left, the "hero" of the Seattle riots, and the guiding spirit of many of the anti–Le Pen protests that are now raging in Paris; he is also the most charismatic leader of the antiglobalization movement in the world.

It was thus alarming to see Bové, after a pro forma denunciation of anti-Jewish violence, informing viewers of the TV channel Canal Plus that the attacks on French synagogues were being either arranged or fabricated by Mossad. "Who profits from the crime?" Bové asked. "The Israeli government and its secret services have an interest in creating a certain psychosis, in making believe that there is a climate of anti-Semitism in France, in order to distract attention from what they are doing."

Since Bové didn't actually say Jews weren't killed in the Holocaust, it may seem excessive to some readers that B'nai B'rith accused him of *négationnisme*, or Holocaust denial. But B'nai B'rith is right. They have simply thought about the roots of Holocaust denial a bit more thoroughly than others. For anyone who inhabits Western culture, the Holocaust made that culture a much more painful place to inhabit— and for any reasonably moral person, greatly narrowed the range of acceptable political behavior. To be human is to wish it had never happened. (Those who deny that it did may be those who can't bear to admit that it happened.) But it did. If there's a will to anti-Semitism in Western culture—as there probably is—then the Arab style of Judeophobia, which is an anti-Semitism without the West's complexes, offers a real redemptive project to those Westerners who are willing to embrace it. It can liberate guilty, decadent Europeans from a horrible moral albatross. What an antidepressant! Saying there was no such thing as the gas chambers is, of course, not respectable. But the same purpose can be served using what Leo Strauss called the *reductio ad Hitlerum* to cast the Jews as having committed crimes identical to the Nazis'. They must be identical, of course, so the work of self-delusion can be accomplished. We did one, the Jews did one. Now we're even-steven.

You can see the attractive force in such an ideology. Author Alexandre Del Valle fears that anti-Semitism could also be a *binding* force, leading to a "convergence of totalitarianisms," of Islamism and the Western antiglobalist left. Elisabeth Schemla, a longtime editor at France's center-left opinion weekly *Le Nouvel Observateur* who now edits the online newsletter www.Proche-Orient.info, says, "The anti-Semitism of the left is more dangerous than that of the right. They have power in the media, the universities, the associations, the political class." Schemla worries that a third of the candidates in the first round of the presidential election were strongly motivated by the conflict in the Middle East. As such, it is not the strong showing of Le Pen that is the most alarming development in the first round of the election, but the record-high score of the three Trotskyite parties on the hard left.

## Bonifacisme

Last August, Pascal Boniface, a top foreign policy adviser to Lionel Jospin, wrote an open "Letter to an Israeli Friend" that appeared in *Le Monde*. The echo of the "Letters to a German Friend" that Albert Camus had written in 1943 and 1944 was not lost on Jewish readers. The lawyer Pierre François Veil remarked that if Boniface had wanted to reach an Israeli friend, he could have written to the *Jerusalem Post*. The letter was, of course, addressed to the Jews of France, and many read a threat in its closing lines: "In France," Boniface wrote, "should it permit too much impunity to the Israeli government, the Jewish community could also be the loser in the medium term. The Arab/Muslim community is certainly less organized, but it will be a counterweight, and it will soon be numerically preponderant, if it is not already."

"I gave my advice not because of the weight of the community but on principle," Boniface said in an interview. The votes of the two communities are about even. Muslims may number as many as 8 million, but only half are citizens. Of the remaining 4 million, 2.5 million are not yet old enough to vote, and of the 1.5 million that remain after they're taken out, over half *won't* vote. But at a time when Jews were being threatened in the streets of France, it seemed that Jews were not being lectured on electoral clout but outright intimidated: *Break your solidarity with Israel,* the deal was, *and we'll leave you in peace; otherwise, you'll be lopped out of the national community.* Boniface is not alone in his opinions; the Coordinated Appeal for a Just Peace in the Middle East (CAPJPO) has asked French Jews to make a "critique of Israeli policy." As Alain Finkielkraut noted, CAPJPO has never asked Muslims to pressure Palestinians to stop their suicide attacks. Boniface was soon being accused of the same thing: making Jews'—but no one else's— membership in the national community contingent on the acceptable behavior of a foreign country. This attitude was given a witty short-hand—*bonifacisme*—in the Jewish press, which condemned it as a form of anti-Semitism.

"I defy anyone to find a single line in any of my work that is anti-Semitic," said Boniface in an interview. He noted that his opinions were fairly generally held. "My students have changed their opinion, too," he said. "Twenty years ago when Israel invaded Lebanon they were evenly divided. Now they are overwhelmingly pro-Palestine." Lionel Jospin followed Boniface's line throughout his campaign, con-demning "communitarianism" and insisting to Jewish, though not to Arab, groups that "we must not import into France the problems of the Middle East." But the "evenhandedness" of Jacques Chirac on communitarian matters was almost worse. During a visit to Paris's grand mosque the week before the vote, Chirac firmly condemned the burning of (many) synagogues (in his own country), but assured the gathered dignitaries that if anyone were to harm a mosque (which no one has done) or a church (like the Church of the Nativity, in another country, where Israeli troops had surrounded Palestinian terrorists holding hostages), it would be equally bad.

And yet "the problems of the Middle East," as Jospin calls them, are all that France wants to think about. It has long alarmed Jews that non-Jews are showing up less and less at their marches. Since October 2000, they have wondered why their fellow citizens were not marching against really existing anti-Semitism in France, the way they used to march against the safely-part-of-history version. ("A demonstration on 13 January 2002 of Jewish leadership assembled in the Créteil syna-gogue—the latest victim of violence—was marked by the sparseness of non-Jewish sympathizers," noted Shimon Samuels of the Wiesenthal Center. "Indeed, the town's deputy mayor used the occasion to publicly revile the Sharon government and was met by jeers from the audience.")

On April 6, pro-Palestinian marches were held across the country. On April 7, the CRIF, the umbrella group of French Jewish organizations, held a march for Israel. They decided also to march against the anti-Semitic attacks of the preceding days. Three of the 21 members of the CRIF board decided to make only the second part of the march. One of those, Olivier Guland of the *Jewish Tribune*, complained, "It's the first time Jewish institutions in France have given the impression that the defense of their own interests is not the same as the defense of the Republic's values." The most commonly held sign—*"Synagogues brûlés, république en danger"*—gave the message that the interests of France and its Jewish community were pretty much identical. But whether that's the message France will *get* is anybody's guess.

"Traditionally," Alain Finkielkraut wrote in the Jewish monthly *L'Arche*, "anti-Semites are those French who worship their identity and love one another against the Jews. Contemporary anti-Semitism involves French people who *don't* like themselves, who have a post-national perspective, who are shedding their 'Frenchness,' the better to identify with the poor of the world. They use Israel to place the Jews in the camp of the oppressors. You have a sort of league between anti-Semitic Islamism and self-disparagement, between repudiation of another and hatred of oneself."

Finkielkraut has for years railed against the dangers of political correctness and the lazy thinking of France's antiracism movement. His writings often seemed merely a necessary means of saving a political movement from sloppy thinking. But now that that movement is "raising a war machine against the Jews in the name of the excluded," such work seems much more important.

The French left has thoroughly assimilated the lessons of World War II. Maybe too thoroughly. After fantasizing for years about how much braver than their parents *they* would have been had they lived in 1938, after waiting stylishly for years for a predictably fogey-ish, Vichy-style anti-Semitism so they could combat it according to their antiracist operator's manual, they suddenly find themselves confronted with evidence that there are at least hundreds of thousands of people in their country who think pretty much as the al Aksa Martyrs' Brigade does, and millions more whose opinions are anyone's guess. The French left may have idealistic reasons for placing its sympathies with the Palestinians, but it has powerful reasons of expedience, too. Thus far its heart lies with the side that has committed the most violence on French soil.

The most dangerous thing about Jean-Marie Le Pen, who loathes the global economy, distrusts the Jews, and practices gesture politics, is not that he'll get elected. It's that he'll serve as the hate object who unites anti-Western Islamists and anti-Western antiglobalists, who march against him night after night over ideological differences that grow harder and harder to discern.

# A YEAR OF FIRSTS
# AND LASTS

## By Matt Labash

*(Originally published in the September 16, 2002, issue)*

*Bronx, New York*

I met Edlene LaFrance on the worst day of her life. Or maybe it was the second worst, or the fifth, there are so many to choose from now. Two days after the Twin Towers fell, her 43-year-old husband, Alan, lay buried at the bottom of one of them. Though the city was awash in acts of unparalleled selflessness, I'd spent that morning thinking selfishly, walking the ash-caked streets of lower Manhattan, trolling for battle-scarred humanity to fill my notebooks before deadline.

On a bum tip, I rushed to the Chelsea Piers, where ambulances full of recovered wounded were rumored to be arriving. When I got there, a bystander scoffed at my naiveté. "There *aren't* any more wounded," he said, letting the thought finish itself. Another reporter suggested the action had moved cross-town to the National Guard Armory on Lexington, and indeed it had. The street outside the building looked like a Third World bazaar. Except instead of merchants peddling trinkets, family members were holding "Missing" posters, begging for whereabouts and clues, as if their loved ones had gone down to the corner for a pack of cigarettes, then had forgotten the way home.

The media were supposedly barred, but I slipped inside the building. In all the chaos, it was about as difficult as crashing Penn Station.

Hundreds of family members sat in rows, their bodies racked with tension and slicked with sweat in the un-air-conditioned hall. Amidst this grimness, I scouted for the most approachable faces, which belonged to Edlene, her son Jody, and his wife, Camille.

I asked if I could follow their family through this process, and they graciously assented. Edlene clutched a photo of her husband in his white wedding tuxedo. Their 21st anniversary was in two weeks, and on the morning of September 11, for some reason, she'd come close to giving him his biggest present early—a new wedding ring.

We made polite chat over grief counselors' incessant offerings of sandwiches and Sprite. Edlene quipped that she could use some tranquilizers instead. She answered all my questions dutifully, but her eyes kept drifting to the archway at the front of the room. It was there, down a staircase, that 20 people at a time were being taken to scour two lists—the first indicating that their loved one had turned up hospitalized, the second that they'd turned up dead.

When it was the LaFrance family's turn, we went downstairs. A Red Cross volunteer offered to look for Alan's name. Edlene, a portrait of poise a few minutes earlier, simply laid her head down on a table while her son sat beside her, stroking her hair. Alan didn't turn up on the deceased list, but he wasn't on the hospitalized list either. A deductive silence enveloped us—the sound of someone's life coming undone.

A year later, as journalists and grief groupies again jostle for a piece of 9/11, it's tempting to sit back and chortle at excesses and opportunism. God knows the last year has seen enough of them: American Paper Optics issuing "Images of 9-11 in 3-D," the commemorative cigarette lighter with a flame flickering right over a picture of a burning World Trade Center, the "9/11: 24/7" mock-tribute album from drag-queen Tina C. featuring songs like "Stranger on the Stairwell" and "Kleenex to the World."

Still, as we congratulate ourselves on how life goes on in all its wretched excess, there are thousands like Edlene. Her life actually has changed, and will remain that way beyond any news cycle.

I catch up with her at an apartment building that sits next to a noisy highway in the Bronx. A landlord's letter posted in the lobby

cautions residents not to let their pets urinate in the stairwell. She wel-
comes me at the door, and says that she has lost some weight and some
hair since we last spoke, but she is just as I remembered: diminutive
and dignified, her warm voice accented with a steel that suggests she is
being brave by necessity, if not by nature. Her rent is $709 per month,
and it seems a bit steep for the two rooms she inhabits. They're the
same rooms she lived in with Alan. And in a way, that $709 is what
got him killed.

An audio/visual technician, Alan freelanced all over the city—
primarily doing conferences at a New York public library and at the
Windows on the World restaurant on the 106th floor of the World
Trade Center. Edlene could always tell to which job her husband was
headed by what he was wearing. If he was off to the library, he'd wear
casual clothes. If it was the World Trade Center, he'd put on a black
suit. The library gig was more lucrative, but his employer took too
long to pay (six to eight weeks).

Though usually Alan just worked nights at the World Trade
Center, he sometimes needed money to make rent. So the morning of
September 11, he went there to set up a breakfast conference. Edlene
can't remember her last words to her husband, but for days after the
towers went down, she tricked herself into recalling that he hadn't
been wearing his black suit.

From the testimonials of his friends and relatives, it is clear Alan
was the kind of guy you'd want to be around. He liked his breakfasts
big and greasy at a nearby diner, and he loved to play drums. At his
Jehovah's Witness congregation's cookouts, he always assumed the role
of grillmaster, perhaps because even at 6'4", he was such a suspect basketball player that friends mockingly called him "Jump Shot." He
took his grandmother shopping almost weekly. He liked to work on
cars, but not for money. Often, Edlene would gaze out the window at
busy Bruckner Boulevard and see a stranded motorist. She'd shoot her
husband a help-those-poor-people look. He'd grumble a little, then
grab his tools and ride to the rescue.

Now, the first face Edlene sees in the morning is Mohamed Atta's.
She keeps a *New York Post* cover photo of the man who killed her husband on the floor next to her bed. Every morning when she wakes up,

she steps on his face. It is a small, desperate gesture, but it's the only revenge she'll ever get. When asked why she'd keep a picture of this murderer in her bedroom (she never calls him a "terrorist" or "hijacker," always a "murderer"), she says, "A lot of times when I don't think it's real, I just turn over and there's his face. Then I know it's real."

It wouldn't seem she'd need any more reminders. Her husband, who had no life insurance, handled the finances, and within the first few weeks after he died—her mind scrambled and her heart literally palpitating—she couldn't even locate her checkbook. Unable to keep track of the bills, she had to get an extension after an eviction notice. Though some charity has found her (Black America Web Relief ended up footing her rent through August), most of the $30,000 or so she's received from victim's assistance funds has gone to offset her son Jody's expenses, since he frequently has to fly back with his family from Chattanooga, Tenn., to help his mom navigate mounds of paperwork and other disasters.

A typical one occurred when Alan's old Volvo had to be retrieved from an impound lot where it was towed from the train station after sitting there for weeks (Alan had the only key). Likewise, her telephone was nearly cut off, but with her nursing job in a "clinic in a bad neighborhood," she has managed to keep it paid up, along with Alan's cell phone account. She doesn't have the cell phone—Alan had it with him when he died—but by keeping the account open, she can still call his messages to hear his voice. "They never found remains," she explains. "It's all we have left of him."

Edlene LaFrance is not a whiner, though she could be forgiven if she were one. She hasn't told her overburdened son that her doctors are worried she has breast cancer. Having switched nursing jobs earlier this year, she has told no one at work besides her boss that she lost her husband on 9/11. Even her own mother, who is senile and who Edlene doesn't wish to traumatize, has no idea her daughter is now a widow. "When she asks where Alan is," Edlene says, "I tell her he's at work."

I ask her if she blames God for any of this. "Why would I?" she asks, out of conviction or convenience or both, "He didn't do it." She says she's been hitting the Scriptures pretty hard lately—not Job, as you might expect, but all the widows 'n' orphans passages. There are a lot more of them than she had noticed before, and she says they present a compelling body of evidence that God won't let her fall through the cracks. So far, she says, He hasn't.

The thing that's changed the most for her is time. She no longer measures it in weeks and months, but in firsts and lasts—the last time she did something with Alan, the first time she must do it without him. She doesn't cry much anymore, but the day before my visit, a light bulb burned out in her hallway. She ended up in a heap on the kitchen floor for 20 minutes. It was a 1,000-hour bulb that Alan had last changed. She has not replaced it.

There are long lists of firsts she is avoiding. She will not go on vacation, and chooses not to go to the movies, since that was Alan's favorite pastime. When she goes to their favorite diner for breakfast, she sits at the counter, since she and Alan used to sit at a booth. She knows she must get over this, and it will be easier to, she reasons, after September 11. Right now, she dreads that date the most. Though she'll be surrounded by extended family, all she really wants to do, she says, "is take some sleeping pills and wake up on September 12th."

After hours of conversation, we set off for the train station on foot, strolling through her neighborhood in a late summer half-light. Another 30 minutes, she says, and she wouldn't be out on these streets. At first, I think she means because they're crime-ridden. But no. "That was the time me and Alan always walked together," she explains. As she says this, I nod understandingly. But I can't under-stand. Not really. We have all grown rather possessive of September 11, taking it out, reexamining it when it suits us, making it mean what we want it to mean. Edlene doesn't have that luxury. I want to make it easier for her, but that can't be done, so I hold my tongue. She thanks me for listening, and I nod some more, as she puts me on a train that will take me back to my wife and son.

# THE FOG OF PEACE

## By David Brooks

*(Originally published in the September 30, 2002, issue)*

EITHER SADDAM HUSSEIN WILL remain in power or he will be deposed. President Bush has suggested deposing him, but as the debate over that proposal has evolved, an interesting pattern has emerged. The people in the peace camp attack President Bush's plan, but they are unwilling to face the implications of their own. Almost nobody in the peace camp will stand up and say that Saddam Hussein is not a fundamental problem for the world. Almost nobody in that camp is willing even to describe what the world will look like if the peace camp's advice is taken and Saddam is permitted to remain in power in Baghdad, working away on his biological, chemical, and nuclear weapons programs, still tyrannizing his own people, fomenting radicalism, and perpetuating the current political climate in the Arab world. And because almost nobody in the peace camp is willing to face the realities that a peace policy would preserve, the peace proponents really cannot address the fundamental calculation we confront: Are the risks of killing Saddam greater or less than the risks of tolerating him? Instead of facing the real options, they fill the air with evasions, distractions, and gestures—a miasma of insults and verbiage that distract from the core issue. They are living in the fog of peace.

When you read through the vast literature of the peace camp, you get the impression that Saddam Hussein is some distant, off-stage figure not immediately germane to matters at hand.

For example, on September 19, a group of peaceniks took out a full-page ad in the *New York Times* opposing the campaign in Afghanistan and a possible campaign in Iraq. Signatories included all the usual suspects: Jane Fonda, Edward Said, Barbara Ehrenreich, Tom Hayden, Gore Vidal, Ed Asner, and on and on. In the text of the ad, which runs to 15 paragraphs, Saddam Hussein is not mentioned. Weapons of mass destruction are not mentioned. The risks posed by terrorists and terror organizations are not mentioned. Instead there are vague sentiments, ethereally removed from the tensions before us today: "Nations have the right to determine their own destiny, free from military coercion by great powers. . . . In our name, the government has brought down a pall of repression over society. . . . We refuse to be party to these wars and we repudiate any inference that they are being waged in our name." The entire exercise is a picture perfect example of moral exhibitionism, by a group of people decadently refusing even to acknowledge the difficulties and tradeoffs that confront those who actually have to make decisions about policy.

Frances FitzGerald recently wrote a long essay in the *New York Review of Books* headlined on the cover "Bush and War." In the piece FitzGerald portrays the Bush foreign policy team as a coterie of superhawks driven by a fierce ideological desire to act unilaterally. This unilateralism leads the Bush advisers, FitzGerald asserts, to see or invent enemies, such as Saddam Hussein. "If one decides to go it alone without allies or reliance on the rule of law, it is natural to see danger abroad."

If you are a writer setting out to evaluate the Bush foreign policy team and its longstanding worries about Saddam, it would seem reasonable to measure whether or not those fears are justified or exaggerated. This is Journalism, or Scholarship, 101. But this is the question FitzGerald cannot ask, because that would require her to enter the forbidden territory of Saddam himself. FitzGerald raises the possibility that war against Saddam might lead to a Palestinian revolt in Jordan, oil shortages, and terrorist attacks. She mentions the daunting cost and scope of an American occupation of Iraq. She approvingly quotes Brent Scowcroft's warning that taking action against Saddam would inflame the Arab world and destroy the coalition that we need to wage war on al Qaeda. But what of the risks of doing nothing? This issue she does not touch. This is the issue that must remain shrouded in the fog of peace.

Reviewing Noam Chomsky, legal scholar Richard Falk, a member of the editorial board of the *Nation*, observes that while he agrees with much of what Chomsky writes, he is troubled by the fact that Chomsky is "so preoccupied with the evils of U.S. imperialism that it completely occupies all the political and moral space."

That is exactly what you see in the writings of the peace camp generally—not only in Chomsky's work but also in the writings of people who are actually tethered to reality. Their supposed demons— Paul Wolfowitz, Richard Perle, Doug Feith, Donald Rumsfeld, and company—occupy their entire field of vision, so that there is no room for analysis of anything beyond, such as what is happening in the world. For the peace camp, all foreign affairs is local; contempt for and opposition to Wolfowitz, Perle, Rumsfeld, et al. is the driving passion. When they write about these figures it is with a burning zeal. But on the rare occasions when they write about Saddam, suddenly all passion drains away. Saddam is boring, but Wolfowitz tears at their soul.

You begin to realize that they are not arguing about Iraq. They are not arguing at all. They are just repeating the hatreds they cultivated in the 1960s, and during the Reagan years, and during the Florida imbroglio after the last presidential election. They are playing culture war, and they are disguising their eruptions as position-taking on Iraq, a country about which they haven't even taken the trouble to inform themselves.

The noted historian and Columbia University professor Simon Schama wrote a long essay for the *Guardian* that was published September 11. He begins by defending President Bush's use of the term "evil." But as he starts to talk about the war on terror and the possible war in Iraq, suddenly all logic is overtaken by his disgust for the Bush crowd:

> *The United States Inc. is currently being run by an oligarchy, con-*
> *ducting its affairs with a plutocratic effrontery which in comparison*
> *makes the age of the robber barons in the late 19th century seem a*
> *model of capitalist rectitude. The dominant managerial style of the*

*oligarchy is golf club chumminess; its messages exchanged along with*
*hot stock tips by the mutual scratching and slapping of backs.*

Schama goes on to attack Dick Cheney for Halliburton, Bush for
Harken Energy, Secretary of the Army Thomas White for Enron, the
proposal to eliminate the death tax, the banality of the architectural
proposals for Ground Zero, Bush's faith-based initiatives, and so on
and so on. It all adds up to one long rolling gas cloud of antipathy,
which smothers Schama's ability to think about what the United
States ought to do next.

This is the dictionary definition of parochialism—the inability to
consider the larger global threats because one is consumed by one's
immediate domestic hatreds. This parochialism takes many forms, but
all the branches of the opposition to the war in Iraq have one thing in
common: Iraq is never the issue. Something else is always the issue.

For Schama and many others, the Bush crowd is the issue. They
stole the election. They serve corporate America. They have bad manners. This is the prism through which Maureen Dowd, Molly Ivins, and
many others view the war. Writing in the *Boston Globe*, Northwestern
University's Karen J. Alter psychoanalyzes the groupthink mentality
that she says explains the Bush crowd's strange obsession with Iraq. The
real problem, you see, is in their psyches.

Among some Democrats in Washington, a second form of
parochialism has emerged. They see the Iraq conflict as a subplot
within the midterm election campaigns. "It's hard not to notice that
the sudden urgency of war with Iraq has coincided precisely with the
emergence of the corporate scandal story, with the flip in congressional
[poll] numbers and with the decline in the Republicans' prospects for
retaking the Senate majority," Jim Jordan, the director of the
Democratic Senatorial Campaign Committee, told the *Washington Post*.
"It's absolutely clear that the administration has timed the Iraq public
relations campaign to influence the midterm elections."

What's fascinating about this wag-the-dog theory is what it reveals
about the mentality of the people who float it. These are politicians (far
from all of them Democrats) who have never cared about foreign affairs,
have no history with the Cold War, have no interest in America's super-

power role. One sometimes gets the sense that these people can't imagine how anybody could genuinely be more interested in matters of war and peace than in such issues as prescription drugs, Social Security, and Enron. If the president does pretend to care more about nuclear weapons and such, surely it must be a political tactic. For them, the important task is to get the discussion back to the subjects they care about, and which they think are politically advantageous.

This explains the strange passivity that has marked much of the Democratic response to Iraq. The president must "make the case," many Democrats say, as if they are incapable of informing themselves about what is potentially one of the greatest threats to the United States. Tom Daschle's entire approach to the Iraq issue has been governed by midterm considerations.

On September 18, as the U.N. was consumed by debate over Iraq, as the White House was drafting a war resolution on Iraq, Daschle delivered a major policy address. The subject? The tax cut Congress passed over a year ago. The speech, the *New York Times* reported, was "the beginning of a party-wide effort to turn attention away from Iraq and back to the domestic agenda." The United States is possibly on the verge of war, and Tom Daschle is trying to turn attention away from it. He's running around Capitol Hill looking for some sand to bury his head in. This is parochialism on stilts.

For a third branch of the parochialists, Iraq is not the issue, America is the issue. The historian Gabriel Kolko recently declared, "Everyone—Americans and those people who are the objects of their efforts—would be far better off if the United States did nothing, closed its bases overseas, withdrew its fleets everywhere and allowed the rest of the world to find its own way without American weapons and troops." For peaceniks in this school, the conditions of the world don't matter. Whether it is Korea, Germany, the Balkans, or the Middle East, America shouldn't be there because America is the problem. This is reverse isolationism: Whereas the earlier isolationists thought America should withdraw because the rest of the world was too corrupt, these isolationists believe that America should withdraw because the United States is too corrupt.

"I Hear America Sinking" is the title of James Ridgeway's recent piece in the *Village Voice*: America is too corrupt and troubled to attempt any action in Iraq. "American foreign policy is like their television," writes John O'Farrell in the *Guardian*. "It has to keep jumping from one thing to another because the president has the remote control in his hand and his attention span is very limited." Writers in this school derive an almost sensuous pleasure from recounting how much people in the rest of the world dislike America; whether those anti-Americans also, by the way, kill homosexuals, oppress women, and crush pluralism is relegated to the background. For these parochials, the immediate priority is hating America.

A fourth form of parochialism is what might be called modern multilateral gentility. For people in this school Iraq is not the issue—the U.N. is the issue. Now, it should be said that there are substantive reasons to care about whether or not the United States has allies. We need friends to help transform the Middle East. But for many of its supporters, multilateralism is purely a procedural matter. They seem to care less whether an action is undertaken than whether it is undertaken according to all the correct and genteel multilateral forms.

Like all forms of American gentility, this multilateralism is greatly concerned with refined manners. There can be no raw bullying around the earth, no passionate declarations of war, no ungentlemanly crusades. Instead, the conflict must be resolved through the framework of the United Nations (which for some reason is seen as a high-toned and civilized center of conflict resolution). Like all forms of American gentility, multilateralism carries a strong aroma of cultural inferiority. We Americans are sadly crude and uncultured. The Europeans are really much more sophisticated and subtle than we are about the affairs of the world. Their ways and manners are more mature.

Multilateral obsessives tend to be more centrist than other people in the peace camp. They are more respectable and more establishmentarian. But like many other members of the peace camp, they simply do not tackle the question of what Saddam might do or what the future might look like. Preferring process over substance, they hold to a multilateralism descended from previous genteel causes, such as civil service reform and campaign finance reform. In their quiet and sober way, they too contribute to the fog of peace.

Now it should be said that within the peace camp, there are honorable exceptions to this pattern. Adam Shatz recently wrote a long piece in the *Nation* surveying left-wing thought on the war. The left is wrapped around its own axle, Shatz noted, because it can't come to terms with American power.

Richard Falk, the left-wing legal scholar, himself has argued that in deciding whether to go into places like Afghanistan and Iraq, "we should look with as much care as possible at the case where the interventionary claim is being made, and consider the effects of intervening and not intervening." This hardly seems like a radical notion, but of course it is precisely this approach that the peace camp, by and large, refuses to take. As Shatz observed in his piece, "Falk has been widely chastised for his vacillations."

Moreover, there are some in the peace camp who are willing to grapple head-on with the risks of preserving the status quo. Madeleine Albright, Bill Clinton's secretary of state, has argued that there is no need to take on Saddam right now because the efforts to thwart him have worked. "Since the administration of former President George H. W. Bush, each time Mr. Hussein has pushed, we have pushed back," she wrote in a recent *Times* op-ed. Furthermore, she argued, "Saddam Hussein's military is far weaker than it was a decade ago. And he must surely be aware that if he ever again tries to attack another country he will be obliterated. All that is grounds for calm, but not complacency."

When you come across the Groundhog Day predictions of what will happen if the United States invades Iraq—the Arab Street will explode, we will create a thousand new bin Ladens, we will become stuck in a quagmire—you're actually relieved. Here are writers who are at least willing to compare the risks of action with those of inaction. Stephen Zunes argues in the *Nation* that Iraq is not a center of anti-American terrorism, international inspectors can ensure that Saddam will not obtain weapons of mass destruction, and the Iraqi people would not welcome a U.S. effort to topple the current regime. Writing in the *New York Times*, author Milton Viorst predicts that if the United States goes into Iraq, Islamists in Pakistan will overthrow the government there and launch a nuclear attack on India. These assertions and predictions may be wrong and far-fetched, but at least

Zunes and Viorst are willing to think about the world and about the future.

They are still the exceptions. For most in the peace camp, there is only the fog. The debate is dominated by people who don't seem to know about Iraq and don't care. Their positions are not influenced by the facts of world affairs.

When you get deep enough into the peace camp you find fog about the fog. You find a generation of academic and literary intellectuals who have so devoted themselves to questioning meanings, deconstructing texts, decoding signifiers, and unmasking perspectives, they can't even make an argument anymore. Susan Sontag wrote a *New York Times* op-ed about metaphors and interpretations and about the meaning and categories of war. It filled up space on the page, but it didn't go anywhere.

Tony Kushner, the fashionably engagé playwright and most recently the author of *Homebody/Kabul*, contributed to a symposium, also in the *Times*. Here is the complete text of his essay:

*Change is not the substitution of one static state for another. The meanings of Sept. 11 continue to be fought over, and the prevailing interpretations will direct future action. Colossal tragedy has made available to America the possibility of a new understanding of our place in the world.*

*Tragedy's paradox is that it has a creative aspect: new meaning flows to fill the emptiness hollowed out by devastation. Are we dedicated to democratic, egalitarian principles applicable to our own people as well as to the people of the world? And do we understand that "our own people" and "the people of the world" are interdependent? Will we respond with imagination, compassion and courageous intelligence, refusing imperial projects and infinite war?*

*The path we will take is not available for prediction. We ought not to believe columnists, think-tank determinists or the cowboy bromides of our president and his dangerous handlers and advisers. We, the citizenry, are still interpreting.*

*Our conclusions will then force our reinterpretation. Urgency is appropriate but not an excuse for stupidity or brutality. Our despair over our own powerlessness is simply a lie we are telling ourselves. We are all engaged in shaping the interpretation, and in the ensuing actions, we are all implicated.*

Tony! We can hear you but we can't see you! You are lost somewhere in the fog of peace.

# LIBERATING IRAQ

## By Stephen F. Hayes

———◆———

*(Originally published in the April 14, 2003, issue)*

*Umm Qasr, Iraq*

The wheels of the four Humvees in our convoy had not stopped turning when Ali al-Ethari jumped out of the back of the second vehicle and sprinted toward the front of the Port Authority building here in Umm Qasr, Iraq. The 15 others in the convoy—11 American soldiers, two Iraqi Americans, and two reporters—knew where he was headed.

Tributes to Saddam Hussein appear everywhere in this southern port town. A smiling, avuncular Saddam hovers over a corner market on a plastic plug-in sign, like the ones that advertise cheap beer in bars throughout America. A few feet later, Saddam the conqueror, wearing a black-brimmed hat and a Western suit, fires a rifle onehanded in a portrait inside one of the U.N. compounds here. Further on, in the middle of the road, a billboard-sized tile edifice depicts a menacing military Saddam, in green fatigues and a black beret, firing a pistol toward the sky. On the flip side, for vehicles traveling in the other direction, is a grinning Saddam in a white naval uniform with gold trim. Each of these monuments had been defaced—one with red X's over the dictator's mug, another with red paint splashed across his face, others simply torn apart. Only the one in front of the Port Authority building was still untouched.

Of the anti-Saddam Iraqis I've met over the past several weeks, Ali al-Ethari is the quietest. In that time, we've spoken twice, and on those occasions, only briefly. In group settings, too, he lets others do the talking.

He said nothing before bailing out of the moving vehicle. As Ali ran towards the unmolested canvas, the other two Free Iraqi Forces soldiers called out.

"Wait for us," said Ali al-Mohamidawi, with a chuckle. "We'll help you." Al-Ethari ignored them, unhitched the long knife on his belt, and began shredding the 15-foot painting. By the time the other two Iraqis joined him, most of the work was done. Several American soldiers from the convoy team joined the Iraqis in front of the few remaining scraps of canvas. They laughed about their friend's uncharacteristic outburst. But when al-Ethari turned around, he wasn't laughing.

The other Iraqis understood, withdrew their playful smiles and, for a moment, said nothing. Everyone has a mission. This was part of Ali al-Ethari's.

Al-Ethari is a member of the Free Iraqi Forces, a program that brings together Iraqi exiles with American soldiers to liberate Iraq. The Pentagon began seeking volunteers for the FIF as early as last August, working through opposition groups and running radio ads in areas with a heavy concentration of Iraqi Americans. The program is not nearly as large as originally conceived—there will be fewer than 100 soldiers who wear the Free Iraqi Forces uniform. The orders for the air base in Taszar, Hungary, where these troops were trained, called for accommodations for 3,000 men. The need for extraordinarily careful vetting, coupled with the slow churning of the vast Pentagon bureaucracy, limited participation. But the numbers reflect no lack of enthusiasm. Thousands of Iraqis in the United States applied to join the FIF—some were rejected, others were bogged down in process and simply never made it to review.

That's a shame. The Iraqis who made it back to their native land are spread throughout the country—from Umm Qasr, to Najaf, to An Nasiriyah—and are contributing to the war effort in valuable ways: reassuring a panicked population in the south that food and water are on the way; helping compile a "blacklist" of Baath party members and Saddam sympathizers who will be prosecuted after the war; describing the underground bunkers that protect the regime; educating military police in the ways of Islam to help them better handle prisoners of war; giving the precise location and capacity of a water plant near Basra. Two Free Iraqi Forces soldiers identified a tattoo on the arm of a captured Iraqi as the mark of the fedayeen—Saddam's death squad irregulars. Another was speaking to a relative on the street in Umm Qasr when two would-be suicide bombers heard him describing his duties in Arabic and, comforted by the presence of a fellow Iraqi, surrendered. The list goes on.

The stories of these Iraqis—each of whom fled Saddam's regime, many with a bounty on their head—are extraordinary. Anyone who wonders what Iraqis think of the war of liberation need only listen to these men.

Base camp for the Free Iraqi Forces is a firehouse near the Iraqi border. The U.S. Marines live in the kitchen. The Iraqis are in a conference room, and the 11 Army reservists and 10 members of the Florida National Guard who helped train them live, two to a room, in small offices throughout the building. Most of our time here is spent waiting for word that we will move forward, deeper into Iraqi territory. The Iraqis huddle outside the front door, chain-smoking, drinking Taster's Choice coffee from the MREs, and following war developments on the radio. The most accurate news, they say, comes from Radio Sawa, a U.S. government outlet that broadcasts in Arabic, and from Kuwaiti Radio. They closely monitor the numerous Arabic-language stations critical of coalition efforts here. Their listening habits mirror those inside Iraq, according to the Iraqis we have met in southern towns such as Umm Qasr and Safwan.

When FIF soldiers finally get word of their assignments, they are sent ahead with a trainer or two—men who have been with them since shortly after they arrived in Hungary in mid-January. They are then integrated with other military units, primarily those whose duties are in civil military operations.

At first blush, the firehouse might seem a collision of two worlds with little in common. Of the 50 or so Iraqis in the first cohort from Taszar, only two are Christian. Some of the Muslims are devout, others less so. But one of the first rules established by Lt. Col. Dan

Hammack, the commanding officer, is that training will stop for prayers. It is not uncommon to be in the middle of a conversation with an Iraqi who abruptly excuses himself to pray.

In contrast, some of the American soldiers do their best to live up to stereotypes solidified in Hollywood. "So what if I nailed Saddam's daughter right in front of him?" asks one Marine, in the company of two Iraqis and a reservist with two grandchildren back home. He follows up with a story about a fellow soldier's sexual encounter in Korea. The tale, extraordinary if true, gets an uncomfortable reaction from the other three. "If you take f— and s— out of the English language," says Saib al-Hamdy, a Free Iraqi soldier, "the Marines wouldn't be able to talk."

"I know," says the Marine. "Your English is better than mine. Most of us guys didn't go to college. You come to Chicago and that's what you hear every other word." After a moment, though, he is contrite. "If it offends you, I won't say it."

Accommodations of that kind are made daily. Major Bret "Huge" Middleton, a banker and former college football player from rural Kansas, keeps an English version of the Koran on the box of water at his bedside. He's made his way through the first several chapters. "It's pretty amazing. It's not much different from the Bible at all."

Hammack, a Special Forces officer now in the reserves, makes the same point at a briefing for 100 U.S. military police officers tasked with handling enemy prisoners of war. "For the Shia, a father and son were martyred in Iraq, in Karbala and Najaf," he says, turning to an Iraqi to doublecheck his pronunciation. "It's Na-jeff, right?"

"Na-jaaf," says the Iraqi.

"Na-jeff," continues Hammack. "In Christianity, it's the same thing. Some leader was brutally murdered. It's still a very painful event. If you really look at it, and get right down to it, there are many beliefs that are very similar to that. It's key that you respect that."

Major Mark "Evil" Green, a reservist from Oklahoma with "three confirmed kills," offers a useful example. "If you ask men to strip nekkid in front of other men, you will be offending everything they are. Give them that respect and that dignity and it will go a long way."

Some of the Americans living in the firehouse say the mission has caused them to lose their prejudices. "After September 11," says one, speaking of Arabs, "I would walk into the Magic Market and give them a glare. These guys here have ruined my life—but in a good way. They've changed my life. Lots of the things I thought I knew before coming here, I don't believe anymore."

The camaraderie among the Iraqis, the Marines, and the Army reservists is genuine. Most of the Iraqis have been given nicknames. Ahmed is "George Michael," because he looks like the British pop star. Another Iraqi is known simply as "Tupac." Before he introduced himself to any of the Americans he flashed gang signs and asked one of the soldiers whether he listens to rapper Tupac Shakur. His brother and father, "Three-pack" and "Six-pack" respectively, are in Iraq now working with American soldiers. Another goes by "Tim." "His name is Tahib or Tabib or something like that," explains one of the Americans. "But no one could pronounce it, so he's 'Tim.'" There's also "Burt Reynolds" and "Robert De Niro." "I don't even know what his name is," recalls Sergeant First Class Curtis Mancini. "Every once in a while we'd get him to say, 'You tawkin' to me?' He was perfect."

The Americans were hand-picked for this assignment, and some of them were ambivalent when learning about the specifics. "It was intimidating," says Hammack. Now, as stories about FIF successes in the field trickle in, units who were not preassigned soldiers from the Free Iraqi Forces are requesting them.

One of those requests came on Thursday, March 27. A civil affairs unit already relatively deep in Iraq needed a Free Iraqi Forces soldier. The soldier would be Hakim Kawy, a soft-spoken but at times garrulous man from San Diego.

A group of Americans and Iraqis gathered in front of the firehouse after lunch that day. We listened to the radio and chatted about any number of things—the water supply in Basra, a huge T-bone with button mushrooms, the Italian Deli in Arlington, Virginia, our wives, fiancées, and girlfriends. Beer. "I'd drink a warm Pabst Blue Ribbon right now," said Gunny Sergeant Randy Linniman, unaware that he was insulting this Milwaukee native.

Hakim seemed distracted. He approached me and began to explain that he hadn't been in touch with his family in weeks. Among those living in the firehouse, I alone had the answer to his problem—a satellite phone. I offered it to Hakim and was surprised when he declined. He told me that he can't talk to his family. It's too emotional. In the two months since he left, he has been in touch with his wife and four children only through intermediaries. I anticipated his next question.

"Will you call my family?" It was just after noon in Iraq and just after midnight in California, when I dialed his home. A groggy voice on the other end paused for a moment when I identified myself as a reporter who had been in touch with Hakim. "Is he okay? Where is he?" I told Hakim's wife that her husband was doing well, was in good spirits, and missed his family tremendously. Hakim paced in the sand about 10 feet from where I stood with the phone. I answered what questions I could and agreed to pass on a message. "Tell him I sold the house," his wife urged. "He'll be so relieved."

The conversation lasted three minutes. Hakim came to me when I put the phone back in my pocket. I reported on my brief chat, and then told him the news. "Your wife asked me to tell you that she sold the house. It's in escrow." Hakim's tense shoulders relaxed and he began to cry—not the loud wailing of the distraught or the muffled sobbing of the overjoyed. Hakim simply wept in silence—cathartic tears of reprieve. He took out several napkins to dry his eyes and offered one to me. "Thank goodness," he said.

When Hakim left home in mid-January, he and his wife discussed the possibility that a prolonged absence from his construction business could cause financial hardship. He urged his wife to sell the house if circumstances worsened. They have. For an average American, having to sell a home under financial pressure would represent something of a life crisis. Hakim Kawy is not an average American.

He arrived in the United States in the mid-1970s after a harrowing escape from his home in northern Iraq. Hakim was serving his compulsory year in the Iraqi military in 1974, after graduating from the university with a degree in mathematics and statistics. He was posted at the top of a mountain, and on leave one day bought some candy to give to Kurdish children who, he says, were left undernourished by civil war raging in northern Iraq. "One of the secret military was watching me, and I went to torture for three months," he says matter of factly.

"You have no idea. I would get so weak I cannot stand on my feet. Psychological and physical abuse. They slam me, and they throw me, and they spit on you. Sometimes they keep your hands like this [he holds his hands up, as if being frisked] for a long time. They punched me in the face while I sleep. They let you bleed and nobody see you. And at the end of the day they throw you piece of bread, old, and water."

In a final effort to make him talk, Hakim says, "they did something to me horrible." His captors loaded him down with a sheet of "light metal," perhaps aluminum—he says it was approximately eight by ten feet—and made him climb the mountain with it, through minefields and hostile Kurdish peshmerga fighters, to reach his former post. They waited for a windy night and sent him on his way.

"The wind threw me twice and I fell on the stone. The second time there is water," he says, indicating a small pool of water among the rocks. As he sat on the rocks, he remembered that he had a small piece of rope in his pocket. He rolled the metal into a tube, tied the rope and made his way to the top.

"They couldn't believe I make it to the top. They were so surprised and also so angry."

Hakim often takes breaks from the chronology to offer his thoughts on the current war. "It's not just ruthlessness they sanction. They enjoy this. They enjoy this. They like to do this—they take you and kick you and try to disturb your dignity." He continues: "Yes, it is a personal tragedy, you know, but I hope the world will look at it as not me only. Everyone should know that Saddam and his terrorism is like a disease—he has no border, nothing to stop it."

Hakim was taken to a judge, accused of being a traitor. He had harsh words in the courtroom in northern Iraq. He knew the judge, who was related to a well-known Shia cleric. "I told the judge in Erbil, 'Sir, I look like your son. I'm just 22 years old. I'm not going to torture for the next six months. This is what I'm saying and it is the truth. If you want to finish it, get your gun and do it.' Like they say, go ahead and make my day—you get to that point. Just finish it." The judge gave Hakim one week to report to another court in Baghdad. His mother told him to leave the country at night. "You are my youngest," she told him. "I cannot lose you all. I need you to go out."

Hakim tells me what his mother meant when she said, "I cannot lose you all." His brother had been jailed earlier that year, and although the government claimed to have released him, he was never heard from again. "And the story keep going like any other Iraqi's story," he says. "It's nothing."

I hadn't talked to Hakim much before this conversation, and I apologized for taking several hours of his time—more than either of us had planned. He didn't mind. "The more I talk, the better I feel. I put a small nail in [Saddam's] coffin. It is my small part."

On Sunday, March 30, shortly after Ali al-Ethari attacked the Saddam portrait in front of the Port Authority in Umm Qasr, he and I ventured inside, accompanied by another FIF soldier, Saib al-Hamdy. The building bore the scars of a battle that had taken place there just a few days earlier. Some of the walls were pocked with fresh bullet holes, paint chips scattered on the floor below. Everything in the building, like everything in the country, was covered with a light dusting of sand.

The scene looked as if it had been frozen in place the day before coalition troops rolled into town. In a glass case in the lobby were routine announcements that revealed no worries about war. One flyer declared that Saddam Hussein had cancelled law 30-999, effective June 1, 2003. "No more taxes will be collected to build the new mosque." Another memo, signed February 23, praised port workers for successfully delivering 218 new cars from an incoming freighter to Warehouse 21 without an accident. The port manager requested bonuses from a government higher-up. Scribbled across the front of the memo, accentuated by a yellow highlighter, was the approval— 7,000 Iraqi dinars ($2) to each worker for a job well done.

Deeper inside the complex, we came upon the personnel office. Stacked neatly on the shelves were three-ring binders with records dating back more than a decade—a treasure trove of information for the civil affairs units here. One of the key elements of the postwar reconstruction is returning Iraqis to their jobs. Early last week, the

British running the relief efforts had already rehired several drivers. The Free Iraqi Forces in Umm Qasr are helping the Brits determine who previously worked at the port, and in what capacity. They've found several Iraqis who were formerly laborers reporting back to work as self-promoted managers.

When we left the port to assess the situation in the town, our convoy was greeted with the kind of reception the White House and Iraqi Americans had long predicted. Iraqis here lined the streets—waving their arms, giving thumbs-up to American soldiers, cheering. "America good, Saddam bad," one elderly man in tribal clothing yelled from the side of the road. Tributes to Saddam Hussein had been defaced. Tile edifices were splashed with red paint. Paintings of the dictator were ripped down from walls. The Baath party headquarters had been vandalized.

Written tributes to the Iraqi tyrant on the crumbling walls and the dilapidated buildings had also been defaced. One, in perfectly stenciled Arabic lettering, declared: "Yes! For the leader Saddam Hussein." The new graffiti sent a different message—"Dun Saddam, Good U.S.A."

But the time we spent in southern Iraq was not all jubilation. Many Iraqis here, unaccustomed to their newfound liberty and the harsh reality it presents, seemed to be fighting their own emotions, lurching unpredictably from gratitude to desperation to apprehension. And the residents of both Umm Qasr and Safwan badly needed water. Even as we circled the town in military Humvees to the cheering of locals, the children were practicing their elementary English. "Mister . . . water," they said, cupping their hands in front of them. "Mister . . . water."

We drove around long enough that we began to pass children we'd already seen, still lining the roadside. Their cries grew more frequent, as if they were calling someone by name. "Mr. Water, Mr. Water, Mr. Water." One prepubescent boy hiked up his shorts and showed a little leg as he pleaded for water. He and his friends were barely old enough to understand the significance of his attempted tease, but they doubled-over laughing anyway.

When our convoy stopped, many Iraqis rushed the soldiers to shake their hands, thanking them for liberating their town. The Iraqis quickly gathered around the three members of the Free Iraqi Forces to talk about the progress of the war. Although they appeared grateful to have Arabic-speaking American soldiers, they immediately began venting their frustrations about food and water. "My children have not had water for seven days," said one man, waving off a reporter trying to snap pictures. "We do not want people to see us like this. We need water."

Although the residents here salted their complaints about life's necessities with an appreciation of coalition efforts to get rid of Saddam, the palpable sense of panic wiped smiles off the faces of these soldiers who moments earlier had been welcomed as liberators. And many Iraqis told us they did not believe Saddam would be eliminated. "How do you expect us to believe that you, the world's two superpowers, can get rid of Saddam, when you can't even get water to a small town on the border?" asked one man.

A man who bore a strong resemblance to Saddam Hussein berated the American soldiers and their Iraqi colleagues. "You have destroyed our town," he said, addressing Ali al-Mohamidawi. "You have destroyed my property. Americans and British go home. No one wants you here. We never had these problems with Saddam Hussein."

His rant drew loud and violent protests from the dozens of Iraqis gathered around us. Without warning, a bearded, middle-aged man in tribal robes lunged at the Saddam defender and grabbed him by the shirt collar. "What property? They did not touch your property! Where's the damage? Do not say these things. We want the Americans. We need the help. You work for Saddam Hussein."

Others joined in, harshly criticizing the Saddam look-alike. A man in his twenties who spoke some English took me aside to assure me that everyone in Umm Qasr supports the Americans and British. They are worried, though, that anyone who rises up will be killed if Saddam survives.

As we talked, the bearded man dashed from the scene and returned 30 seconds later. His younger brother, carrying a heavy metal pipe, accompanied him. The man's wife came too, wailing loudly and begging him to walk away. Ali, backed up by Dan Hammack, tried to settle the group, at one point reminding them that the Americans had powerful guns that would have to be used if the situation worsened. The crowd, now in the hundreds, struggled to keep the combatants apart. After several tense minutes, the pro-Saddam man left—alone— walking slowly back to his house. The others quickly reported that he was a well-known Baath party official, one of a handful remaining in this section of liberated southern Iraq.

Even as we left, the bearded man told us that he would exact his revenge that night. "I will kill him," he said.

We returned to our Humvee but were unable to leave for several minutes. The crowd from the town square had followed us. They wanted to know more from Ali. One man who couldn't make his way to the front of the cluster ran around to the back seat, behind the driver, where I was sitting. He leaned far inside the vehicle, over my lap, and grabbed Ali by the shoulder.

"How do you know Saddam Hussein will be gone?"

*"He will go. I promise you. 100 percent."*

"But how can you be sure? He will live."

*"I promise you with my life, he will go. 100 percent."*

The four of us in the Humvee rode away in silence. Ali wondered why the coalition couldn't get water to the town. On the trip here, he noted, we passed several semi-trailers filled with water. Why was it taking so long? His frustration grew when we returned to the port. On a quick tour, we were stunned to see boxes upon boxes of bottled water lying around. Ali spoke up again. "Why is this water sitting here? What can we do?" One of the soldiers—a mid-level American officer who had not been part of the Free Iraqi Forces group—offered an answer meant to be reassuring. "It's being taken care of," he said. "All of the stuff is being taken to warehouses for storage."

The Pentagon reported early last week that a water pipeline between Kuwait and southern Iraq had finally been opened. The 610,000 gallons of water it will pump daily should wash away the concerns Iraqis here have about their own survival, and allow them to focus on the survival of the dictator in Baghdad.

As we set out for Safwan, our convoy came upon a group of Iraqis along the side of the road. They were up to something, but it was hard to tell what. The convoy came to a sudden stop, and the Americans jumped out of the vehicles, guns drawn. They immediately, almost reflexively, formed a perimeter around the Iraqis, who dropped to the ground as commanded. One man waved a white T-shirt. Col. David Blackledge and an FIF soldier approached the group. One of the Iraqis pointed to a makeshift coffin and explained that their friend had been killed in fighting in nearby Al Zubayr. They had come to bury him where he was born. Blackledge told them to do it quickly to avoid arousing further suspicion.

Ahmed, known to everyone at the firehouse as "George Michael," moved forward last week. When I talked to him shortly before he left, he told me his mission began in 1991. Twelve years ago last month, on March 18, 1991, he put on a business suit and sunglasses and walked the road between Basra and Nasiriyah, in southern Iraq, to surrender to the U.S. Army.

For weeks he had hidden at his sister's house. A local Baath party leader had seen Ahmed agitate against the regime and notified Iraqi intelligence. They had his name, and they knew where he lived. When the authorities came to Ahmed's house, they asked his father where he was hiding. His father pleaded ignorance. Being less concerned with punishing the actual revolutionary than with simply inflicting punishment on someone, they took Ahmed's brother, Ali. He was tortured for a week—hung from the ceiling with his arms tied behind his back. One of his arms was broken. Days later, Ali was taken to the front of a local government building that functioned as a site for public executions. As he was led to the tall, wooden post where he would be tied, he stared at horrific reminders of his ill-fated predecessors: Directly behind the support pole, the wall was painted with several coats of dried blood and clumps of human hair.

As his captors were tying his hands behind the post, Ali made a strange request. "Please shoot me in the back," he pleaded. The six gunmen, three standing and three lying on the ground, howled with laughter. Their commander, also amused, asked him what crime he had committed. "I did nothing," Ali told them. "They took me because of my brother."

"You did not participate in the uprising?" the commander asked. "You are innocent?"

"Yes."

With that, the commander motioned for his assistants to untie Ali, and told him, "Go home."

Ahmed calls this the "miracle." "He did not do this because he is a nice man," the Free Iraqi soldier says. When his brother returned home, Ahmed left. He turned himself in to the U.S. forces in southern Iraq, setting in motion a process that would see him bounce from nation to nation, and from one refugee camp to another, for the next two years. If he had stayed, he would almost certainly have been caught and killed by Saddam's regime. When he left, he didn't know if he would ever see his family again.

After living and working for 10 years in Portland, Oregon, that moment is at hand, perhaps within days. It will not be a perfect reunion. His father died in 1999. "My father made me a cassette, and he's singing to me and crying, and he says he knows he won't see me again. He says that he's not worried about me, though. He says he's proud of me."

After his father died, he planned a trip to Syria to see his mother, two brothers, and sister, and to meet for the first time several nephews and nieces. They stayed for five weeks, trading stories and remembering their times together in Iraq. Ahmed learned then that his family had deliberately spread rumors about his fate when he fled in 1991. They told everyone that he was killed in action—not wanting to risk further retribution from the local Baath party and Saddam's henchmen.

Ahmed is grateful today that he saw his mother in 1999. Shortly before he left for training in mid-January, he received word from his sister that his mother has cancer. "Lung cancer," he explains. "The bad kind, not the good kind. How you say it?" Malignant? "Yes, malignant."

He has had plenty of opportunities to check on his mother, but he's not sure he wants to hear how she is doing until he returns home. He has relatives in Umm Qasr, the town likely to be the first official stop on his mission in Iraq. "When I get to Umm Qasr, maybe I call from my aunt's or my cousin's. If I'm there, I'm doing big thing. I could die too and could be killed in action. I don't know how I'm going to act. It's going to be the happiest day of my life if I call and they say 'Here, talk to your mom.'"

His family is expecting him. They don't know exactly what he's doing, but they know he's coming to see them. In Syria, in 1999, he devised a way to communicate with his brother. They spoke in code for years, worried that the government was eavesdropping. Whenever Ahmed mentioned the name of the local Baath official who ratted on him in 1991, the brothers agreed, it meant he was talking about Saddam's regime.

On January 14, 2003, Ahmed called his brother. "I'm going to get my money from [the Baath party leader]," Ahmed said, referring cryptically to the training he was to receive in Europe. "And [Ali] knew exactly what I mean. I asked him, 'You got it?'"

"We'll help you get the money," said Ali. "That guy owes everyone lots of money."

So Ahmed moved forward last week with a civil affairs unit. He is now working to calm the Iraqi people, to explain the mission, to lay the groundwork for the humanitarian effort to come, and to reassure small pockets of a frightened population. It is a job he takes very seriously.

"I'm going to be proud if they think I am an American soldier," Ahmed says. "I have no fear to go there. I believe we live one life and we die one time. And if I die, I die for a good cause. For my family and for my people."

Ahmed carries a picture of his girlfriend around his neck. He showed it to me with evident pride and recalled seeing her for the first time at a Starbucks in Beaverton, Ore. He shared his memories of the giddy days of new love—playing pool and bowling, making eggs at 2:00 a.m., seeing *My Big Fat Greek Wedding*.

Before the war, he and several Iraqi friends gathered regularly at the Starbucks to talk about life, politics, and the coming war. Some of them didn't share his enthusiasm for the mission. None of them wants to keep Saddam in power, but several of his friends don't approve of his willingness to fight with U.S. troops. As he explained their arguments, he became very animated. He called them "cowards."

"Let me ask you a question—why the American people, why the American soldier have to die in our homeland? I say, we have to die there. So I said to them, [he points] you and you and you, you have to volunteer so less American people go. If you are American soldier, you go to Basra, why you have to die there?"

Many of the American soldiers here with the Free Iraqi Forces, men who could die here, as Ahmed puts it, have already made significant sacrifices. Major Bret Middleton, whose brother was killed in the first Gulf War, left a wife and four little girls back in Kansas. Master Sergeant Frank Kapaun, also a former Special Forces soldier, got his orders just three days before being deployed. He left his job as a telephone line installer and an occupied apartment in Columbus, Georgia, that friends and relatives would have to clean out. No one I spoke to complained. They have gotten as much as they have given, something few of them expected when they first met the Iraqis in Taszar, Hungary.

Sergeant First Class Curtis Mancini, a soldier's soldier and a 17-year veteran of the police force in suburban Fort Lauderdale, Florida, jotted his first impressions in a notebook. "They are rough looking, some look disheveled, but not unclean. They think nothing of talking to each other during lectures and loudly, oblivious to the instructors and the class. They are eager and intelligent above my expectations. Many with advanced degrees, many want to engage in intellectual conversations."

The Iraqis went through two weeks of improvised basic training in Hungary. Since many of them are older, out of shape, and accomplished professionals, they did not always take kindly to the rigorous regimen. The training had to accomplish two potentially conflicting goals: preparing men for possible combat and not alienating them.

The second part of their instruction was precisely tailored to the work they are doing now in Iraq. Kapaun was one of the first American soldiers to go forward with Free Iraqis. He was air assaulted into An Nasiriyah with one of the FIF soldiers and assigned to a military police unit, and then a counterintelligence subunit, responsible for interrogating enemy prisoners of war.

"You could not buy the assets and intelligence, all the benefits we were getting out of them," says Kapaun, who recently returned from Nasiriyah, leaving his FIF soldiers with their new unit. "Their interviews yielded time-sensitive, real-world intelligence." The Free Iraqis helped American military intelligence sort out real enemies from Iraqi civilians caught in undefined "battle space."

Kapaun, like many of the other Americans here, has become emotional about his men. "It tore my heart out to say goodbye to them," he says. "I made plans to see them back in the States, but hopefully I'll see them in Baghdad first."

The moment the war began with 40 Tomahawk missiles in a "decapitation" attempt, Ali al-Mohamidawi ran to the road in front of the base and began flagging down buses heading north. None of the Free Iraqi Forces soldiers I met was as eager to fight as Ali. And with good reason.

Ali told me his story in three separate sessions in the supply room that also serves as my bedroom and office. He was dressed in the "chocolate chip" desert camos given to the Free Iraqi Forces. Emblazoned across the left front pocket of his shirt were the letters "FIF." A patch on his right sleeve gave the same identification. He sat on a box of water and began talking.

Ali lives in Alexandria, Virginia, and works at a nonprofit foundation that helps international refugees. He knows their situation better than most, having come as a refugee to the United States in 1994 after three years in a Saudi refugee camp for displaced Iraqis. When he finally arrived at National Airport, no one showed up to get him. He was supposed to have been picked up by someone from the nonprofit that now employs him. But they forgot.

Ali didn't panic. He asked the cashier at one of the restaurants for a cigarette and waited, just happy to have finally made it to America.

Ali was one of the instigators of the 1991 Shia uprising near Basra. He and several friends had begun stockpiling weapons and ammunition months before American forces started military operations in January of that year. Ali graduated from the university in 1989, just as the Iran-Iraq war came to an end. Like all young men his age, he was compelled to "serve the flag." His Iraqi army unit deployed in northern Iraq for three months, taking the place of soldiers discharged after the war.

A disabled man from his unit had been assigned to guard a lot of old cars outside of Basra, and when that man had acquired enough points to retire, he came to Ali, knowing that he had family nearby, and offered to recommend that Ali take his place. Ali got the assignment, to the great displeasure of the other soldiers in his unit. One, from a wealthy Basra family, was particularly frustrated that he hadn't heard of the opening in time to bribe his commanders for the plum position.

Ali reported to his new post in early 1990. The cars there were mostly old and beat up, and no one ever came to check on him. "First fifteen days, I go eight and go home at five. Then nobody come look at me, and finally I go 11 and come home 12. My dad, he says, why you not go and be so lazy?" The entire time he worked there, his commander took his salary, which was around $30. So he stopped reporting to work. On August 10, 1990, Ali's unit was sent to Kuwait—part of the invading army. A fellow soldier visited him and told him to leave the cars and report to his unit immediately. Ali refused.

"I said, 'If I don't see any document I'm not going to leave, because the cars are my responsibility.'"

The documentation came several days later, hand-delivered by a small delegation from Saddam's regime. "One officer from my unit came to my house with three soldiers, one of them with intelligence. I remember that day very clear, because I was helping my father fix the water pump on his car, and I remember seeing their car. It was a military car. They said, 'Ali you need to go to Kuwait. You are going to be on the border with Saudi Arabia.'"

They wanted Ali to return with them. But Ali's father begged the men for two extra days, to help prepare the family for his son's departure. "I can give you my word that he will join you," Ali's father told the soldiers. They relented.

But Ali had no intention of reporting for duty. He and his father had discussed the invasion and agreed that Americans were not likely to let it stand. Ali went first to his sister's house, and then to his uncle's. For months, he kept in touch with his friends planning the uprising.

In late February, one friend, Ahmed, who now lives in Iran, had made arrangements to obtain bullets for the weapons they had stored. He asked Ali to retrieve the ammunition and take it to Al Kebla, a town 15 minutes away. Ali avoided main roads, taking side streets and rural roads where there are no checkpoints.

"That day, unfortunately, there are checkpoints. And I saw six people about a mile away. I'm driving and they have a motorcycle. If I stop and turn back, they are going to follow me. I said let me continue what I'm going to do, the bullets are in the trunk. And the idea come to me, I give them high beam, and I speed up so they can't recognize I'm speeding. About 100 meters they are telling me to slow, and they wasn't prepared for me. And one guard stayed in the middle and when I got close to him he go to the sidewalk. And I heard from behind me the [gun]fire and I speed up between the streets and I got away from them."

Ali returned to his house with the bullets, and his brother Karim, who had heard the shots, quickly helped him cover the car. "And he take it from the trunk and we take it to the roof. And we put it somewhere where they cannot find it. And we went to the roof—watching, watching—and no one come find us."

Two days later, on March 2, 1991, with the Iraqi Army in rapid retreat from Kuwait, their supplies depleted and morale low, Ali was awakened by gunfire. "I heard the shooting and I know the guys have started. I went to them with my father, and my three brothers—Karim, Mohammed, and Rahim. When we got there we saw our guys."

Ali paused to collect his thoughts.

"And I'm saying to history now, that Ahmed and some other 15 guys start the revolution in the south at two o'clock in al-Jamhoria. And when they finish call the people and shooting at the air, and the people, they listened to them. It's an uprising, and some people they come with them." The rebels quickly set up checkpoints of their own. They stopped each car that passed, trying to convince everyone they saw to join them in the uprising. One man, driving a red car, was a well-known officer in the Iraqi army. He refused to join and accused Ali and his friends of belonging to a "mafia."

They warned the officer against moving through the checkpoint. "And he just ignored them and he left," says Ali. "And they shoot him with an RPG and they kill him."

Ali and his friends waged a fierce battle with Baath party members and Saddam's intelligence service, the Mukhabarat. The rebels kept their captives in a small mosque at al-Husseinia. Iraqis began pouring out of their homes to participate in the uprising. In less than 24 hours, the rebels had taken Basra. They assigned neighborhoods different military functions—one would serve as the mess hall, another as the ammunition depot. The rebels would control Iraq's second-largest city for 15 days.

Things turned bad quickly. The rebels were running out of ammunition. The help they had been promised—from the United States and Iran—never materialized. Saddam dispatched his Republican Guard to the south, and ordered Ali Hassan al-Majid, better known as "Chemical Ali," to put down the rebellion. Some of the leaders fled to Iran, others turned themselves in to American soldiers.

On March, 20, 1991, Ali and his friends staged one final battle in al-Jamhoria. He begins the story of that last conflict with the most important detail. "And that time my older brother, Karim, gets killed. He was with me. We saw the Republican Guard. He was killed with my cousin, named Ali. He was killed and other guys, too. I don't remember their names right now, 15 guys from my small area. A lot more, actually—70, 80 guys, maybe."

Ali and the remaining rebels took the dead bodies and piled them inside a nearby house belonging to Abdul Khalik, one of the earliest instigators who now lives in Iran. "The army getting close to us—very close, very close, very close," says Ali, pacing in between rows of boxes in the storage room. "My responsibility that time—how we going to get my brother's body and my cousin's body from the house before the army burn all the area."

They agreed to split up. Ali's father and an uncle would take care of the deceased. When Saddam's soldiers confronted them, they would place blame equally on the Iraqi army and the uprising. The two men loaded the bodies in a truck. They were stopped twice by the regime. Each time, they gave the same explanation. "We don't know who killed them, the bombs from you guys or from the uprising." They took the bodies to Al Zubayr. Although each of the dead men was Shia, they were buried in a Sunni cemetery.

Meanwhile, Ali gathered the women and children in three families—all related—and began to move them to his home in Al Kebla, several miles away. After walking for 35 minutes, they were stopped by Iraqi soldiers. "A general called me over to him. When he talk to me, I know from his accent he is from Tikrit. And he asked me, what is my name. And I said Ali. And he said where is your ID, where is your unit. And he said where is your unit—and I said Kuwait. And he kept looking at me and he says, you are one of them, you are one of them. You are uprising against us.

"And my aunt, she is very brave, and she says to him, 'We are all women and we have lost all of our family and he is our only man. And you are stopping us.' And that time she gave me her daughter to carry her in my hand, so that I can avoid the people who doubt me. So it look like I'm helping, holding the daughter. She said, 'We look like we came from Israel? No. Do we look Iranian? No. Why you stop us? Why you investigate us? You don't have any children?'"

"And he said, 'I'm talking to him, man-to-man.' And he said to me, 'Tell me the truth and it's okay.' And I said, 'I'm not with [the uprising].' And he hit me, in the face like this. [Ali makes a violent slapping motion.] And he said, 'You are lying.' And that time when he hit me, my aunt and other girls start crying and shouting and cursing him. And telling him a lot of stuff, 'You are not a man, you are not a brave man and you let us go.' And he said, 'I'm going to leave you because you are with the family.' And he said, 'If I see you again, I'm going to kill you.'"

They walked for another 30 minutes, until a man in a truck stopped to ask if they needed a ride. The entire group—15 people—piled into his Toyota pickup truck. The drive to Al Kebla took 15 minutes.

When his father returned from the cemetery, they discussed their options. His father told Ali to wear his army uniform and "just get lost somewhere." Ali followed his father's advice, walking to the nearby town of al-Jammyet, to visit a friend named Sajad. When Ali arrived, Sajad took him to a small river behind the house, where they hid under a bridge. Sajad pointed to the corner. About 75 Iraqi soldiers were arranging many locals in some sort of line. Each person was linked to the next with a rope, tied around his waist. Ali recognized some of them as uprising participants, from Al Faw Island, south of Basra. The others, he said, were innocent. The captives were led to the desert directly behind the University of Basra and executed.

With no chance of another uprising and Baath party members looking for him, Ali decided he would retrieve his brother Rahim from their home and surrender to American soldiers. But first he would have to make it back to his home, and then to the Americans. This wouldn't be easy. Coalition forces had mostly withdrawn from the cities and, as they left, Saddam's soldiers and Baath party members filled the void. Basra and its suburbs were crawling with Saddam loyalists, patrolling the city with guns drawn, looking to kill rebels like Ali and his brother.

Because he was wearing his Iraqi Army uniform, he was stopped only once. He lied and said he was looking for his unit. The Iraqi Army officer let him pass. Ali saw many others who weren't as fortunate. "When they see civilians walking—oh my god—they stop him, and if he don't stop, they shoot him right away."

Ali kept walking. "At each corner of a block, seven to ten guys blindfolded, lied next to each other and they are already dead. And I asked one soldier—he said, 'You know why they don't move them? Because they want to show the people that abuse, or that miserable.' It was a big disaster, oh my god. I remember. And behind the university they just kill them."

Among the men and the pain and the killing, Ali saw an old lady pushing a cart full of vegetables. From a distance, she seemed to be going about her business untroubled by the bodies strewn about the streets. They walked toward one another and soon Ali could see her eyes. The woman offered Ali some food and some water.

"My son, why you walking by yourself?"

"I'm going to my unit." It was clear from her expression that she didn't believe him.

"And she said, 'Okay I'll ask God to keep you.' And I said, 'You don't need to cry.' And she wants to bring some water and some food to her family.

"She saw the people at the corner laid down next to each other, and she told me some of them still move and they're bleeding. And she looks like she lost her mind. And she said, 'Be careful, maybe they going to kill you.' And she saw the disasters. And she saw also guys who belong to [Chemical Ali] force these guys to drink the gasoline, and then they shoot them. They had a special kind of bullet—at the front it's red, and when it's shooted at the night, it's not going just the bullet, it's going with the fire. And when it takes the bodies of the people filled with gasoline it makes the people explode—like a bomb. Each corner, goddamn it."

Ali made it home. He stayed there for three days and reviewed his options with his brother Rahim and their father. He remembers his father's advice. "They have your name. Maybe it's better you're going to disappear." There was an American checkpoint on a highway north of Basra. It was too far to walk. On April 1, 1991, Ali called a friend, Nakeeb Karim, a high-ranking officer in the Iraqi army who had— with a cloth over his face like a bandit—anonymously participated in the uprising. He had once again taken his position in the army, and therefore could deliver Ali and Rahim through most Iraqi checkpoints without arousing too much suspicion.

Nakeeb Karim drove them to the town of Al Zubayr. The Iraqis had shut down the highway leading out of town, toward the Americans. Ali and Rahim thanked Nakeeb Karim for the ride and began to walk across the desert. They walked for perhaps one hour when they saw a tent in the distance. It made them nervous, but then, everything made them nervous. They each had a gun. As they approached the tent, they were confronted by a Bedouin and his family.

"Who are you?"

"We are Iraqi soldiers, and we looking for our unit."

"Tell me the truth."

"We are uprising, and the government take over everything."

The Bedouin gave them water and goat's milk, and told the brothers that he had received two visits from American soldiers in recent days.

"They came from that direction and they left in that direction," he told Ali and Rahim, who started walking.

As they approached the highway, they ditched their guns in the desert. Ali says he could make out six or seven tanks in the distance and took off his shirt to wave it as a sign of surrender.

"And I separate with my brother and they can check on us and they know that we have nothing. And they knew that people come to them. We are not the first case or the last case, so they take it easy with the people. Most of the soldiers over there sympathize us and they help us. We told them we are the uprising. They wasn't hard with us, they gave us a little conversation and they brought a Kuwaiti interpreter and they help us. And they gave us food and water. There is black, big guy, lieutenant, and they give us that food [Ali holds up an MRE], and he says, 'I know you eat halal meat and we don't have it here.'"

Ali spent several years in refugee camps before making it to the United States in 1994. He has grown accustomed to living in America, and is likely to return after the war is over. He is open to the possibility, however, that he won't return at all.

"I left Iraq by fighting and I come back with the fight, and maybe I'm going to dead with the fight and I have no problem with that."

Each of the Free Iraqi soldiers I spoke to expressed that same thought. Theirs is clearly—and always has been—a war of liberation. But the same is true for the Americans here. Yes, they are well aware of the more immediate reasons for this war—weapons of mass destruction, eliminating threats of terrorism, stabilizing a region. But the Iraqis' "fight is our fight," says Hammack.

Standing outside the firehouse one night last week, one of the FIF soldiers asks Major Mark "Evil" Green about his tattoos. Like many soldiers, he has several. One is a sword with lightning bolts. Another wraps around his left biceps, barbed-wire with three drops of blood, representing his kills. The third takes up most of one side of his chest. It shows a grim reaper holding crossed pistols—above it, "Death before Dishonor."

The other side of his chest is blank. For now. With one finger, he traces an outline of the one he plans to get when he gets home. "Free Iraqi Forces."

Says Hammack: "We're all FIF now."

# THE HOLOCAUST SHRUG

## By David Gelernter

*(Originally published in the April 5, 2004, issue)*

I HEAR AND READ all the time about Democratic fury; evidently, enraged Democrats are prepared to do whatever it takes to rid the country of George W. Bush's foul presence. Somehow Republican rage doesn't seem quite as newsworthy (and when it does show up, the story-line is usually "Republicans Angry at Bush"). To be fair, Republicans *do* control the presidency and both houses of Congress, and ought to be far gone in euphoria. But they are not. There are lots of unhappy and quite a few furious ones out there, and they are not *all* mad at the president. Some reporters will find this hard to believe, but quite a lot of them are actually mad at the Democrats.

Consider Iraq. By overthrowing Saddam, we stopped a loathsome bloody massacre—a hell-on-earth that would have been all too easily dismissed as fantastic propaganda if we hadn't seen and heard the victims and watched the torturers on videotape. Now: There is all sorts of latitude for legitimate attack on the Bush administration and Iraq. A Bush critic could allege that our preparation was lousy, our strategy wrong, our postwar administration a failure, and so on ad infinitum . . . so long as he stays in ground contact with the basic truth: This war was an unmitigated triumph for humanity. Everything we have learned since the end of full-scale fighting has only made it seem *more* of a triumph.

But Democratic talk about Iraq is dominated not by the hell and horror we abolished or the pride and joy of what we achieved. Many

Democrats mention Saddam's crimes only grudgingly. What they really want to discuss is how the administration "lied" about WMDs (one of the more infantile accusations in modern political history), how (thanks to Iraq) our allies can't stand us anymore, how (on account of Iraq) we are shortchanging the war on terror. But *don't you understand*, a listener wants to scream, that Saddam's government was ripping human flesh to shreds? Was consuming whole populations by greedy mouthfuls, masticating them, drooling blood? Committing crimes that are painful even to *describe*? Don't you understand what we achieved by liberating Iraq, what *mankind* achieved? When we hear about Saddam and his two sons, how can we help but think of the three-faced Lucifer at the bottom of Dante's hell?—"with six eyes he was weeping and over three chins dripped tears and bloody foam," *Con sei occhi piangea, e per tre menti / gocciava 'l pianto e sanguinosa bava*, as he crushes human life between his teeth.

I could understand the Democrats' insisting that this was no *Republican* operation; "we were in favor of it too, we voted for it too, and then voted more money to fund it; we want some credit!" Those would be reasonable political claims. But if you talk as if this war were one big, stupid blunder that we are stuck with and have to make the best of—you are nowhere near shouting distance of reality; people would suspect your sanity if you were not a politician already. Instead of insisting that the war belongs to them, too, Democrats are running top speed in the other direction. Howard Dean led the way on this flight from duty, honor, and truth, but it didn't take long for most of the nation's prominent Democrats (with a few honorable exceptions) to jump aboard the Dean express—which is now, absent Dean, a runaway train.

People ask, why this big deal about Saddam? "Isn't X evil too, and what about Y, and how can you possibly ignore Z?" But we aren't automata; we are able to make distinctions. Some evil is beyond our power to stop. That doesn't absolve us from stopping what we can. All cruelty is bad. Yet some cruel and evil men are worse than others. By any standard we did right by overthrowing Saddam—and do wrong by denying or belittling that fact.

The Democrats' refusal to acknowledge the moral importance of

the Coalition's Iraq victory felt, at first, like the Clinton treatment—
more relativistic, warped-earth moral geometry in which the truth
gradually approaches infinite malleability. Overthrowing vicious dictatorships and stopping crimes against humanity were no longer *that*
big a deal once Republicans were running the show. It seemed like the
same old hypocrisy, sadly familiar. (I will even concede, for what it's
worth, that Republicans can be inconsistent and hypocritical too.)

But as we learned *more* about Saddam's crimes, and Democrats
grew *less* convinced that the war was right and was necessary . . . their
response took on a far more sinister color. It started to resemble the
Holocaust Shrug.

I suggest only diffidently that the world's indifference to the
Coalition's achievement resembles its long-running, well-established
lack of interest in Hitler's crimes. I don't claim that Saddam resembles
Hitler; I do claim that the world's *indifference* to Saddam resembles its
indifference to Hitler.

The Holocaust was unique—"fundamentally different," the
German philosopher Karl Jaspers wrote, "from all crimes that have
existed in the past." Hitler's mission was to convert Germany and
eventually all Europe into an engine of annihilating Jew-hatred. He
tore the heart out of the Jewish nation. There is nothing "universal" or
"paradigmatic" about the Holocaust, and next to Hitler, Saddam is a
mere child with a boyish love of torture and mass murder.

Yet Saddam, like Hitler, murdered people sadistically and system-
atically for the crime of being born. Saddam, like Hitler, believed that
mass murder should be efficient, with minimal fuss and bother; it is no
accident that both were big believers in poison gas. Saddam's program,
like Hitler's, attracted all sorts of sadists; many of Saddam's and
Hitler's crimes were not quite as no-fuss, no-muss as the Big Boss preferred. Evidently Saddam, like Hitler, did not personally torture his
prisoners, but Saddam (like Hitler) allowed and condoned torture that
will stand as a black mark against mankind forever.

Hitler was in a profoundly, fundamentally different league. And
yet the distinction is unlikely to have mattered much to a Kurd
mother watching her child choke to death on poison gas, or a Shiite
about to be diced to bloody pulp. The colossal scale and the routine,

systematic nature of torture and murder under Saddam puts him in a special category too. Saddam was small compared with Hitler, yet he was *like* Hitler not only in what he wanted but in what he did. When we marched into Iraq, we halted a small-scale holocaust.

I could understand people disagreeing with this claim, arguing that Saddam was evil but not *that* kind of evil, not evil *enough* to deserve being discussed in those terms. But the opposition I hear doesn't dwell on the nature of Saddam's crimes. It dwells on the nature of America's—*our* mistakes, *our* malfeasance, *our* "lies." It sounds loonier and farther from reality all the time, more and more like the Holocaust Shrug.

Turning away is not evil; it is merely human. And that's bad enough. For years I myself found it easy to ignore or shrug off Saddam's reported crimes. I had no love for Iraq or Iraqis. Before and during the war I wrote pieces suggesting that Americans not romanticize Iraqis; that we understand postwar Iraq more in terms of occupied Germany than liberated France. But during and after the war it gradually became impossible to ignore the staggering enormity of what Saddam had committed against his own people. And when we *saw* those mass graveyards and torture chambers, heard more and more victims speak, watched those videotapes, the conclusion became inescapable: This war was screamingly, shriekingly *necessary*.

But instead of exulting in our victory, too many of us shrug and turn away and change the subject.

Young people might be misled about the world's response to the Holocaust by the current academic taste for "Holocaust studies" and related projects. It wasn't always this way.

In the years right after the war, there was Holocaust horror all over the world. The appearance of such books as Elie Wiesel's *Night* and Anne Frank's diary kept people thinking. But after that, silence set in. In 1981 Lucy Dawidowicz, most distinguished of all Holocaust historians, wrote of "this historiographical mystery of why the Holocaust was belittled or overlooked in the history books." I remember the 1960s (when I was a child growing up) as years during which the Holocaust was old stuff. On the whole, neither Jews nor gentiles wanted to think about it much. I remember the time and mood acutely on account of travels with my grandfather.

He was a rabbi and a loving but not a happy man. His synagogue was in Brooklyn, at the heart of an area that was full of resettled Holocaust survivors. He would visit them often, especially ones who had lost their families and not remarried. Naturally they were the loneliest. But what they suffered from most was not loneliness but the pressure of not telling. Pressure against their skulls from the inside, hard to bear. They needed to speak, but no one needed to listen.

Old or middle-aged men with gray faces and narrow wrists where the camp number was tattooed forever in dirty turquoise, living alone in small apartments: They would go on for an hour or more, mumbling with downcast eyes as if they were embarrassed—but they were not embarrassed; they were merely trying to keep emotion at bay so they could finish. Not to be cut down by emotion was the thing; they wanted to make it through to the end. So they would mumble quickly as if they were making a run for it, in Yiddish or sometimes Hebrew or, occasionally, heavily accented English. My Hebrew was inadequate and my Yiddish was worse, but I could get the gist, and my grandfather would fill me in afterward. Once an old man wanted to tell us how one man in a barracks of 40 had stolen a piece of bread (or something like that), and in retaliation the whole group was forced at gunpoint to duck-walk in the snow for hours. He didn't know the right word, so he got down on the floor to show us—an old man; but he *had* to tell us what had happened.

Steven Vincent went to Iraq after the war and reported in *Commentary* about Maha Fattah Karah, an old woman, sobbing. "I look to America. I ask America to help me. I ask America not to forget me." Saddam murdered her husband and son. That story takes me back.

My grandfather was driven. He spent years at one point translating a rabbi's memoir from Hebrew, then more years trying to find a publisher—any publisher; but no one wanted it. Holocaust memoirs were a dime a dozen, and (truth to tell) had rarely been hot literary properties in any case. Then he shopped the "private publishers" who would bring out a book for a fee. He tried hard to raise the money. He was a good money-raiser for many fine causes. But this time he failed. No one wanted to underwrite a Holocaust memoir. The book never did appear.

The Holocaust Shrug: To turn away is a natural human reaction. In 1999 (Steven Vincent reports) the Shiite cleric Sadeq al Sadr offended Saddam—whose operatives raped Sadeq's sister in front of him and then killed him by driving nails into his skull. Who can grasp it? In any case, today's sophisticates cultivate shallowness. They deal in cynicism, irony, casual bitterness; not in anguish or horror or joy.

Lucy Dawidowicz discussed the unique enormity of the Holocaust. It destroyed the creative center of world Jewry and transferred pre-meditated, systematic genocide from "unthinkable" to "thinkable, therefore doable." Mankind has crouched ever since beneath a black cloud of sin and shame.

Nothing will erase the Holocaust, but it is clear what kind of gesture would counterbalance it and maybe lift the cloud: If some army went selflessly to war (a major war, not a rescue operation) *merely* to stop mass murder.

That is not quite what the Coalition did in Iraq. We knew we could beat Saddam (although many people forecast a long, bloody battle); more important, we had plenty of good practical reasons to fight. Nonetheless: There were many steps on the way to the Holocaust, and we can speak of a *step toward* the act of selfless national goodness that might fix the broken moral balance of the cosmos. The Iraq war might be the largest step mankind has ever taken in this direction. It is a small step even so—but cause for rejoicing. Our combat troops did it. It is our privilege and our duty to make the most of it. To belittle it is a sad and sorry disgrace.

# ON TYRANNY

## By William Kristol

*(Originally published in the January 31, 2005, issue)*

A social science that cannot speak of tyranny with the same confidence with which medicine speaks, for example, of cancer, cannot understand social phenomena as what they are.
                                    —Leo Strauss, *On Tyranny*

Tyranny, like hell, is not easily conquered. Yet we have this consolation with us, that the harder the conflict the more glorious the triumph.
                                    —Thomas Paine, *The Crisis*

INFORMED BY STRAUSS and inspired by Paine, appealing to Lincoln and alluding to Truman, beginning with the Constitution and ending with the Declaration, with Biblical phrases echoing throughout—George W. Bush's second inaugural was a powerful and subtle speech.

It will also prove to be a historic speech. Less than three and a half years after 9/11, Bush's second inaugural moves American foreign policy beyond the war on terror to the larger struggle against tyranny. It grounds Bush's foreign policy—American foreign policy—in American history and American principles. If actions follow words and success greets his efforts, then President Bush will have ushered in a new era in American foreign policy.

That era will of course build on the efforts and achievements of his predecessors—especially Harry Truman and Ronald Reagan. The invocation of Truman is clear. Here is Truman, in his address to a joint session of Congress on March 12, 1947, announcing what came to be known as the Truman Doctrine: "I believe that it must be the policy of the United States to support free peoples who are resisting attempted subjugation by armed minorities or by outside pressures." And here is Bush: "So it is the policy of the United States to seek and support the growth of democratic movements and institutions in every nation and culture, with the ultimate goal of ending tyranny in our world."

Truman's basically defensive formulation of the doctrine of containment was appropriate at the beginning of the Cold War. Reagan was able, two decades later, to go further and to talk of transcending or overcoming communism. So we did, and Bush claims we are in a new and more hopeful era: "America's vital interests and our deepest beliefs are *now* one." Our previous victories allow a more expansive embrace of America's "ultimate goal."

Expansive does not mean reckless. Bush avoids John Kennedy's impressive but overly grand, "pay any price, bear any burden" formulation. Bush states that military force will of course be used to "protect this nation and its people against further attacks and emerging threats," and that "we will defend ourselves and our friends by force of arms when necessary." But he explains that the task of ending tyranny around the world is not "primarily the task of arms." The goal of ending tyranny will be pursued through many avenues, and is the "work of generations."

And Bush makes careful distinctions among the nations of the world. There are democratic allies, to whom he reaches out for help. There are "governments with long habits of control"—in Russia, or China, or the Arab dictatorships—whose leaders Bush urges to start on the "journey of progress and justice, and America will walk at your side." But he also makes clear to these leaders that we will pressure them and hold them accountable for oppression, and that we will support dissidents and democratic reformers in their countries.

Then there are the "outlaw regimes." It is their rulers who call to mind Lincoln's statement: "Those who deny freedom to others deserve it not for themselves; and, under the rule of a just God, cannot long retain it." So for those nations we intend to promote regime change—primarily through peaceful means, but not ruling out military force in the case of threats to us.

If the critics of the speech who have denounced it as simple-minded were to read it, they would find it sophisticated. They might even find it nuanced.

Still, sophisticated and nuanced as it is, it does proclaim the goal of ending tyranny. And just as Truman's speech shaped policy, so Bush's will. As he implicitly acknowledges, his presidency will be judged not by this speech but by his achievements. The speech, by laying out a clear and compelling path for U.S. foreign policy, will make substantial achievements easier. There will be vigorous debates over how to secure these achievements—debates over defense spending and diplomacy, over particular tactics and operational choices. We will at times differ with the president on some of these matters, as we have at times in the past. But on the fundamental American goal, President Bush has it right—profoundly right.

# A REALIGNING ELECTION

## By Robert Kagan and William Kristol

(*Originally published in the February 14–21, 2005, issue*)

THE DAY AFTER IRAQIS went to the polls, the London *Independent* commented, "In the long term, it is possible that yesterday's elections in Iraq may be seen as marking the start of great change across the whole region." Needless to say, the editors hastened to add that it would be "utterly wrong, now or in the future, for President Bush or the prime minister to claim that Iraq's elections vindicate their invasion." But the first statement was by far the more striking, both because it came from an antiwar, anti-Bush newspaper and because it was undeniably true.

Let's set aside for the moment President Bush's two recent speeches, and all the doctrinal debates they have spurred, and simply focus on what has actually happened, in the real world, over the past year. First, there were the elections in Afghanistan last October. Despite predictions of disaster, eight million Afghans voted for the first time in their war-savaged lives. Afghan women, who but three years before were among the most oppressed people on earth, were able to cast ballots as full-fledged citizens. As one Afghan told a *New York Times* reporter, "In the whole history of Afghanistan this is the first time we come and choose our leader in democratic process and free condition. I feel very proud and I feel very happy." The *Times* reported that the man, a Tajik, had voted for Hamid Karzai, a Pashtun.

Then, in December, came the crisis and democratic triumph in Ukraine. Elections stolen by a corrupt Ukrainian government with the connivance of Russia's ruler, Vladimir Putin, were reversed by a mas-

sive display of "people power" in the streets of Kiev and other
Ukrainian cities. A new round of elections brought some 27 million
Ukrainians out to vote—roughly three-quarters of those registered—
in what will go down in history as the "Orange Revolution." "This is
the people's victory," one man told a *Washington Post* reporter. "Ukraine
will finally achieve what it wanted when it got its independence from
the Soviet Union. Democracy will finally reign in this country. It
won't happen overnight, but it's begun."

Then, last month, the Palestinian people held elections for a new
prime minister, the first in nine years. There, too, turnout was huge,
and the new Palestinian prime minister, Mahmoud Abbas, received an
overwhelming majority of the votes. As one senior Fatah leader told
the *Washington Post*, "This is a historic vote for us. The most important
thing is not the winner. The most important thing is to see the
Palestinian people committed to the principle of democracy."

And this commitment has improved the chances for Israeli-
Palestinian peace. Israel is beginning the process of withdrawing from
West Bank towns, as well as from the Gaza strip, and has released hundreds of Palestinian prisoners. Prime Minister Abbas appears to be
taking serious steps toward ending Palestinian attacks on Israel. And
on February 8, Israeli prime minister Ariel Sharon and Prime Minister
Abbas will hold a summit in Egypt, the first such summit in nearly
two years.

The elections in Palestine were critical to this progress, as was the
death of Yasser Arafat. President Bush had all along insisted there
could be no progress toward peace so long as Arafat remained in
power, and that any progress would come as a result of new, democratic elections in Palestine. The president was pilloried in Europe, and
by some in the United States, for holding to that position over the past
two years. Now, it appears, he has been proven right.

Finally, there were the elections in Iraq. We don't need to add to
the stories that Americans already know well, of millions of Iraqis
risking their lives to cast votes, defying the terrorists who threatened
to kill them and in some cases succeeded. But it is worth contemplating whether, as the *Independent* suggests, the Iraq elections may mark
"the start of great change across the whole region."

Not so long ago, indeed right up until the day of the elections, this kind of thinking was treated as delusional. The vast majority of the American foreign policy establishment—Democrat and Republican, left, right, and center—ridiculed the whole notion that "democracy" should be America's goal in Iraq, not to mention across the broader Middle East and Muslim world. Even the community of professional democracy "experts" cluck-clucked at the Bush administration's "childish fantasies." Larry Diamond, perhaps the dean of that community, flatly declared several weeks before the elections in Iraq that they would "grease the slide to civil war."

Indeed, even as millions of Iraqis were casting their votes, we were being told, in *Newsweek*, in the *New Republic*, and elsewhere, that their votes were essentially meaningless. The "wrong" people would be elected, because the Iraqis are not decent enough, "liberal" enough, to elect the right people. "Elections are not democracy!" we were reminded. True enough. Nor does one election guarantee "liberalism." But, the fact is there can be neither democracy nor liberalism without elections.

And then there is this simple point: How can anyone living in this flourishing democracy tell the people of Iraq that they should not vote for their own leaders, that they are not "ready"? President Bush is sometimes accused of arrogance, but the true and appalling arrogance consists of telling the Iraqi people that they are not capable of electing the right kind of people. And are we so afraid of letting the Shia, who make up more than 60 percent of the Iraqi population, or the Kurds, who make up about 20 percent, win their fair share of votes in a free election? Are we really willing to deny these people the right to choose their own representatives?

Thankfully, President Bush never accepted the notion that Iraqis or other Arab or Muslim peoples are not "ready" for democracy. As a result millions of Iraqis (and Afghans) have now voted. How will this remarkable exercise of democracy affect the rest of the Arab and Muslim world? We remain confident that progress toward liberal democracy in Iraq will increase the chances that governments in the Middle East will open up, and that the peoples of the Middle East will demand their rights. And the chances increase every time the president singles out nations like Egypt and Saudi Arabia, or Iran and Syria, for special mention, as he did in the State of the Union. Words do matter, especially against the backdrop of deeds in Iraq and Afghanistan. There will, for example, be elections in Lebanon this summer, where an opposition victory could spell the beginning of the end of Syria's imperial role in that country. As for Egypt, Jordan, and Saudi Arabia, you don't have to take our word for it. Jordan's King Abdullah put it best: "People are waking up. [Arab] leaders understand that they have to push reform forward, and I don't think there is any looking back."

Here in the United States, the partisan reaction to the recent successes has been truly stunning. Never have so many been so miserable in the face of such good news. The Middle East experts who predicted disaster have not been able to bring themselves to acknowledge that it wasn't a disaster after all. Instead, they have simply shifted to predicting disaster in the future, or to falsely claiming that Iraqi Shia, who follow Ayatollah Sistani's lead, are tools of Iran. The democracy experts have been particularly egregious as well. Has their hatred of Bush made it impossible for them actually to applaud democratic elections when they occur?

We also have to admit being disappointed at the reaction of Democrats. We have no naive expectation of bipartisanship. We recall perfectly well how many Republicans refused to give Bill Clinton credit when he deserved it, in Bosnia and Kosovo. Nor is there anything surprising in Ted Kennedy's monotonous counsel of doom: In Kennedy's world, as in John Kerry's, the dream will never die, and the Vietnam war will never end. But where are the other Democrats, even a handful of them, to stand up and applaud the gains of democracy around the world?

There was a time when the spread of freedom was a foreign policy ideal Democrats cherished. In 1984, when El Salvador held its own round of miraculous elections in the midst of a bloody civil war, many prominent Democrats threw their support behind Ronald Reagan's policies in that country—not because they liked Reagan but because they cared about spreading democracy, and fighting communism, in Central America. And in 1999, while many Republicans attacked

Clinton's intervention in Kosovo, some stood by the president and even criticized their colleagues. This magazine supported Clinton throughout the Kosovo conflict, not because we were exceptionally fond of Clinton, and not because we had complete confidence that he was prosecuting the war effectively, but simply because, at the end of the day, we thought he was doing the right thing. Is it so hard for Democrats, with the next presidential election still almost four years off, to overcome their Bush-hatred just for a moment in order to join in supporting the cause of freedom and democracy?

The next steps in Iraq will of course still be difficult. In particular, the brave Iraqi voters deserve the commitment of the United States to remain fully engaged in the struggle to defeat the terrorists. And even as the security situation improves, as we trust it will, the political process will remain messy. No one should expect miracles. But the fact remains that it is today more possible than ever before to envision a future in which the Middle East and the Muslim world truly are transformed. For this, no one will deserve more credit than George W. Bush.

# THE MIDDLE EAST GEORGE W. BUSH HAS MADE

## By Reuel Marc Gerecht

⟨⟩

*(Originally published, as "What Hath Ju-Ju Wrought!,"*
*in the March 14, 2005, issue)*

HAVE THE IRAQI ELECTIONS PRODUCED a democratic earthquake that has changed forever the fundamental political dynamics in the Muslim Middle East? Only the culturally deaf, dumb, and blind—for example, Michigan's Democratic senator Carl Levin—can't see what George W. Bush's war against Saddam Hussein has wrought. The issue is not whether the basic understanding of contemporary Muslim political legitimacy has been overturned—it has—but how forcefully the regimes in place will resist the growing Muslim democratic ethic.

And the crucial question for the United States is whether the Bush administration will realize that the most consequential regimes in place—Hosni Mubarak's in Egypt, the Saudi dynasty in Arabia, the military junta in Algeria, and the theocracy in Iran—probably won't evolve without some internal violence. The Bush administration ought to be prepared to encourage or coerce these regimes into changing sooner, not later. What the United States should fear most is not rapid change—the specter of the fallen shah of Iran will surely rise in many minds—but the agonizing, dogged resistance of dictatorship. (Would that the United States had understood in 1971, after the shah's delusional and obscenely expensive celebration of 2,500 years of Persian kingship, that Washington had an increasingly sclerotic, corrupt autocracy confronting perhaps the most intellectually dynamic and angry society in the Middle East.)

Although it is now beyond doubt that President Bush is philosophically a Reaganite—holding, that is, that the United States' self-defense is inextricably connected to the expansion and protection of democracy—many within his administration share Europe's overriding concerns about "stability" in the region. And even among Reaganites, it's not hard to find those who are profoundly anxious about Muslim fundamentalists becoming potentially powerful players if free elections were actually held in the Arab world. The Bush administration has not yet worked out a grand strategy of democratization: Clear, simple principles applied with as much consistency as practicable would be an entirely adequate approach. Events are likely to make Elliott Abrams's democracy-promotion job on the National Security Council perhaps the most critical office to President Bush. Iraq has unleashed a wave of pent-up frustration and anger against the status quo throughout the region. The clever dictators, like Mubarak and Tunisia's Zine el Abidine ben Ali, will try to preempt it by fixing multiparty elections and adopting pro-American/pro-Israeli foreign policy initiatives. The Bush administration will likely get hit from several directions at once, as the peoples of the Middle East and their rulers continue to react to what started on January 30, 2005.

Let's take a quick *tour d'horizon* and see where we are.

LEBANON—This may be the most promising—though it may not be the most important—aftershock of the January 30 elections. Syria is obviously in trouble in Lebanon. The assassination of former prime minister Rafik Hariri, coming so soon after Arabic satellite television beamed astonishing pictures of Iraqis risking their lives to vote, ignited long-simmering, anti-Syrian animosity among the Lebanese Christian and Sunni communities. (There may well be a Lebanese who doesn't believe Hariri was murdered by Syria's ruler Bashar al-Assad, but what is striking about the Lebanese rumor mill—one of the most energetic in the Middle East—is how unified the view is on Syrian culpability.) The most urgent question now is whether the Lebanese Shiite community, specifically the Amal and Hezbollah political movements, will back the Sunnis and the Christians in their call for Syria's ejection. Both organizations have substantial ties to Iran—Hezbollah is revolutionary Iran's only true child and remains the clerical regime's only foreign-policy success—and would be petrified of completely losing Tehran's support. It remains unclear what the Lebanese Shia are going to do, but if one had to bet, the odds are decent that Amal and Hezbollah will not break from the Lebanese Christian and Sunni communities.

As the Lebanese-American scholar Fouad Ajami regularly points out, the Lebanese Shia as a people do not want to be left behind in the country. If the vast majority of the other Lebanese have decisively broken with Syria—and they have—then the Shia will not separate themselves from their countrymen. This is even more true if clerical Iran, Hezbollah's mother ship, does not ride to the rescue of Bashar al-Assad. And there is reason to hope this will not be the case.

First and perhaps foremost, Bashar is inept. The cool, calculating rule of his father, Hafez al-Assad, has given way to the blundering of a young leader who has galvanized anti-Syrian sentiment even among traditionally pro-Syrian Lebanese. Say what you will about Iran's ruling clerics—they are a nasty collection of highly ideological power politicians willing to deploy terrorism at home and abroad whenever necessary—they are not fond of expending their own prestige and power on behalf of juveniles, especially when the odds are they would lose. Iran probably wouldn't mind seeing Bashar al-Assad fall from power in a palace coup—not an unlikely possibility if Syria gets forced out of Lebanon. As long as the Alawite clan (a heretical branch of Shiite Islam that has dominated Syria's Baath party) stays in power, the Iranians aren't likely to become too worried.

And the events in Lebanon don't necessarily spell disaster for the Syrians. What Thomas Friedman called the "Hama rules"—the willingness to slaughter regime opponents by the thousands, as Hafez al-Assad did in the town of Hama in 1982—still hold, and the Alawite regime appears cohesive enough to do this without hesitation. The Syrian Sunni desire for revenge against the minority Alawites is easily enough to ensure Alawite solidarity. The Sunnis, who believe they have always had the historic right to rule Syria, would probably not show the same consideration that Iraqi Shia have so far shown their former Baathist tormentors. It is possible that the democratic ethic may be growing among Syria's Sunni Arab population—Syria's awful tyranny, like Baathist Iraq's, can teach well the benefits of restraining state power—but that won't matter much against a savage regime with a ferocious internal security service and elite military units capable of artillery barrages against civilians.

Also, Lebanon has seen some form of democracy. Lebanon has never been fully of the Arab world—it is historically, religiously, culturally, and geographically a special place—and the idea of a democratic Lebanon probably isn't nearly as scary to the Middle East's despots as is the idea of a democratic Iraq or Egypt. (A Palestinian democracy has a bit of the same quality about it—Palestinians have existed in a surreal world for decades, where their triumphs and tragedies don't relate well to the day-to-day lives and local political frustrations of most Arabs.) Iran's clerics, or Syria's Alawites, or the Saudi princes, or the Mubarak family in Egypt don't necessarily view the return of Lebanese democracy as a dagger aimed at them. It is something they could live with—a price worth paying to eliminate from among them a damaging, Paris-Washington-uniting incompetent like Bashar al-Assad.

Unless Iran's clerical regime views the liberation of Lebanon as a lethal defeat for Hezbollah—and the organization's chief, Hassan Nasrallah, has been rhetorically fence-sitting about joining or damning the Christian and Sunni opposition to the Syrians—then the odds are good that the Syrians will withdraw. One can appreciate why the Lebanese youth cannot stop praising "Ju-Ju," an affectionate Arabic take on "George." They are willing to admit easily what comes much harder to many in Congress and in Washington's Democratic think tanks.

SYRIA—Drive them out of Lebanon but don't spend much time or effort trying to tighten the noose around the Baathist Alawites. The state is not as Orwellian as was Saddam Hussein's, but the ethnic and religious dynamics of its regime will make regime solidarity very difficult to overcome. However, if the Syrian Baathists are aiding the Iraqi Baathists to the extent that the Bush administration alleges—and the allegations appear solid—the United States ought to strike militarily. If American and Iraqi lives are being lost because of Bashar al-Assad's support of Iraqi Baathists in his country, then the Bush administration is being tactically and strategically negligent in not retaliating. This doesn't mean the United States should invade Syria. But Syrian intelligence and military bases—and any locales where Assad is hosting Iraqi insurgents—are legitimate targets for air and special-ops raids. It is possible that such limited military strikes could threaten the stability of the Alawite dictatorship, allowing an opportunity for a Sunni civilian and military opposition to gain ground.

But the administration shouldn't bank on the democratic aftershocks of Iraq shaking Syria itself. It might happen. A good rule of thumb is that an appreciation for democracy has become more widespread in the Arab world than the American and European "realist" crowd would have us believe. But the best we should hope and plan for is an eventual cracking of Alawite power, allowing for a return of Sunni rule. With the Sunnis in charge, political evolution has some chance, particularly if the United States starts to focus its democratizing attention on the country in the Arab world that matters most—Egypt.

EGYPT—Democratizing Egypt is what President Bush's post-9/11 "forward strategy of freedom" is all about. End the nexus between tyranny and Islamic extremism in the lower Nile Valley—the perverse pattern of Egyptian dictators Gamal Abdel Nasser, Anwar Sadat, and Hosni Mubarak fueling anti-American Islamic militancy through both suppression and support—and one of the two most important intellectual breeding grounds for bin Ladenism (Saudi Arabia is the other) will turn into a laboratory where both secular and fundamentalist Sunni Muslims can make their case democratically. If Egypt doesn't democratize, bin Ladenism will not end. The hatred for American-supported dictators will continue to grow; the Muslim Brotherhood, the fount of all Sunni fundamentalists, will not be able to evolve politically further, moving devout Sunnis from Koranic shibboleths to democratically derived legislation that will be seen by most Islamic activists as both legitimate and at odds with the Muslim Holy Law.

Since Iraq, President Mubarak, who used to equate democracy with "chaos," sees a need for "more freedom and democracy" in Egypt. The odds are excellent he is actually trying to devise a system whereby, with less friction, he continues in power and the chances of succession for his son increase. But that doesn't mean the United States shouldn't take advantage of Mubarak's opening. If Mubarak thinks Egypt is ready for more democracy and freedom, then far be it from the United States not to take him at his word. Now is the time to announce that American aid to Egypt is henceforth conditioned on democratic progress. Mubarak cheats, the aid is cut. Mubarak cheats a lot, the aid ends. We should not allow Mubarak to scare us again with the specter of Islamic extremism. Fear of another Ayatollah Ruhollah Khomeini— who, by the way, didn't come to power democratically—has too long paralyzed our thinking about Egypt.

As has Egypt's peace treaty with Israel. Many supporters of Israel in the United States have become de facto backers of dictatorship in Egypt because they fear the Islamist boogeyman. They want to believe that the system in Egypt can liberalize—even though nowhere in the Arab world have we yet seen an Ataturkist evolution. Indeed, the evolution of Arab dictatorship has been in the opposite direction, toward the nexus that gave us bin Ladenism and 9/11. They really don't want to give fundamentalists a chance to compete in elections; they want "progressive Muslims" to somehow be nourished—to spring more or less full-grown from the head of Mubarak, as did Athena from Zeus. They would rather not reflect too long on the history of democratic Christendom—that you don't get to arrive at Thomas Jefferson unless you first pass through Martin Luther. But Natan Sharansky is right: Democracy, sooner not later, is the only way out. The liberal critics of the Bush administration's democracy promotion have usually been cranky and unfair—it's pretty hard to envision the region's democrats, particularly those on the front lines in Iraq and Lebanon, cheering John Kerry—but they may soon have a point. In the not too distant future, Washington is going to have to break with the Mubarak regime. If we don't, bin Laden's jihadist call, and not the shouts for "Ju-Ju," will be the summons with lasting appeal.

SAUDI ARABIA—Continue to push the democratic agenda publicly in the Arabian peninsula. The rather pathetic Saudi attempt to defuse democratic ferment at home and the Bush administration's growing anti-Saudi attitude by holding highly restricted municipal elections is likely to do the opposite of what the royal family intended. The Shiites of the Eastern Province—where most of Saudi Arabia's oil is located—may, as the Arab Shiites of Iraq continue to advance democratically, become more inclined to protest. The turnout for the municipal elections clearly showed that the Shiites in the Eastern Province didn't consider the exercise a joke (as was the case among many Sunnis).

The Wahhabi clerical establishment, the religious backbone of Saudi power, may become more inclined to use older, violent means to oppress the Shiites. Washington should rhetorically preempt the issue, by declaring loudly and often that it favors modern democracy in Saudi Arabia, where minority rights are protected. We would be wise not to assume that the Saudi royal family is more "modern" than the people of the country. It may well be more "modern" than the average Wahhabi in the Najd region, the heartland of Wahhabi power. But Saudi Arabia is much larger than the Najd.

It is possible that a variation of the Iranian experience has been at work in Saudi Arabia, that Saudi-Wahhabi power has distanced an ever greater number of people from the Saudis' rigorous fusion of religion and state. Saudi Arabia is an odd place, with a large number of people permeated with Western ways. Sometimes that fuels Islamic militancy. Sometimes it does the opposite. Both may be happening in Arabia. In either case, we know for certain that Saudi Arabia was the cradle of bin Ladenism. There is scant evidence to suggest that the Wahhabi establishment has changed its spots (philosophically it can't). The Wahhabis should have to compete for their flock. Inside the country and out, the United States should be relentlessly pushing for democracy. As in Egypt, we should increasingly tie government-to-government relations and joint programs directly to Saudi progress with real national elections.

ALGERIA AND TUNISIA—North Africa has traditionally been ignored by the Americans. It shouldn't be. It would be a good test of France's desire to advance democratic change in the Middle East to see if Paris would rhetorically join the United States in energetically encouraging democracy in both countries. Tunisia has an increasingly lively democratic culture developing on the Internet in the form of blogs and virtual publications, both inside and outside the country. Stealing a page from Hosni Mubarak's playbook, President Zine el Abidine ben Ali recently invited Israel's prime minister Ariel Sharon to Tunisia in a crude (but with the Egyptians, effective) effort to get on America's good side through the Israelis. Ben Ali read the tea leaves after the January 30 election and decided to preempt.

The Bush administration should relentlessly thump ben Ali—criticize his dictatorship whenever and wherever possible. Since ben Ali, like Mubarak, has recently discovered the damage the lack of democracy has done to the Arab world, Washington could begin simply by using his words against him. Tunisia, like Algeria, is hardly a strategically critical country for the United States. There are no airfields there that we absolutely must use to continue the war on terror. All official dealings with these two states should be premised on their governments' support of democratic reform. And the Bush administration would be wise to revisit the position of Algeria in the Arab world. Scarred by the civil war of the early 1990s, Algerians are probably a much wiser people than they were when Islamists first began to challenge the corrupt military dictatorship.

Algeria's highly Westernized young would probably embrace the chance to remove the military dictatorship over them—if they had some possibility of doing so without confronting the official black ninjas who rival the throat-slitting Islamic militants in savagery. Algeria's failed experiment with democracy in the 1990s was closely watched in the Muslim world, particularly among fundamentalists. If Algeria were to get back on track and follow through with democratic reforms, the impact on the region, and on the millions of Algerians who live in Europe, would likely be significant.

IRAN—Don't compromise the democratic future of the country by trying to buy the mullahs' nuclear goodwill. Democracy in Iran is the key to ending that country's long embrace of terrorism. And if there is a nationalist desire in Iran to have nuclear weapons (we only know for sure there is a clerical will to have these arms), then talking with a democracy about them is entirely different from trying to appease a dictatorship, which is what the French, British, Germans, and certain quarters at the State Department and the National Security Council would like to do. One can live with a nuclear-armed democracy. The Bush administration should realize that the American policy of containment has helped create the most pro-American Muslim population in the Middle East. We should be patient. Let Iraq's Shia, in particular Grand Ayatollah Ali Sistani's democratic opinions and actions, have their effect.

IRAQ—Remember that January 30 was only the first democratic wave to come out of that country. If Iraq doesn't go off the rails—and the odds are very good that the Shiites, Kurds, and Sunnis will find workable democratic compromises—there will be more election aftershocks issuing from Mesopotamia, probably of a magnitude greater than January 30. The trial of Saddam Hussein is coming. There are many things the administration should do to exploit the people power of Iraq, but first and foremost is an Iraqi C-SPAN *controlled by Iraqis.* There is a large audience in the Arab world, especially in Egypt and Saudi Arabia, just waiting to see the next episode from Baghdad. Let the millions watch Saddam's trial live. Republicans and Democrats who believe in spreading democracy in the Muslim Middle East shouldn't disappoint these hungry viewers. Perhaps Osama bin Laden will also watch and see the end of his dreams.

# IV

## BOOKS AND ARTS

# THE BRILLIANT SHOW THAT KILLED BROADWAY

## By John Podhoretz

*(Originally published in the October 30, 1995, issue)*

TWENTY-FIVE YEARS AGO, the Stephen Sondheim musical *Company* opened on Broadway, and made a sensation. *Company* has now returned to Broadway for the first time in a revival at the Roundabout Theater. But something interesting happened in the years between the two productions: Broadway died. And one of the causes of its death was this extraordinary piece of theater.

In 1970, *Company* seemed like a message from the future—an entirely new style of musical, one that would liberate a form that had grown stale and formulaic. Indeed, despite the Roundabout's awkward and second-rate staging, the revival seems entirely contemporary (despite some jarring references to "busy signals," "answering services," bubble-headed stewardesses, and kooky free spirits). That is due primarily to Sondheim's songs, amazingly compact and precise depictions of modern American life—as when an aging matron sings of her life in "The Ladies Who Lunch": "Another long, exhausting day, / Another thousand dollars, / A matinee, a Pinter play, perhaps a piece of Mahler's / I'll drink to that. / And one for Mahler." It would take John Updike or John Cheever pages to capture what Sondheim captures in just those 25 words: the self-loathing and self-congratulation of a decaying cultural elite.

*Company*'s emphasis on "relationships" at the expense of plot prefigured the success of such 1990s sitcoms as *Seinfeld* and *Friends*, as did its caustic take on its central character, a 38-year-old charmer named Robert who cannot bring himself to wed despite (or because of) the imprecations of "these good and crazy people, my married friends." Those friends are four couples whose own flawed marriages are portrayed in the broad strokes of a brilliant *New Yorker* cartoonist in George Furth's sharp libretto.

A series of vignettes, *Company* isn't really about Robert or his friends—its subjects are marriage, commitment, fear of commitment, New York, men, women, and sex. And through it all, Sondheim offers a view of adulthood that is sophisticated, depressing, and sobering. About the best he can offer in praise of marriage is this lyric, sung by the four husbands to Robert: "You're always sorry, you're always grateful. You think of things that might have been. Then she walks in."

People enter marriage with dread and foreboding. A woman lets loose with a torrent of words and emotions in a showstopper called "Getting Married Today," at the relentless, literally breathtaking pace of the "Flight of the Bumblebee": "Pardon me is everybody here because if everybody's here I want to thank you all for coming to the wedding I'd appreciate your going even more I mean you must have lots of better things to do and not a word of it to Paul you 'member Paul you know the man I'm going to marry but I'm not because I wouldn't ruin anyone as wonderful as he is. . . ."

Nor are unmarried relations any better. In Central Park, one of Robert's girlfriends sings the first anthem to New York that actually describes the place well: "They meet at parties through the friends of friends whom they never know. / 'Do I pick you up or shall I meet you there or shall we let it go? / Did you get my message cause I looked in vain / Can we see each other Tuesday if it doesn't rain, / Look, I'll call you in the morning or my service will explain. . . .' "

This is a show in which the audience-pleasing production numbers are all intended ironically. The high-stepping "Side by Side by Side" is, you realize slowly, a terrifying salute to life as a third wheel: "Year after year, older and older, side by side . . . by side. / One's impossible, two is dreary, three is company, safe and cheery." And in "You Could Drive a Person Crazy," Robert's three girlfriends make like the Andrews sisters as they attack him for his flaws in the catchiest tune of Sondheim's composing career: "When a person says that you upset her, that's when you're good. / You impersonate a person better than a zombie should."

It's startling to realize that *Company* made its debut at a time when Broadway was still dominated by shows that seem a million years old now—lumbering behemoths like *Hello, Dolly!* and *Fiddler on the Roof* and *Man of La Mancha* and *1776*, all of which were going strong when Sondheim and his collaborators came along and smashed their genre to bits.

The show was a turning point in the American theater, as notable in its way as *Oklahoma!* had been when it premiered in 1943, seamlessly merging song, story, and dance and thereby ushering in the socalled golden age of the Broadway musical. *Company* was the first wholly successful break from the rules of the golden age, which were pretty simple. First, musicals were generally set in the past. Not the past as anyone had ever known it, but an idealized past, a time more innocent than when the show was written. *The Music Man*, which opened in 1957, is set in 1911 Iowa. (1956's *My Fair Lady*? Edwardian London. 1945's *Carousel*? Turn-of-the-century New England. And so on.) Second, with few exceptions (like *Carousel*), they concluded happily, with a marriage or the possibility of a marriage. And finally, they were designed to rouse their audiences, to knock them for a loop with an explosive combination of comic patter, ballads, dance routines, broad and obvious jokes, and a big finish—an "11 o'clock song" designed to send the audience home with a buzz of excitement.

The form worked, and worked brilliantly, for about 25 years. Indeed, for about five decades altogether, the Broadway theater (especially in its pre-*Oklahoma!* days) was the single greatest source of the American popular song, which history will record as one of the glories of American creativity in the 20th century. Cole Porter, Irving Berlin, Richard Rodgers and Lorenz Hart, Jerome Kern and George and Ira Gershwin—these were theater people, writing show after show, year after year, with standard after standard in each score.

Alas, like most art that is produced unconsciously by those who think of themselves primarily as craftsmen, the musical was soon forced to contend with the disease of ambition—the pretentious desire of people who are rich, famous, and successful to be celebrated for their artistry, not for making people laugh or sing or cry. Theater folk ached to prove themselves deserving of the title of artist, and actually succeeded in winning converts to their cause with shows like *West Side Story*—a supposedly daring and adult 1957 musical about street gangs based (may God help us) on *Romeo and Juliet* that today seems as quaint as *The Music Man*.

*West Side Story* was the first show Sondheim worked on. Later, he would write his own music, but in *West Side Story* he was writing words to Leonard Bernstein's soaring melodies and gave few hints of the rhythmic and linguistic intricacies he was later to display: "A boy like that, / Who kill your brother! / Forget that boy and find another! / One of your own kind, stick to your own kind . . ." was about the best he could do.

So at the beginning of his career, Sondheim had already been party to a movement to change the Broadway musical, to make it more adult, more serious, more capable of addressing important issues. Only who wants serious issues addressed in a musical? Why would you even want to try? It would be like attending *Madame Butterfly* because that opera has a lot to say about suicide, or *The Magic Flute* because you are interested in exploring Masonic philosophy.

After all, an absurd thing happens every 10 minutes in a musical: Somebody bursts into song. This, needless to say, happens rarely in real life, and if it happened more frequently it would be cause for deep alarm. The reason that musicals traditionally concerned themselves with trivialities of plot and incident was because its makers knew they could not fool audiences into taking seriously the antics going on in front of them. Any resemblance to real life was strictly coincidental. The greatest stars of the musical stage—Ethel Merman, Mary Martin— were bold and brassy caricatures of women, every gesture overdone and over-deliberate so that it could be seen in the third balcony rear. When Merman was required to break down in tears during the bitter and powerful 11 o'clock number called "Rose's Turn" that closes 1959's *Gypsy*, Sondheim's second show, the original cast album records her for posterity sputtering and muttering like a dinner-theater amateur.

So along come the 1960s, with their mania for self-expression and celebration of the young, and at decade's end along comes *Company*. Staged with all the glitter and dazzle that Broadway could muster— which was all the glitter and dazzle of the world then—*Company* had a gasp-inducing set, striking costumes, hammy crowd-pleasing performances, and legendary choreography by the late Michael Bennett—all deployed in the service of a show that was, at root, a devastating criticism of the Broadway musical itself. (This would become more explicit in Sondheim's next show, *Follies*, which takes place as a great old Broadway musical theater is literally falling down around the characters' heads.)

Even the audience came under attack: The aforementioned "Ladies Who Lunch," so memorably skewered by Sondheim, were none other than the women sitting in the audience at the Alvin Theater night after night, paying $25 to hear themselves belittled.

No longer would the Broadway musical be shackled to the formal demands of plot, story, character, happy ending. No longer would they require a chorus (the original production had four extra singers, called "the Vocal Minority"; the revival dispenses with them entirely). No longer would they offer only a song, a dance, a laugh; no, now they were to deal with the humiliations and traumas of dancers and homosexuals (*A Chorus Line*). Or life on the verge of a nuclear war (*Dance a Little Closer*). Or the traumas of having a baby (*Baby*). Or discovering your father is gay and his lover has AIDS (*Falsettos*). Broadway musicals could now be about "concepts."

This artistic impulse, this desire to free the musical from the strictures of its past, proved a devastating, horrifying failure. The "concept" musical didn't liberate Broadway; it destroyed it. There was no new intimacy between audience and show as *Company* had promised. Instead, the concepts themselves changed—shows ceased being about marriage, relationships, imperialism and death, and increasingly concerned themselves with singing cats (*Cats*) and singing trains (*Starlight Express*).

Instead of brilliant shows like *Company* and the Sondheim shows of the 1970s that followed it—*Follies, Pacific Overtures, A Little Night Music,* and *Sweeney Todd*, each with a score more impressive than its predecessor's—Broadway audiences today are assaulted by garish, overproduced, earsplitting examples of showbiz excess in the form of Europop bilge like *Miss Saigon* and *Les Misérables*. They aren't works of theater; they're magic shows, Vegas revues, in which the singing, the dancing, the acting, *everything* is secondary to the spectacle. Instead of the indescribable uplift of an infectiously tuneful Broadway show, they pummel audiences into mute submission and allow them to leave the theater thinking they've gotten their $75 worth.

Still, audiences come; old theaters are being renovated; Disney is making a mint on its stage version of *Beauty and the Beast*. *Cats* and *Les Misérables* will run forever—and that may be the literal truth; *Cats* has been at the Winter Garden for 13 years, *Les Miz* at the Imperial for 9. But Broadway is dead—or rather, the *idea* of Broadway is, the idea that Broadway is a show-business culture that is not simply a stepchild of Hollywood's but one with its own stars, its own legends, its own stories—and that, moreover, offers its fans and spectators a high they can find nowhere else. The only two musicals scheduled to open this season on Broadway are *Victor/Victoria* and *Big*, both adaptations of movies. In 1970, when *Company* opened, 22 new musicals vied for attention. Sondheim and his collaborators made the old Broadway archaic and replaced it with something far worse.

The idea of Broadway is now as remote and dated as the depressing fact that this month on the Great White Way, the 74-year-old Carol Channing will open a revival of *Hello, Dolly!*—in which she must play a desirable woman of 40. Apparently, like the body of Vladimir Lenin, Broadway will be there, entombed and embalmed, right in front of us, for many decades to come. And each year, the tickets will get more and more expensive as the body slowly crumbles to dust inside its well-preserved sepulcher.

# EDWARD R. MURROW: INFOTAINMENT PIONEER

## By Andrew Ferguson

*(Originally published in the July 22, 1996, issue)*

JOURNALISTS LIKE TO THINK of themselves as skeptics, but when it comes to their own trade they lean toward romance. TV journalists are especially susceptible, since their particular subclass of the profession, where the reach can be so vast, the depth so shallow, and the pay so large, carries with it a certain insecurity; TV people are often considered the spoiled, slow-witted kid brothers of the Big Boys in newsprint. (And with reason: Consult the card catalogue under Broadcast Journalism and you'll find entries like *Connie Chung— Broadcast Journalist.*)

Through every indignity, TV journalists hold tight to the mythic figure of Edward R. Murrow—"the spiritual leader," in Jim Lehrer's phrase, "of everyone in serious broadcast journalism." Stanley Cloud and Lynne Olson's *The Murrow Boys: Pioneers on the Front Lines of Broadcast Journalism* is yet another attempt to burnish the legend. The third Murrow biography in a decade, at least the fifth since its subject's death in 1965, it concentrates on Murrow's relationship with the team of journalists he recruited to cover Europe for CBS News during the Second World War, and tells the tale well. By the end of the book, though, the pull of sentiment proves too great and the moral of the story is uncorked.

Most of Murrow's correspondents moved from radio to TV after the war, Cloud and Olson write: "Commercial television at once trivialized and corrupted what they did. Then it tired of them and tossed them aside. . . . It is discouraging . . . that CBS tilted the balance so quickly and now so completely away from a commitment to news and public affairs and toward lowest-common-denominator programming."

There it is, straight up: the creation myth of TV news, an idyll brought low, a profundity trivialized, a purity defiled by the dollar. In this telling, as in all previous, Murrow looms as the exemplar of a vanished Golden Age, a long ago Eden of TV and radio news, where "standards" were higher and everybody knew it, and profits were lower and nobody cared. Murrow remains the paragon: objective in his judgments, unsullied by commerce, immune to the lures of squalid showbiz, unstinting in his commitment to the principled presentation of the news. All correspondents and anchors have learned to invoke his name, for like most mythic figures carved from the recent past, Murrow is meant to stand as a rebuke to the Fallen present. He is offered as final proof that "serious broadcast journalism" need not be an oxymoron.

Do I exaggerate? So deep is Dan Rather's devotion that he even gets his suits made at Murrow's old tailor in Savile Row. "To this little boy," Rather once wrote, "Murrow was a hero right out of the adventure books. Risking his life for the *truth* [emphasis, amazingly, in the original]. His work heightened my sense, even then, that being a reporter was a kind of vocation: demanding sacrifice, needing courage, requiring honor." Here, in a brief three sentences, is almost everything that people find annoying about contemporary journalism. Rather, of course, personifies the sanctimony, the self-flattery, the missionary zeal, and the Olympian remove America has learned to expect of its TV news stars. But so, alas, did Edward R. Murrow, even back in the Golden Age. When his friend and contemporary Howard K. Smith called Murrow "the most influential journalist of our time," he was righter than he knew. The next time you see Gunga Dan Rather huffing through the Khyber Pass, with a camera crew in tow and a bathroom towel wrapped around his head, or Peter Jennings congratulating himself for having the guts to "take on" the tobacco companies, or even Barbara Walters asking a stuttering movie star about his favorite tree, you know whom to credit.

Murrow foreshadowed it all, was father to it all. This is owing less to the man's own professional weaknesses than to the nature of broadcast journalism itself. Broadcast journalists need to believe that their trade wasn't always so silly and meretricious. Unfortunately for them, the real moral of Murrow's life is that broadcast journalism was always silly and meretricious.

Disinterested observers will have some trouble fixing the Golden Age of broadcast news in chronological time. It doesn't help to work backward. Shortly after Charles Kuralt retired in 1994, he lamented the lost era, implying that it had closed not too long ago. "The bean counters are really in control now," he said. "I decided to leave before they could invite me to leave." (Thanks.) But several years earlier, in the late eighties, Dan Rather was lamenting the lost Golden Age also, the "tragic transformation from Murrow to mediocrity" that had recently been accomplished. To the early eighties then? No, for Walter Cronkite himself had announced that by then the "Murrow continuum" "had really come to a terminal point." Cronkite may have placed the Golden Age in the years leading up to his retirement in 1981. He would have gotten an argument from Eric Sevareid, who in the mid-1970s said CBS News had "degenerated into show biz."

Perhaps Sevareid was referring to the golden time as the glorious sixties and early 1970s, the period leading up to his own retirement. Alas, no. For in 1969 Alexander Kendrick, himself a Murrow Boy and author of the first gargantuan Murrow biography, announced that "the Murrow window on the real world had been shrunk to a peephole. . . . Controversy, with its pros and cons, had given way to compatibility. . . . Emotion replaced editorial perspective." Fred Friendly, one-time president of CBS News, agreed, although Kendrick was apparently off by a few years. By Friendly's account, CBS had wholly succumbed to worldly forces by 1966, the year, coincidentally, of *his* retirement.

And so the Golden Age recedes and recedes, until we reach its first autopsy, performed in 1958 (!) by Murrow himself. In a widely noted speech he declared TV news to be trivial and soporific, given over at last to "decadence, escapism, and insulation." No matter what day it is, the Golden Age of Television News always ended the day before yesterday.

It shouldn't surprise us that this long tradition of breast-beating and bogus nostalgia, two generations' worth, was begun by Murrow. By 1958 his own career at CBS was winding down, and a final snap at the hand that fed him would have seemed almost obligatory. He had come to the network 20 years earlier, hired as "director of talks" for CBS in Europe. He was 27; he told the company he was 32. He had majored in speech and drama at Washington State but claimed to have majored in international relations and political science at the (slightly) more prestigious University of Washington. In applying for the position he even awarded himself an M.A. from Stanford. A year earlier he had lost a job because of his resumé-padding, but he neglected to mention that, too.

Whatever his early prevarications, they were ever after swallowed up by his achievements in Europe, which were considerable. Foreign radio coverage at the time was similar to C-SPAN today. As director of talks, Murrow was to schedule live, unedited broadcasts of speeches and other events and assemble roundtables with newsy guests. But as war approached he began to do much more. He began to report the news, live: Munich, Sudetenland, the Anschluss. With CBS's mounting enthusiasm, he hired his Boys to extend the network's reach. They were mostly young and hungry wire-service reporters, among them several who would be elevated into the carefully tended pantheon of broadcast journalism: William Shirer, Charles Collingwood, Howard K. Smith, and Eric Sevareid.

The young men Murrow assembled were, like him, tireless and possessed with enormous physical courage. They were pushed to their utmost by the circumstances of war. Although as reporters they were often scooped by the competition, the mythology remembers them as masters of language, the written and spoken word. This is in keeping with the vanity of today's broadcast journalism, whose practitioners will tell you their "craft" requires "good writing" above all. Not surprisingly, the quality of Murrow's writing, and that of his Boys, swung widely—from the tersely eloquent to the deep purple to the banal. Murrow himself favored the plain style of most good newspaper reporters: an Ernie Pyle of the airwaves. On the page it can wear thin. Here is Murrow reporting the Blitz:

"The antiaircraft barrage has been fierce but sometimes there have been periods of twenty minutes when London has been silent. Then the big red buses would start up and move on till the guns started working again. That silence is almost hard to bear. . . . You know the sound will return—you wait, and then it starts again. That waiting is bad."

Murrow himself understood the limitations of his writing. "It ain't no false modesty," he once told an interviewer, "to say that I don't know enough to write a book. I can write the language of speech, but that's totally different. When I write a book review or a preface to somebody else's book, Janet [his wife] has to go through it and scatter the commas." To really gauge the power of Murrow's war reporting, the basis of his lifelong reputation, you have to hear it. War brought out Murrow's greatest gifts, but they were not a journalist's gifts so much as those of an actor and showman. "He understood the value of silence—of no words—better than any of us," his colleague Larry LeSueur said not long ago. His famous radio sign-on—"This [pause] is London"—was shaped according to the advice of his college drama coach, who continued to send him pointers throughout the war. His great live broadcasts, particularly those during the Blitz when he spoke from rooftops and street corners as the bombs fell around him, draw their authority not from his words or even from his skills of observation, the traditional tools of a journalist. When you hear Murrow's most memorable work you're hearing the voice of a man in mortal danger. It's hard to forget that voice.

And what a voice—a smoky rumble, so grave and biblical it could make the funny papers sound like a threat to civilization. Murrow's voice was essential to his success, as, say, Basil Rathbone's voice was essential to his. Murrow demonstrated in its infancy that broadcast journalism, unlike print, was utterly dependent on the elements of show business for its effect. This is a truth that broadcast journalists always deny in theory, though they observe it in practice. And it was true even before the "tragic transformation from Murrow to mediocrity."

Murrow's work during World War II is still accepted as the nonpareil of broadcast journalism. It probably is. But its standing is curious even so, given the peculiarities that surrounded it—anomalies that could only shock those who idealize the standards of the Golden Age. Murrow broadcast under strict censorship, for one thing; as he spoke, a British information officer stood at his side, ready to tap the correspondent on the wrist the minute he got out of line. And beyond their CBS salaries, the Boys would receive direct payments from the sponsors of their broadcasts, huge corporate behemoths like Sinclair and Chesterfield (an oil company and a tobacco company!). The payments were viewed as an incentive for the correspondents to file more stories; today the purists would view them as symptoms of an unacceptable coziness with thine enemy, the sponsor. Charles Kuralt would have an aneurysm.

Before and during the war, CBS strictly forbade the use of taped material in its broadcasts. There were good commercial reasons for the ban, but the company also wanted to assure its listeners that they were getting the story straight, unmanipulated by technological wizardry. Murrow found the ban inconvenient and violated it often. Unknown to CBS, he sent a sound truck around London early in the Blitz to record the noises of the bombardment. He used the sounds as background, presumably for verisimilitude, when he wasn't broadcasting from the scene himself—a forerunner of the "dramatic recreations" that have so alarmed critics of TV news in the 1990s.

It wasn't long before Murrow and his Boys were caught up, quite happily, in another bête noire of today's purists, the "star system." The Boys became very famous very quickly. They hired agents and signed quickie book contracts. They made trips stateside, to be greeted at the dock by Movietone camera crews and interviewed in nightclubs by Walter Winchell. Lucrative speaking tours filled their brief sabbaticals. No one was so ripe for stardom as Murrow, whose great good looks were as arresting as his voice, and no one took to it with quite the same ardor. "He's much too much the showman not to have fun in the part he's playing," a colleague said a few years after the war. "When he gets out of a plane, wearing a trench coat and a hat with the brim pulled down, he's Ed Murrow, the big correspondent. Maybe one reason he enjoys acting like a newspaperman so much is that he never was a newspaperman."

Much of this stardom depended on the immediacy of radio itself, which Murrow skillfully exploited. He always spoke directly to the listener: "Last night you'll remember I was telling you about . . ." But the stardom was goosed along by the vast public relations apparatus of CBS. Even by the mid-1930s it was the best in the business, and it went to work for Murrow early. He was badly beaten by NBC on the story of the Anschluss, for example, but the company's sales-promotion staff was having none of it. The flacks quickly got out a glossy and expensive brochure, "Vienna, March 1938," which, as *Scribner's* magazine later reported, "artfully arranged the calendar of news flashes so as to [convey] the impression without saying so that CBS had been omnipresent and omnipotent throughout." The Headliner's Club presented a medal to Murrow, too: "the first time," a rival said, "a reporter ever got an award for not getting a story." Sitting on the Headliner board at the time was Paul White, Murrow's backer at CBS.

It's interesting but probably fruitless to speculate how much of Murrow's stature today is really the residue of CBS's remorseless effort to make him a star. Most people, most broadcast journalists, recall Murrow for his television presence rather than his radio celebrity, in any case. He and the Boys came to TV reluctantly, deeming it an inferior medium to radio, but they came to it nonetheless. In fact, the differences between radio and television journalism were less pronounced than Murrow pretended. The value of both lay in technology, in the capacity to convey news at once, as it happened. The broadcast journalist could be on the air with the story while the newspaper publisher was still trying to get his printers out of bed. Then as now, however, journalists in the electronic media tended to be second-tier as gatherers of news. Radio got the soundtrack, television got the pictures, but newsprint got the scoops. In his memoirs David Schoenbrun, one of Murrow's Boys, told the story of a scoop he got in the fifties. Schoenbrun was CBS Paris bureau chief at the time, a giant of broadcast journalism. But his editor in New York refused to run the scoop, because, the editor said, he hadn't seen it in the *New York Times* and so couldn't confirm it. For Schoenbrun, the story illustrated the difference between print journalism and its simulacrum on TV.

But TV was kind to Murrow, as radio had been. The impression he

made on the tube was indelible: the ever-present Camel with its curl-
ing smoke, and the squint, and the eyebrows that rose and lowered
with ominous portent—these combined with the already famous voice
to create an almost unbearable gravity. "No one's brow furrows," went
a doggerel of the time, "like Edward R. Murrow's." As with radio, the
essential gifts were cosmetic, pieces of showbiz. Among other innovations, Murrow pioneered the technique, favored today especially by
Jennings and Tom Brokaw, of looking away, to the left and slightly
below the camera, as he paused in reading his lines. The impression is
one of a thoughtful fellow, caught in unhurried rumination. But of
course, with Jennings and Brokaw, it is only an impression.

Murrow differed from his handsome successors in one important,
and now quite endearing, respect. His presence on the screen was not
particularly . . . cozy. His sign-off, at the end of each TV appearance,
was "Good night, and good luck"—the suggestion being (as Reuven
Frank noted) that you're sure as hell gonna need it. Sometimes his face
and voice together could convey an unmistakable contempt for whoever was stupid enough to be watching television at the moment. He
once devoted an episode of his documentary show, *See It Now*, to a trip
he had made to the front lines in Korea. He closed the show staring
into the camera. The troops, he said "may need blood. Can you spare a
pint?" There are many ways to read this line: imploring, cautionary,
chummy. Murrow's curled lip and cold eyes expressed something else:
"Can you spare a pint, *you soft, pampered, pasty-faced little wetsmacks?*"
The attitude worked in the fifties, apparently. We were a different
country then.

*See It Now* was one of two Murrow series on CBS TV, the other
being a weekly half-hour celebrity interview called *Person to Person*. (In
its quest to keep the legend alive, CBS not long ago issued a four-vol-
ume set of video cassettes, "Good Night and Good Luck," that provides a useful overview of Murrow's TV work.) *See It Now* was a "broadcast," *Person* was a "show." Among broadcast journalists the distinction
is crucial. Broadcasts are news, documentaries, chin-wags. You will
never hear Dan Rather call the *CBS Evening News* a "show." Shows are
entertainment. Comedians, singers, dancers, actors . . . *entertainers* do
shows.

Murrow did both. Legend tells us that Murrow was embarrassed by *Person to Person*, considering it beneath the dignity of a broadcast journalist to be joshing around with the likes of Marilyn Monroe and Jayne Mansfield. But josh he did, and the surviving tapes show him enjoying himself. They also show him to be a genial but not terribly skilled interviewer, perhaps because, as was later revealed, most of the questions and answers were scripted. "What's been your biggest thrill in show business, Frank?" he asked the great crooner. Murrow was father to Larry King, too.

Purists complain that the iron wall between news and entertainment has been breached in our present benighted era. But it looked pretty permeable in Murrow's Golden Age, as well. The documentary half-hours of *See It Now* were often devoted to interviews; and it is not at all clear why Carl Sandburg, a guest on *See It Now*, should be considered highbrow and Fred Astaire, on *Person to Person*, should be low. Other broadcasts dealt with the issues familiar to viewers of the "newsmagazine" shows of today: education, race relations, nuclear power, war and peace. Here, too, Murrow knew to use the tricks of technology to make the broadcasts more seamless and—well, entertaining. Interviews conducted by other correspondents, for example, would be redubbed to make it appear that Murrow was the questioner. In the very first *See It Now*, Murrow showed footage of soldiers in Korea, digging trenches, building bunkers, even firing their weapons, with the clear implication that they were at the front. In fact they were nowhere near the enemy. Murrow's correspondent, Robert Pierpont, objected to the dodge, to no avail.

*See It Now* was famous then, and now, for losing money—further evidence, it was said, of a commitment to news above all else. Only Murrow's own high standards kept it on the air, according to Murrow. To those corporate "vice-presidents" who wanted him, glug, to show a profit, he talked tough, he said. "I say, 'If that's the way you want to do it, you'd better get yourselves another boy.' " CBS was happy to let this impression stand, probably because the show wasn't losing all that much money, if any. Its sponsor, Alcoa, underwrote expenses up to $2,300 a week, which was usually sufficient to cover costs. CBS picked up any overage; not so great a burden, since Alcoa was also paying an additional $34,000 for the airtime. Murrow bragged often of his hostility to the "money men," and especially his sponsors. But at the same time he was willing to film promotional documentaries for them, as he did for Alcoa. By 1953, his disdain for profit was earning him more than $200,000 a year, more than $300,000 by 1960.

Murrow's most celebrated moment on television—the summit of his career—came in March 1954. This was the broadcast broadside against Joe McCarthy. The mythologists point to it as the blow that brought down the terrorist Tailgunner. The claim calls to mind Mark Twain's remark about his military service in the Civil War: "I left the Confederate army in 1865. The South fell."

The show was indeed masterfully done. It consisted almost entirely of clips of McCarthy himself—a compendium of every burp, grunt, stutter, nose probe, brutish aside, and maniacal giggle the senator had ever allowed to be captured on film. Murrow, it was said, allowed McCarthy to hang himself, but in truth McCarthy had been hanging himself quite efficiently in the several months before Murrow offered him more rope. By the time the show aired, a mutiny was underway on his own subcommittee to relieve McCarthy as chairman. Prominent Republicans had joined Democrats in publicly denouncing him, even, gingerly, his former comrade Vice President Richard Nixon. In the mainstream press, anti-McCarthy feeling was endemic. Among those routinely critical were *Time* magazine and Col. Robert McCormick's *Chicago Tribune*. If Col. McCormick and Henry Luce were denouncing a right-wing icon, you could feel pretty safe in firing away. "Ed didn't want to get too far ahead of public opinion," Fred Friendly said. And he didn't.

It is revealing that Murrow is most honored today for broadcasts—1960's "Harvest of Shame," about migrant farm workers, is another—that were works of unbridled advocacy. He had always been as much advocate as reporter. In the late 1930s Murrow saw himself as a subtle propagandist for bringing America into the war. His advocacy was always artful. Cloud and Olson quote a CBS executive in the 1930s, advising a reporter to "disguise his own opinions by attributing them to others." "Don't be so personal," the executive said. "Use such phrases as 'It is said . . .' and 'There are some who believe . . .'"

This is how Murrow did it, he said, and it's a trick still widely employed. Other times, Murrow's advocacy was completely behind the scenes. In 1956, the newsman secretly rented a studio to coach Adlai Stevenson, then running against Dwight Eisenhower, in hopes of improving the candidate's television performances.

The *New Yorker* once ran a profile of Murrow called "The World on His Back." It was an image he carefully cultivated. In time he began to believe his own legend—not surprising, since he had done so much to concoct it. By 1958, when he delivered his famous denunciation of television's preoccupation with revenue, he had fully evolved into a national scold, the last honest man in a world twisted by commerce. This is the image that has survived, the one so desperately invoked by the scolds of our own day. Murrow left journalism for good in 1961, to head up the United States Information Agency. All those cigarettes finally killed him in 1965, at the age of 57.

But his relationship with CBS effectively came to an end much earlier, in 1959. In the aftermath of the quiz-show scandals, the president of CBS, Frank Stanton, announced that the network would no longer tolerate any technological "hanky-panky" that might deceive its viewers. He said he had in mind, specifically, Murrow's practice of scripting *Person to Person* interviews in advance. People who tuned in, Stanton said, needed to know that "what they see and hear on CBS is exactly what it purports to be." And so the top "money man" censured Murrow publicly—for violating the high standards of broadcast journalism. Murrow never forgave him for it.

# THE *NEW YORK TIMES* MAGAZINE AT 100

## By Philip Terzian

⟞⟝

*(Originally published, as "100 Years of Turpitude,"*
*in the April 29, 1996, issue)*

WHOEVER IT WAS WHO SAID that journalism is the first rough draft of history was, presumably, a journalist. For no historian would ever suppose such a thing. And what better proof is needed than the periodic spectacle of journalists attempting to behave as historians?

Comes now the *New York Times Magazine*, which observed its centennial on April 14 by publishing a lush, slightly thicker than usual edition, gleaning (in editor Jack Rosenthal's words) "bursts of passion and energy" now bequeathed to posterity. To recapture those "moments of memorable writing," Rosenthal explains, excerpts were presented in chronological order; but, he goes on, "we were struck as we kept reading to see how much powerful writing was provoked by enduring subjects like women, civil rights, Vietnam, and South Africa." Enduring? More like incessant. So here, then, we know we are in the realm of the journalist—even worse, the baby boom journalist. In a hundred years of history, what matters is the world since the baby boom began; what endures is the product of the past quarter century.

Most striking, in all its pages, is what cannot be found. The world before 1960 or so can scarcely be discerned; the world before 1940 is practically invisible.

There is one brief, quirky piece set in World War I; there are eight dispatches from Vietnam. From the founding of the magazine in 1896 until the eve of World War II—the "Early Years," as the editors call them—nearly half the 20th century is artfully dismissed with an airy visit to the Hamptons, four paragraphs about state senator Franklin Roosevelt, two bird's-eye views of Adolf Hitler, vignettes of Joe Louis and Henry Ford, and nothing whatsoever about the Bolshevik revolution, psychoanalysis, the modernist schools in any of the arts, the Progressive movement, Prohibition, the crash, the Depression, Charlie Chaplin, the talkies, the Armory show, the Armistice, or Aimee Semple McPherson.

Of war, there is nothing about Manchuria, or Korea, or the Marne, or the Somme, or the Bulge, or the Chinese or Spanish civil wars. But there are Bosnia and Beirut and Frances FitzGerald and David Halberstam and "Ho Chi Minh on the Move" against the French.

Of literature, no mention is made of F. Scott Fitzgerald, W. H. Auden, T. S. Eliot, Sherwood Anderson, Evelyn Waugh, William Faulkner, Jean Cocteau, Robert Frost, Thomas Hardy, Franz Kafka, George Orwell, Alberto Moravia, E. M. Forster, Vladimir Nabokov, H. L. Mencken, Boris Pasternak, Ezra Pound, Edith Wharton, James Joyce, Ernest Hemingway, Luigi Pirandello, Wallace Stevens, Virginia Woolf, Henry James, Paul Valéry, Willa Cather, William Butler Yeats, or Rainer Maria Rilke.

Space is found, however, for Kurt Vonnegut Jr., Norman Mailer ("The emotional meat of the heart might be free of the common bile"), Joyce Carol Oates, Joyce Maynard ("My generation is special because of what we missed rather than what we got"), and Delmore Schwartz on Marilyn Monroe. Meanwhile, Leo Tolstoy, who once wrote an essay for the *New York Times Magazine*, is banished to make room for monologist Spalding Gray ("So everyone takes their clothes off and lines up like this huddling mob of naked refugees") and articles editor Gerald Marzorati, trailing Salman Rushdie around London.

Of science, there is nothing about the theory of relativity, DNA, antibiotics, the fight against cancer, or splitting the atom—not even about the Bomb, a startling omission that would not have occurred in the nuclear-minded 1980s. And in this century of medical science's greatest triumphs, the cynical reader may guess in advance which topics are included: toxic shock syndrome, breast cancer, and AIDS. No classical music, of course, or jazz, or opera, but a tribute to Woodstock ("It only made the world a little less uptight") and a plastered Janis Joplin ("Such a strange, unsettled mix of defiance and hesitancy, vulnerability and strengths").

Where contemporary feminism is concerned, no dissenting voices may be heard on any page: It's all Vivian Gornick, Bella Abzug, Susan Brownmiller, Kate Millett, even Susan B. Anthony II and a locker roomful of sweating high-school athletes. One woman named Mrs. Arthur M. Dodge makes the case against votes for women (1915); but otherwise, it's forever 1972 and the Equal Rights Amendment is marching toward passage. Katha Pollitt is enamored of abortion on demand; Gloria Steinem is enamored of herself.

Indeed, once the distant past is thankfully behind them, the editors descend upon familiar ground, retooling the recent past to present-day perspective. From the 1950s, we are once again acquainted with Sen. Joe McCarthy, Jack Kerouac, and John F. Kennedy; you would never have known that Dwight D. Eisenhower had lived, or Bishop Sheen, Adlai Stevenson, Helen Keller, Dean Acheson, Norman Rockwell, Charles de Gaulle, David Riesman, or Robert Taft.

You would learn, however, about the courtier spirit, which the *Times* seems to nurture. Here is Herbert L. Matthews on Fidel Castro: "Apparently it is hard for some to understand how otherwise he can work so feverishly for 20 or 21 hours a day, every day without a break. But the answer is simple; he has the build of a professional football player and the strength of a bull." This note of adoration is frequently struck: From Robert Lipsyte on Muhammad Ali ("Youth and light and magic, a Technicolor genie in a bottle-green world"); from Mel Gussow on Meryl Streep, and Joanne Stang on Woody Allen ("What Allen projects . . . is wistful futility"); from Anne Taylor Fleming on Truman Capote ("He sees everything and can make stories out of everyone"); and from William Serrin on Jesse Jackson ("How he can preach. He is perhaps the finest preacher in the country . . . as good as, perhaps better than, Malcolm or Martin. Oral Roberts? Billy Graham? They are run-of-the-mill honkies compared to Jesse").

For the 1970s, there is J. Anthony Lukas on the Watergate scandal to tell us all we need to know of Richard M. Nixon. There is a somber exploration of the soul of Jimmy Carter, and Merle Miller's declaration of homosexuality. The 1980s are slightly more varied, but not by much: There is greed, Donald Trump, the Gary Hart affair, a guffaw or two at Dan Quayle's expense, and poor Andy Warhol on the threshold of death. What, no essay on the homeless? No mention of Ronald Reagan? Mere oversights, no doubt.

And yet, toward the end, a pattern seems to form. By the 1980s and the first half of the 90s, as the century is waning, there's a mournful, disturbing, almost elegiac tone to the picture Jack Rosenthal is piecing together. The battles of the culture war are starting to be lost, or so James Atlas seems to think. Salman Rushdie faces death. The Communist collapse seems to agitate the world. Chinese dissidents are shot. The Argentine writer Jacobo Timerman is tortured. Nazi Klaus Barbie taunts the French. Bosnia is drenched in blood. The specter of AIDS stalks the land. Right-wing survivalists retreat into the woods, and Howell Raines stages a strategic withdrawal: fly-fishing for a while, then riding poor Grady, his family's old servant, to a Pulitzer prize, refreshing himself for the challenge of conservatism.

Maybe history, for these journalists, is coming to an end: The baby boom editors are starting to slow down, and the last half of the century is slipping away. For the *New York Times Magazine*, all those four-color graphics and self-congratulations cannot hide a painful truth: The sixties are over, and it's all downhill from here.

# THE TORTURE OF WRITER'S BLOCK

## By Joseph Epstein

*(Originally published, as "Music Without Words,"
in the November 4, 1996, issue)*

THE PAST FEW YEARS HAVE SEEN the deaths of Ralph Ellison and Joseph Mitchell, two of America's most remarkable writers. The one a novelist, the other a journalist, each was thought by many people the best at his respective trade.

*Invisible Man*, published in 1952, may well be the most solidly made and most intelligent novel produced by an American in the past half century. Wildly comic, philosophically deep, socially significant, it is a book of a kind that, if one had written nothing else, would be enough to give one a strong reputation for the rest of one's life, and perhaps beyond. And in the case of Ralph Ellison, who wrote no further novels, it did just that.

As for Joseph Mitchell, he represented, at its highest power, the urban tradition at the *New Yorker*, the tradition of John McNulty and A. J. Liebling, as opposed to the small-town tradition of James Thurber and E. B. White. Mitchell was fortunate to live long enough to see his own reputation revived in 1992, when *Up in the Old Hotel*, a compendium of all his earlier books, appeared with heavy jeroboams of praise.

Ellison and Mitchell both had famous—perhaps the country's most famous—cases of writer's block. Block is the supreme torture for a writer. When a writer is "blocked," he cannot write; his craft and tal-

ent and energy suddenly flee him. The condition can last a week, a
month, a year, a lifetime. And like inspiration, writer's block can show
up utterly without warning. It is a condition that seems inexplicable,
and is painful in the extreme.

I believe most cases of writer's block do have an explanation, and
one not necessarily to be found down in that dark psychic disco where
the superego is doing the tango with the id. Sometimes a writer may
be blocked because he really isn't prepared to write what he has
promised to write; he just doesn't have the knowledge, experience, or
wit to carry out the job. Or sometimes, midway through a lengthy
piece of writing, he discerns the falsity of all that he has written up to
now and hasn't the stomach or stamina to return and begin again.
Sometimes he simply cannot bring the introspection and honesty to
the job that it requires, and sometimes contemplating the serious consequences of what he is writing—in loss of friendships, status, or
future earnings—may be more than he can bear.

The great fear of a writer is that he will find himself locked into
the kind of writer's block that afflicted both Ellison and Mitchell. The
one question you didn't ask Ralph Ellison was, "Hey, kiddo, how's the
new novel going?" For after the splendid success of *Invisible Man*,
Ellison was never able to produce that novel. There were stories about
his having lost nearly an entire manuscript to a fire. After *Invisible
Man*, forty years passed, during which Ellison produced two books of
essays, collected a vast number of honorary degrees, served on endless
editorial and other boards, and kept his cool and courage at a time
when it was easy for a black writer to make a serious jerk of himself.
But the thing he was put on earth to do, write more beautiful novels,
was precisely what he was unable to do.

Joseph Mitchell's block was of a different order, though it lasted
thirty years. Mitchell's last book, *Joe Gould's Secret*, was published in
1965, and its first sentence shows his perfect touch: "Joe Gould was an
odd and penniless and unemployable little man who came to the city
in 1916 and ducked and dodged and held on as hard as he could for
over thirty-five years." Many of his admirers, myself among them,
longed for more such sentences, though they were never to come.

Mitchell once described to me, in his deep North Carolinian

accent, the book he said he was working on. It was about "double exile," which, he went on to explain, was the peculiar condition of feeling a stranger wherever he was: In New York, he felt himself a southerner, while in the South he felt himself a pure New Yorker. The book seemed one of the purest of "enchanted cigarettes," the term Balzac gave to those books one dreams about but almost are certain never to get written.

Another of the things Ellison and Mitchell had in common is that each treated me to a single memorable afternoon, the better part of both of which I spent in the same chair on the second floor in the ample room facing onto 43rd Street at the Century Club in New York. I met each man once, felt I had made a friend, and afterward never had another contact with either of them.

I spent the afternoon of January 26, 1978, with Ellison. He had earlier published a fine essay in the *American Scholar*, the magazine I edit, called "The Little Man from Chehaw Station," which I had read in its first-draft form as a commencement address and which he added onto and greatly improved for publication. My journal entry for the day notes: "He is a smaller man than I had imagined him to be, though, as I had imagined, well turned out sartorially (the only flaw here being too large a wristwatch) and with the manner of a courtly gent." He nicely broke through this formal manner by using the phrase, ten or so minutes into our lunch, "f—ing distinguished" to refer to a pompous figure in publishing who came up to our table. Our conversation was desultory—we told jokes, he told Depression stories, we discussed literature, personalities, politics—and unrelievedly wonderful. I arrived at 12:30 and left, in the dark of a Manhattan evening, at 5:00.

I thought I had a friend for life. When I returned to Chicago, I wrote Ellison a brief note, thanking him for the lunch and for the fine afternoon. No answer. A month or so later, I wrote again, this time proposing an essay for him to write for the *American Scholar*. Again no answer. Perhaps a year later, I wrote yet again, and again no response. All very strange. It was as if you had gone out with a very attractive woman, and thought you had both had a swell time, except that she refused afterward to take any of your calls. What was going on?

Five or six years later, I corresponded with a man who, in the course of one of his letters, asked me if I knew Ralph Ellison. The reason he asked was that he and his wife were once on a cruise with Ellison and his wife, and during that cruise they were nearly inseparable. The four of them seemed to have a perfectly lovely time. Yet when they returned to the United States, Ellison failed to answer any of this man's letters. Did I have any idea what was going on?

The only reason I can come up with to explain Ralph Ellison's odd behavior is his writer's block. He was a naturally friendly, happily gregarious man, I think, and yet he must have worried about the cost of his sweet openness of spirit. With that uncompleted second novel hanging always before him, like surgery that he knew he couldn't postpone forever, though he somehow did, what he least needed was lots of new friends: insistent, responsibility-exacting, time-consuming friends. Friends may have been fine things, but it was the consequences of friendship that he couldn't afford, the consequence of time above all—time that would have to be taken from that damn unfinished novel to be a friend. That the novel wasn't getting written anyhow still didn't mean that one could take time from it. Such is the nightmare of a writer's block—one can't for more than a moment take pleasure in the leisure it imposes or think of anything except the writing one cannot do.

My one meeting with Joe Mitchell was less lengthy but no less enjoyable. We had corresponded years before, and had even spoken over the telephone a time or two. From his few letters to me, I was surprised to learn that Mitchell, whose specialty as a writer was observation of the common life, was a regular reader of intellectual journals and knew all about "the boys on the quarterlies," as his old friend Joe Liebling used to say. He wrote to me, who was nobody if not one of the boys on the quarterlies, and said that he had read me over the years in *Encounter, Commentary,* the *New Criterion,* the *TLS,* and elsewhere.

In print, Mitchell was all cool objectivity, without opinionation, self-effacing, benevolent in his views of human nature. In private he was sly, opinionated, witty in a way that his writing didn't quite reveal. He turned out to be greatly interested in visual art and apparently spent much time going to exhibitions and galleries. When I told him that the art critic Hilton Kramer was a dear friend, he expressed admiration for him, especially for the courageousness of his views. Straightforward expression of views was never part of Mitchell's own modus operandi, at least not as a writer. That didn't mean he didn't have views, quite strong ones. He talked a good deal about missing his friend Liebling. He was critical of E. B. White, the preachiness of whose writing he couldn't abide.

Much in the current scene put him off, not least its liberationist tendencies. "You know, Joe," he said, "I am of a generation that can never consider sex a trivial act. When I was a young man, growing up in the South, if you did ugly to a girl, her brother would shoot you." *Did ugly to a girl* is a phrase I am not soon likely to forget.

I had no sense that Mitchell suffered greatly from not writing. My best explanation for his writer's block is that his subject matter had disappeared on him, and my guess is that he knew it. Mitchell had made his reputation writing about characters in New York, but characters, interesting idiosyncratic characters, had long since been replaced by cases, some of them quite dangerous. Joe Gould today would be viewed as a slightly menacing homeless person; the amiable drunks at McSorley's, the "wonderful saloon" which is still there on East 6th Street owing in good part to the fame Mitchell gave it in various *New Yorker* essays, would now just seem hopelessly lost. Mitchell was left as bereft of a subject as Hogarth might have been under communism.

Writers who have written something substantial and then are blocked are one thing; writers who are blocked long before they hit their peak quite another. The toughest trick, and one of the greatest causes of writer's block, may be that of following one's own strong opening act. Writing a good or financially successful book the first time out can be filled with peril. This appears to have been not only Ellison's problem. It was also, at a lower level of literary creation, the problem of Thomas Heggen and Ross Lockridge, the authors, respectively, of *Mr. Roberts* and *Raintree County.* Both men had enormous commercial successes and each killed himself before producing a second work.

Frank Conroy took nearly twenty years between *Stop-Time*, his fine autobiographical book, and his next work, a collection of stories not many people remember. When I knew him, Conroy seemed in no hurry to produce a second book; having gotten it right the first time, perhaps he felt there was no rush. As the author of eleven books, with a twelfth in press, I often wonder if I would have written less if I had got it right on the first try—or, for that matter, on the eleventh. Sometimes quite good writers, usually highly productive ones, will suddenly go silent. When inquiries are made of people who know them, one learns that they are blocked. This apparently is the case with Michael J. Arlen, who wrote some excellent television criticism for the *New Yorker* and a fine book about his Armenian forebears and who hasn't been in print for a number of years. Renata Adler, a key writer at the *New Yorker* and at the *New York Review of Books*, also has a block that has caused her to close up shop for the better part of a decade.

The saddest case of writer's block I know was that of my friend, the late Marion Magid, for many years the managing editor of *Commentary*. Marion began brilliantly, in her twenties, writing winningly about such varied subjects as Tennessee Williams for *Commentary* and hippie life in Amsterdam for *Esquire*. She straightaway had a style and a point of view—perhaps they are the same thing—and could, as they say about the best infielders, really pick it. You have to imagine a young Joan Didion, but smarter, more amusing, without the depression.

And then, for no good explanation I ever heard, the flame went out. Through her thirties, forties, fifties, up to her death at the age of sixty, Marion never really broke out of her block. All that I can remember her publishing those many years was a single review, in *Commentary*, of a book about American communism by a woman named Vivian Gornick. I made the mistake of complimenting her by calling that piece "a nice little review"; she rejoined by telling me that if I knew what effort went into it, I would never call it "little." On another occasion, when a letter she sent to me failed to arrive, she all but groaned and said, "God, talk about being blocked—now even my letters aren't getting through."

I don't think I can hope to understand the suffering Marion Magid went through. She was the real thing, a true writer, and the early evidence shows a brilliant one, but unable to work at the trade she loved above all others. How immensely frustrating! She could hear the music, but never find the words.

One of the signs of a real writer is the need to write almost all the time; and along with this need goes the feeling of self-loathing when one isn't writing. To what can one compare the pain of the blocked writer? Perhaps to a fine athlete, still in his prime, banned from playing the game he most loves. It has to be hellacious.

When I lived in New York, in the early 1960s, a time when psychotherapy was a dominant force among artists and intellectuals, lots of writers seemed, if not altogether blocked, then highly costive. If in the course of a year these people wrote a single book review, or an essay, or short story, or two or three poems, it seemed production enough. Wallace Markfield, in his novel *To an Early Grave*, captures the spirit of the blocked writer in a character, a literary critic named Holly Levine, whom he shows moving the word "certainly" in about six or seven recastings of the same sentence, until he is saved by a phone call that allows him to abandon the effort. Another day of nothing accomplished.

So endemic was writer's block in New York in the '50s and early '60s that there was a shrink named Edmund Bergler, one of whose specialties was unblocking writers. In a Teutonic accent, or so I have been told, he all but yelled at his patients, telling them that they were immature, they must knock off this nonsense, return to their typewriters, pay his fee. It may well be that some blocked writers derive a perverse pleasure from their blocks. But I doubt that pleasure had much to do with it.

Perfectionism is yet another important reason for writer's block. For a writer stuck with perfectionism, writer's block just about comes with the territory, and such a writer figures to be blocked fully half the time. A one-book novelist named Hannah Green—her one book is a novel that she finished at the age of forty-six and that I have never read titled *The Dead of the House*—died a week or so ago, and her obituary in the *New York Times* noted "her almost obsessive pursuit of a perfection that always seemed just one rewrite away." The real wonder, her husband tossed in, was that she finished a book at all. "She never wanted to let go," he said. (One never truly finishes a poem, Paul Valéry said, one merely abandons it.) Miss Green had been working on a second book for the last twenty-five years when she died at the age of sixty-nine.

I write all this in trepidation, lest I incur the wrath of the furies and they strike me with a block of my own. Fifty years in the business, Zero Mostel moaned in *The Producers*, and I'm wearing a cardboard belt. More than 30 years in the business, say I with merely a grin and a great thumping knock on wood, and I have had the good luck to write pretty much what I have wanted without any hint of block. Perhaps a few people who have felt themselves unkindly treated by me are even now sticking pins in my books and ringing their own ever-so-slight change on the old football chant, "Block that kick."

I suppose that every writer for whom writing is not the painful drama it is often made out to be, but is instead an intense delight, worries a little about exhibiting this delight too frequently in public. Edith Wharton coined the term "magazine bore" for those writers who appear too often in too many magazines. You pick up a magazine and there they are, like the fellow in the orange fright-wig and the Jesus Saves T-shirt who used to show up at all major sporting events.

It is probably a mistake to make writing look too easy. Anthony Trollope's having done so by recounting his writing regimen in his autobiography—he averaged 10,000 words a week, and some weeks wrote as many as 28,000 words—caused his reputation to suffer for many years. He made writing seem altogether too mechanical an activity. A writer, Trollope believed, must approach his task as if he were working at the post office, which Trollope, in fact, for many years did: "He should sit down at his desk at a certain hour." He should eliminate the *Sturm und Drang* aspect of writing: "He need tie no wet towels round his brow, nor sit for thirty hours at his desk without moving,— as men have sat, or said that they have sat." Writing cannot be done in one's sleep, but for Trollope it could be done on the edge of sleep: "A man to whom writing well has become a habit may write well though he be fatigued." Trollope prided himself on having published twice as much as Carlyle and even more than Voltaire, and having accomplished this while holding a second job much of the time.

Writing quickly does have its own odd satisfactions, among them the delusion that one has mastered one's craft. Usually, one hasn't. More likely one has only had a lucky good day. I once had an essay reprinted in a college reader, and the editor offered me an additional $600 if I would write a thousand or so words explaining any difficulty encountered along the way of composing the original essay. I told myself I would do it if I could complete the job in less than an hour. And— yippee ti yo!—I did. Not only was this gratifying in itself, but I could now tell myself that I was a $600-an-hour writer. Unfortunately, the next thing I wrote, a long essay on the Austrian writer Robert Musil, took me so long to write that it returned me to thinking of myself as a minimum-wage man.

I once asked the critic F. W. Dupee, at that time freshly retired from the English department at Columbia and a finely polished prose writer, to write for the *American Scholar*. He wrote back to thank me for my invitation, but told me, in a sentence that sent a little chill through me, "I have stopped writing."

"I have stopped writing." That is the admission that Ellison could not make, nor Mitchell, who went into his office at the *New Yorker* every day for thirty years before his death without ever submitting a word to his editors. I cannot imagine myself, short of a knock-out stroke or severe illness, ever saying that. My guess is that Dupee, good as he was, didn't really *have* to write. He must have been among those perhaps fortunate people who can write, and write extremely well, but can get through life quite nicely without having to write. A true writer, for better *and* worse, needs to write.

Needs to—and, despite all the widely advertised agonies, loves to. Writing has given me pleasure like nothing else I have ever done, and I mean the very act of writing itself. Raymond Chandler, when he learned that the detective-story writer John Dickson Carr disliked writing, speaks for me when he writes:

> *A writer who hates the actual writing, who gets no joy out of the creation of magic by words, to me is simply not a writer at all. The actual writing is what you live for. The rest is something you have to get through in order to arrive at that point. How can you hate the actual writing. . . . How can you hate the magic which makes of a*

*paragraph or a sentence or a line of dialogue or a description something in the nature of a new creation?*

The only argument I would have with that passage is the use of the word "creation." Best, when thinking about writing, to keep the pretension level as low as possible. The first rule in avoiding writer's block is never to think of writing as in any way a creative activity, with its own dramas and tensions. For as soon as one does think about writing as "creative"—a bogus word in any case—one thinks about all that can go wrong with it. Much better, I have always found, to demystify writing as completely as possible. I frequently remind myself that formulating sentences remains one of the most amusing of all pastimes— and, besides (though I shouldn't want this to get around) it beats working all to hell.

# THE TERRIBLE BEAUTY
# OF W. B. YEATS

## By Christopher Caldwell

——◆——

*(Originally published in the April 28, 1997, issue)*

THE LATE CRITIC RICHARD ELLMANN thought William Butler Yeats the most important poet to have written in English since Wordsworth. Ellmann also admitted that, had Yeats died in 1917 and not in 1939, "he would have been remembered as a remarkable minor poet who achieved a diction more powerful than that of his contemporaries but who, except in a handful of poems, did not have much to say with it." So Oxford historian R. F. Foster has quite a task to fulfill in the first of a planned two-volume biography of Yeats, because volume one only takes us up through 1914, before Yeats really hit his stride.

Foster would agree that Yeats's major poems were still to come—but Yeats was more than a poet. He was the most ambitious and active member of modern Ireland's first generation, and his role in founding Ireland's cultural institutions arguably had consequences as profound as did his poetry. "Most biographic studies of WBY are principally about what he wrote," Foster admits. "This one is principally about what he did." It is a wholly defensible endeavor—and one that Foster pulls off with depth and panache.

At the heart of Foster's biography is Yeats's lifelong involvement with the occult. It gives shape to the whole narrative and has implications for Yeats's politics and character. From his early enthusiasm for Irish folklore to the bizarre "system" of cycles he would write about in *A Vision* in 1925, Yeats was consumed with the idea that powerful forces beyond our comprehension were controlling the course of history. He went from talking about leprechauns and banshees in the period covered by this volume to the idea of a "rough beast" out of his own mystical imagining that was about to take over Western civilization.

Foster sees Yeats's obsession with the occult as largely a source of metaphors for his poetry. If so, then this volume gives one a sense of the price that Yeats paid in a wanton search for poetic inspiration.

Born outside Dublin in 1865, a poor student who lacked the classics or math to get into Dublin's Trinity College, Yeats arrived in London in the late 1880s and immediately sought out Madame Blavatsky, Europe's leading clairvoyant, whose "theosophist" empire had been exposed as a fraud by a freelance investigator only months before. Yeats also joined a more radical supernaturalist group called the Hermetic Order of the Golden Dawn and researched in cabalism, black magic, Rosicrucianism, and a variety of Eastern religions he only vaguely understood. Even in 1912, at 47, Yeats was consulting with Etta Wriedt, an American medium who introduced him to "Leo Africanus," the shade of a 16th-century Spanish Arab explorer Yeats would come to see as his alter ego.

We are thus almost immediately at a central issue of Yeats's poetry: Did he actually believe any of this nonsense? Yeats was asked the question constantly, of course, and succeeded in evading it by citing Socrates' words in Plato's *Phaedrus*: "I want to know not about this but about myself." Why, then, was the system he wrote about in *A Vision* based on images his wife saw during bouts of automatic writing? What are we to think of a poet who loudly proclaims his disaffection with established religion and instead bases his metaphysics on quite literally the first thing that comes into his, or his wife's, head?

That question is best answered by examining Yeats's youthful subject matter: Ireland itself. "Mad Ireland hurt [him] into poetry," Auden said, but in truth, the vein of poetry into which Ireland "hurt" Yeats was not a particularly rich one. His first major dramatic poem, *The Wanderings of Oisin* (1889), was "an azure-and-gold tonal arrangement of islands, caverns, basaltic castles, painted birds, milky smoke and grass-blades hung with dewdrops," as Foster describes it. Yeats was clearly keen to write in the folkloric mode. He took an early interest in Irish tales during his childhood summers in County Sligo, but according to his sister Lily there was something bogus, too, in his notion that Sligo people believed in fairies and talked about them all the time. ("So they did, of course. To *children*.")

Yeats assembled the Irish elements in his poems as if he were a professional folklorist, and exploited them as expertly as a mau-mauing modern-day ethnic novelist. There was a ferocious canniness in his business practices at odds with the disarming rusticity of his poems. He was ruthless in dunning subscribers to his poems. And, in setting up a favorable review of his early novel *John Sherman*, Yeats wrote his confidante Katharine Tynan, "You might perhaps, if you think it is so, say that Sherman is an Irish type. I have an ambition to be taken as an Irish novelist not as an English or cosmopolitan one choosing Ireland as a background."

Foster even sees "something curiously self-conscious in his immediate idealization of" Maud Gonne, the English political agitator who came to visit Yeats in 1889. Although Yeats would not have his first love affair until age 29, and would not marry until age 52, the "*fin-de-siècle* beauty," as Foster calls Gonne, was the love of Yeats's life. (Until about twenty years ago, this love was thought to have been unrequited, but Foster adds his voice to the growing consensus that the two slept together sometime in late 1908.) Yeats's delicate and allusive love poems to Gonne are the writings of the period that stand up best against his later work, indeed against any love lyrics in the language. The emotions of love were something not even Yeats could completely smother under a contrived value system—although he tried.

Foster gives us a considerably more nuanced view of what it means to be a mystic, a holy man, a seer in modern times than Yeats biographers before him. He shows that Yeats was as much a striver as a seeker—that the poet cannot be understood except as a man on the make, in pursuit of fame, love, and revelation. This is where Foster's focus on "what he did" rather than "what he wrote" is most appropriate, for it is to that very side of his personality that 20th-century Ireland owes most of its literary institutions.

Yeats launched both the Irish Literary Society of London and the National Literary Society in Dublin, spending a good deal more time tending to the former than to the latter. But it was in 1904, when he established the Abbey Theatre, that Yeats not only changed the cultural face of the country but transformed himself into a genuine national leader.

In 1907 the Abbey produced John Millington Synge's masterpiece *The Playboy of the Western World.* It caused a sensation, in part because of descriptions of women that were considered pornographic. Riots broke out on opening night. Yeats returned from a lecture tour to announce that "so far as he could see the people who formed the opposition had no books in their houses."

The rioters were Catholic; Yeats was a Protestant. What Catholic nationalists saw as a striving for common values and a fear of the wages of immorality, the Protestant Yeats saw as mob psychology and rank philistinism. The *Playboy* conflict drew Yeats into real politics and away from the romantic pose of his early folkloric nationalism. The spiritual *raison d'être* of the modern Irish state—Irish Catholic fears both of persecution and the condescension to which Yeats was given— left him in the uncomfortable position of professing Irish nationalism while being wholly out of sympathy with the myths that gave rise to it. "This endless war with Irish stupidity," he wrote to Tynan, "gets on my nerves." In essence, Yeats was seeking to carve out a place for himself and other Protestants in a country that no longer particularly wanted them.

What is romantic when applied to leprechauns is dangerous when applied to religious and ethnic strife, and Yeats's mystical flights led him to a view of history that didn't correspond to anything he *thought* he was writing about, and which was racist to the extent that it did. Yeats supported Ireland's abortive "Blue Shirt" movement of 1933, and critics have long believed that his mystical nationalism drew him close to fascism. As Conor Cruise O'Brien has put it, "Yeats the man was as near to being a Fascist as his situation and the conditions of his own country permitted." Exhibit A for this claim is Yeats's poem of the early 1930s "Blood and the Moon":

> *Blessed be this place,*
> *More blessed still the tower;*

*A bloody, arrogant power*
*Rose out of the race . . .*

But the sentiment was already present even in such poems as his
1922 "The Fisherman":

*All day I'd looked in the face*
*What I had hoped 'twould be*
*To write for my own race*
*And the reality . . .*

Had this been written by a German, it would make us shudder.
But "The Fisherman" doesn't because it is such a strange kind of
poem: racial bragging that relies on the continued subjection and fecklessness of the race in question. Yeats chose to identify himself with
the Irish peasantry in order to hold a moral high ground—the supposed high ground of the Irish nationalist victim—that would otherwise be indefensible. Which is merely another way of saying that Yeats
was the first poet of identity politics.

Three years after Yeats's death, the critic Randall Jarrell wrote:
"When people who admire Yeats's poetry ridicule or deplore his 'crazy
system,' they do not realize that it was the system which enabled him
to produce the poetry. . . . However wrong that system is for you and
me, it was magnificently right for Yeats: it made his last poetry the
fulfillment of his whole life, it made him write about our times as no
other poet has."

Jarrell is right that Yeats's "crazy" system was "magnificently
right" for him, and this is as true of Yeats's political poems as it is of
his mystical ones. He turned out poetry of a raw, new beauty, and
Ellmann's estimation of Yeats as the most important poet since
Wordsworth is a fair one. But does it matter whether the systems that
spawned Yeats's poetry were violent or dishonest, or whether he
believed in any of them at all?

Yeats used to say that a poet should be able "to say he believes in
marriage in the morning, and that he does not in the evening." To the
extent that he is merely guarding himself against those who would

read a poem as if it were a political manifesto, that view deserves to be defended. But it can't be ignored that Yeats's poetic vision was dangerously nihilistic in some respects, leading him into a cul-de-sac of fascism and paganism. He was one of many poets of the time—Pound, Rilke, and Eliot among them—who wound up fashioning beautiful poetry at the expense of poetry itself.

Sometimes he made his poetry dependent on an ideology that would make it too hot for future generations of poets to handle; sometimes he contrived a poetic voice that (as in much of the work Foster focuses on in this first volume) seems almost willfully fraudulent. Either way, poetry, always a fragile art, was unlikely to long survive having such additional burdens placed on it.

As, indeed, it has not.

# WHAT WORLD WAR II
# WAS FOUGHT FOR

## By David Tell

————

*(Originally published, as "Americans at War,"*
*in the December 29, 1997–January 5, 1998, issue)*

TODAY IT IS WIDELY UNDERSTOOD that, for front-line troops, the fundamental experience of warfare is fear—fear of an intensity that seems nearly beyond human endurance. And yet, as often as not, soldiers manage to endure it. So the obvious questions are how they do it, and why.

These have become central questions of combat only recently. Until the second half of the nineteenth century, Western military science taught that the leadership of generals and the collective discipline they imposed on entire armies were what determined combat effectiveness. It was Charles Ardant du Picq, a French infantry officer, who first seriously argued that armed engagement is inevitably sealed by an outbreak of existential panic among one side's privates—regardless of the training and direction they have received from above. Individuals in the rank and file are "the first weapon of battle," his 1871 treatise *Battle Studies* proclaimed. "Let us then study the soldier in battle, for it is he who brings reality to it."

But another seventy years went by before anyone thought to pursue such a study in systematic, eyewitness fashion. Throughout World War II, the American War Department, eager to discover how troops might overcome the terror of combat, sent Information and Historical Service teams into every theater of the war. The researchers collected mountains of paper evidence and conducted extensive oral interviews with the troops. Then, shortly after the war, they wrote a pioneering work of quantitative sociology, based on after-action questionnaires completed by the GIs themselves.

This study, *The American Soldier*, concluded that once the shooting starts, a man on the front line quickly restricts his mental energies to simple self-preservation. That soldier's sense of mission only rarely extends further than his foxhole buddies, who, he is convinced, represent his only available and dependable support. Larger war aims and patriotism matter little to him. Indeed, at the front there is "a taboo against any talk of a flag-waving variety."

This view of combat attitudes and behavior was given most extreme expression by S. L. A. Marshall, the Army's highest-ranking historian during the war. Marshall's *Men Against Fire*, published in 1947, reported that their instinctive recoil from violence was so powerful that at least 75 percent of American GIs in Europe and Asia could not bring themselves to use their rifles even once. Those few who did shoot back, the book insisted, were motivated exclusively by feeling for the men in their immediate company: "I hold it to be one of the simplest truths of war that the thing which enables an infantry soldier to keep going with his weapon is the near presence or the presumed presence of a comrade."

*Men Against Fire* was profoundly influential. Its lesson about the primacy of small-group dynamics ("unit cohesion," in contemporary parlance) was quickly and permanently absorbed into the tactical doctrine of every Western military establishment. And its portrait of the American combat soldier—bonded by fright to his platoon; heedless, even scornful, of cause and flag—assumed an unshakable place in the scholarly literature on the Second World War.

All of which, of course, is wildly paradoxical. American fighting units in World War II could not have been particularly "cohesive": the turnover in their membership was too rapid. Thirty-seven U.S. divisions spent at least a hundred days in European combat, and more than half of them suffered losses, counting replacements, that exceeded their original strength. The 4th Infantry had a 252 percent casualty rate. The average lifespan of an American platoon leader in Europe, from the moment he took command, was thirty days. You can't cohere with a dead man.

If, just for the sake of argument, these units were somehow cohesive, and cohesion was so vital a battlefield motivation, how is it that only a small fraction of GIs could summon the will to fire their weapons? If, for that matter, the vast majority of GIs were routinely too paralyzed by dread to help out in combat, how did the Allies ultimately prevail? Either way, can it really be true that all these GIs served their time insensible to the fact that something even more important than their own lives was at stake—in this, the most fateful and least morally complicated of wars? And does not this last suggestion dishonor the dead, the survivors, and the country in which they were born?

Yes, it does. American historians have been ignoring or evading this problem for decades. They are ignoring and evading it still.

Stephen E. Ambrose, one of America's leading popular historians of World War II, has recently published *Citizen Soldiers*. His 1994 book, *D-Day*, was an account of the Normandy campaign's first twenty-four hours. This latest work, a sequel, pushes forward to V-E Day, eleven months later. Ambrose again relies upon memoirs, oral histories, and interviews with the junior officers and enlisted men who did the fighting. And he claims to be addressing the same issues raised in his earlier work: "Who they were, how they fought, why they fought, what they endured, how they triumphed."

But except where the soldiers' endurance is concerned, Ambrose doesn't really have much new to say. He is a storyteller, not an original analyst. And the story he tells—though in the voice of the grunts—is consistent with the general conclusion of previous histories that focused on officers in the rear: It was principally the American logistical achievement that defeated Germany in the Second World War, the overwhelming weaponry and manpower we were able to send across the ocean and into the enemy's lines.

Six weeks after D-Day, for example, the German front in northwest France was held by General Fritz Bayerlein's Panzer Lehr Division, spread along the N-800 highway between St.-Lô and Périers. On the morning of July 25, Panzer Lehr was attacked for twenty minutes by 550 American fighter bombers. Then it was attacked for an hour by eighteen hundred B-17s. Then it was attacked for another hour by a thousand U.S. ground guns while 350 P-47s dropped napalm and 396 Marauders did the mopping up. In all, sixteen thousand tons of explosives were released on a target twelve kilometers square. Panzer Lehr was obliterated. The next day, American troops began their unimpeded race east to the Seine.

By the end of the Normandy campaign, Germany had lost more than four hundred thousand men. It had recovered only twenty-four of the fifteen hundred tanks it had thrown into battle. Its fewer than six hundred remaining aircraft faced an Allied fleet of fourteen thousand. By the time the U.S. First Army captured the Ludendorff Bridge at Remagen in March 1945, and Americans began pouring into Germany, Allied pilots were flying eleven thousand sorties a day, and the Luftwaffe had ceased to function. The war ended two months later.

But between Normandy and the Rhine had come the winter of 1944–45, and it is in Ambrose's gripping coverage of these months that his book makes an invaluable contribution. War is hell, we know to say, but the hell we have in mind is usually a Gettysburg photograph by Matthew Brady or the mud and idiocy of the Somme in 1916. The suffering of American combat forces before their epochal triumph in World War II has assumed a casually stipulated—and consequently disrespectful—quality in our popular imagination. Already a bestseller, *Citizen Soldiers* cannot help but restore the grim record.

They did not fight at night during the Civil War and routinely spent weeks marching or camping unmolested. Months sometimes went by without incident in World War I, and even during fire, Allied battalions were removed from the front for rest on regular sixteen-day rotations. Soldiers during World War II, by contrast, fought at the front until they were wounded or killed. They fought round the clock, on maybe two hours sleep, for as long as sixty days at a time.

In the Hurtgen Forest and the Ardennes, there was winter light only eight hours each day. The weather was the coldest in fifty years, frequently well below zero. The GIs had neither warm clothes nor snow boots. Trenchfoot took mens' toes, frostbite took their fingers, and thousands of German "Bouncing Betties"—mines that sprang two and a half feet in the air, spraying a curtain of razor-sharp scrap metal—took their genitals. American foxholes and command posts, tenuously established in frozen mud, were subjected to continual artillery bombardments whose concussive force alone could break a soldier's bones and whose noise was enough to send blood streaming from his ears. When he was able to march on actual roads, his feet slipped on the slime of dead bodies crushed by tanks. GIs wet themselves or wailed for their mothers or vomited from fear. One-fourth of all U.S. battlefield evacuations in Europe were for cases of nervous collapse.

How, then, did they endure such stress and continue fighting? Was S. L. A. Marshall right or wrong? Ambrose circles around this question, awkwardly and quickly, in both his introduction and epilogue, and winds up answering it both ways. He reveres the GIs too much to accept the obviously insulting judgment that they operated without any consideration for national objectives and ideals. "At the core," Ambrose writes, our troops were patriots; it's just that they were modernists, too—uncomfortable with public displays of passion and "embarrassed by patriotic bombast" about the war from combat-ignorant home folks.

At the same time, Ambrose is clearly daunted by the ironclad historical consensus about his "citizen soldiers." "In general," he concludes, "in assessing the motivation of the GIs, there is agreement that patriotism or any other form of idealism had little if anything to do with it. The GIs fought because they had to. What held them together was not country and flag, but unit cohesion. It has been my experience, through four decades of interviewing ex-GIs, that such generalizations are true enough."

In other words: Our troops held together because they held together, not because they were patriotic. And they were patriotic. This is an explanation that doesn't explain.

If Ambrose's book is ultimately unsatisfying, however, Gerald F. Linderman's latest work is ultimately unacceptable. In *The World Within War*, Linderman, a history professor at the University of Michigan, announces the same purpose as Ambrose—wishing to "see World War II through the eyes of those American combat soldiers." But he has done it upside down and backward. *World Within War* is a work of pure theory—viewing American combat soldiers through the eyes of an elaborate, prefabricated diagnosis of warfare's general psychosocial effect on the individual. The book watches American combat soldiers at such close quarters it winds up going blind.

The GIs enlisted, Linderman reports, eager to fight, confident of their prowess, and certain of their personal invulnerability. Once they had seen combat, however, they were first amazed that anyone might mean them harm, and later silent, tense, and narrowly obsessed with their own security.

Standing alone, these observations are commonplace enough, and unobjectionable. But Linderman is not finished. He describes battle-hardened vets as almost feral. Stories of eerily fulfilled foreboding—a soldier is overwhelmed by the objectively inexplicable notion that a bomb is about to drop *right there*, and then it does—are ubiquitous in the literature of warfare. But Linderman treats his World War II examples with alarming seriousness, as though some GIs might literally have become animals of instinct. Bomber pilot John Muirhead, Linderman recounts, once broke radio silence over Italy to say "Group Leader, I smell flak." "Yes, I smell it, too," came the reply. Then flak actually appeared.

And the GIs' transformation into brutes, in Linderman's account, grew deeper still. They became fully inured to the presence of death. Frank Mathias, whom Linderman identifies as an "Army machine gunner," sat eating K-rations next to a Philippine-island ditch filled with Japanese corpses. "I absentmindedly watched bubbles of gas and liquid moving around under their tightly stretched skins as I munched my crackers," Linderman quotes from Mathias's memoir. "The June sunlight was bright and hot. They were in their world and I was in mine. I had to eat, didn't I?" One half suspects this story is a piece of super-macho apocrypha; Mathias's memoir, a buried footnote indicates, is called *GI Jive: An Army Bandsman in World War II*. But it suits Linderman's purpose.

That purpose seems finally to be a denial that the Second World War contained *any* value or meaning for its soldiers. Even the legendary recourse to "unit cohesion" was at some point abandoned by these troops, Linderman says. After over a hundred days on Guadalcanal, a

Marine corporal wrote his dad: "My best buddie . . . was caught in the face by a full blast of machine gun fire and when the hole we were laying in became swamped by flies gathering about him and [he] being already dead, I had to roll him out of the small hole on top of the open ground and the dirty SOBs kept shooting him full of holes. Well anyway God spared my life and I am thankful for it." From such ambiguous evidence Linderman concludes that long service left GIs "bereft of any broad emotional support that might have checked some of combat's denaturant effects." In combat, at bottom, American troops could find "nothing of sufficient worth to justify their presence."

This comes perilously close to a claim that America's overall presence in World War II was unjustified. Gerald Linderman deeply pities the American soldier's plight. But he does not much respect the work they did; indeed, he appears revolted by it. This is a sin. They were working to destroy Hitler.

And they were working pretty well. Trucks and planes and gasoline and explosive shells might have tipped the scale in World War II, but the conflict remained always one for territory, which ground troops alone could capture. The GIs on whom this mission fell were constantly frightened, of course. But they were never immobilized by this fear, not in any dangerous numbers. Marshall's famous statistical contention that 75 percent of American troops never fired their weapons was thoroughly debunked in 1989 when Roger Spiller of the Army's Command and General Staff College at Fort Leavenworth went public with the results of his research among Marshall's private papers and official records. The old man had never conducted company-level "fire-ratio" interviews; he had made the whole thing up, as his wartime assistant later acknowledged.

Why it took forty years for someone to figure this out is a mystery. The historical record is quite clear: The GIs were excellent soldiers. During the Battle of the Bulge, two thousand of them from the 28th Division held off a ten-thousand-man Panzer force for an entire day before falling back a few miles to regroup. Fifteen German divisions attacked Bastogne during the same battle. The town was successfully defended by a single American division, the 101st Airborne, and parts of another.

More important, the historical record is dotted with intriguing pieces of evidence that contradict the other half of received sociological wisdom about the war—that circular business about "unit cohesion" which portrays the GIs as bored by or deaf to or even contemptuous of the great civilizational crisis whose vortex they occupied. True enough, American soldiers had little time for fancy philosophical pronouncements. Their deeds, they surely felt, were sufficiently eloquent. At the war's conclusion, Dwight Eisenhower issued a one-sentence statement from headquarters: "The mission of this Allied force was fulfilled at 0241 local time, May 7, 1945."

But further down the line, the privates and corporals were sometimes more explicit about their overriding convictions and motivation. In the rivers of mail they sent back home to the States, full of unselfconscious tributes to America as "the best country on this Earth." In their mumbled reactions to the German slave-labor camps they liberated. In a thousand other places, most likely. Gerald Linderman cannot bring himself to believe in such transparent patriotism. Stephen Ambrose, who might be expected to seize upon it, hesitates instead, and looks no further.

Linderman tells the story of a woman on a Kentucky public bus who was making good money off the wartime economy and was overheard to say she hoped the fighting might continue at least until she'd paid off her refrigerator. An old man on the bus became enraged. "How dare you!" he bellowed, and began crowning her on the head with his umbrella. No other passenger intervened. Eventually, someone with a spirit like that old man's will write a history of the GIs in World War II. And then, finally, the citizen soldiers will get their due.

# UPDATING GEORGE WASHINGTON

## By Andrew Ferguson

*(Originally published in the April 5–12, 1999, issue)*

THE REGENT OF THE MOUNT VERNON LADIES ASSOCIATION is a woman named Mrs. Robert E. Lee IV—which I think is the most satisfying piece of information anyone could ever hope to come across. Notwithstanding her title, however, and the name of her organization, and indeed the powerful resonance of her own married name, Mrs. Lee is a thoroughly modern woman who favors elegant jacket-and-pants ensembles and brightly colored turtlenecks and outsized eyeglasses that are, curiously, shaded violet.

I met her not long ago, in a tastefully appointed pastel sitting room in the association's administrative building. The offices are hidden from public view behind a towering hedgerow several hundred yards beyond George Washington's mansion, which sits in turn on a bluff on the Virginia side of the Potomac River, sixteen miles south of the nation's capital.

This has been a busy year for Mrs. Lee and the Ladies, who are chartered under the Commonwealth of Virginia as proprietors of our first president's estate. It is the 200th anniversary of Washington's death, and to mark the occasion the Ladies have embarked on a flurry of activities far surpassing anything they have ever undertaken before. They have launched no fewer than three touring exhibits, renovated the estate's museum and restored Washington's tomb, installed a "multimedia mood theater" that dramatizes the moment of Washington's death, and added new thematic tours of the grounds. For the first time professional "re-enactors" have been hired, to roam the estate in period costume and give little lectures about everyday 18th-century planta-tion life to inquisitive tourists. More than one hundred Washington artifacts have been borrowed from collections across the country and placed in the house itself, so that, as the press release says, "the estate will resemble as never before the beloved private retreat that Washington knew at his death two hundred years ago."

So now seemed as good a time as any to ask Mrs. Lee about the teeth.

Or dentures, rather—George Washington's dentures, which are the centerpiece of one of the traveling exhibits, "Treasures from Mount Vernon: George Washington Revealed," now at the Huntington Library in San Marino, California, and soon to travel to Atlanta, Richmond, and Chicago. The Ladies acquired the dentures from a descendant of Martha Washington in 1949. For fifty years, however, they have declined to put them on display, for a reason that explains a lot about the Ladies.

"It just seemed so personal, so—I don't know—so private," said Mrs. Lee, when I brought up the subject. She fingered her gold neck-lace. "Washington was a very dignified man, you see. Very proper, very reserved. I can tell you he would not have approved of having his den-tures on display."

So why now?

"It was something we thought long and hard about, I can assure you. And there was some resistance, and it was something we had to agree on as a group. But"—and here Mrs. Lee leaned forward from her wingback chair, suddenly animated—"we are in a crisis! The man is fading from the history books! People come here and know so very lit-tle about him—very nice people, hard-working people who bring their children here, trying to teach them about history. But there is a huge lack of knowledge about what an incredible man he was.

"You should see the children flock around the dentures. 'Wow,' they say. It lets them see Washington as a man. It makes it relevant and fun for them, and then maybe they'll want to know more, do you see? George Washington has always been the example for all citizens to emulate, the man who embodied our Founding Principles, and maybe this will help us get that across."

Mrs. Lee has resigned herself to doing whatever it takes to make George Washington fun. She went on: "And now sometimes I think, 'Well, maybe he wouldn't mind the dentures.' I think, I really do, that he would approve of us doing what we needed to do so that we can make this the country we should be proud of—a country that needs him."

Mrs. Lee looked aside for a moment. "There's so much to do, to make people realize the essence of the man," she said. "And no one else is going to do that. It's our responsibility. If we don't do it, who else will?"

Washington has not, as Mrs. Lee claims, faded from the history books, not entirely anyway, but he has receded over the past generations—a remote figure who grows ever more distant. "We impute coldness to him," Richard Brookhiser, a recent biographer, wrote, "and we respond to him coldly." The Ladies now endeavor to correct the misapprehension. The association is 146 years old. In many respects it remains an artifact more of the 19th century than the 20th, much less the 21st, and there's something bittersweet in the Ladies' attempt "to meet people where they are," to use the cant phrase of the day. For where people are now is a long way from where the Ladies have been, by tradition and resolve. Their effort to revive Washington, two hundred years after his death, tells us something about them, and something about him, and even more about ourselves.

Even before he died his home was a place of pilgrimage. Washington was, of course, the most famous man in America, certainly the most revered, and the richest, too. He had inherited Mount Vernon from the widow of his half-brother Lawrence and had steadily expanded both the house and the grounds, until the property stretched across 8,000 acres, down the Potomac River and back deep into the Virginia woodlands and then up again to the southern tip of Alexandria. At his death it comprised five separate, self-sufficient farms, numberless outbuildings, housing for three hundred slaves, and a twenty-two-room house designed by Washington himself, resting on a promontory above a bend in the river.

He was never without visitors—a steady stream of fellow politicians, foreign dignitaries, old friends, and unknown well-wishers who felt compelled to see the great man in the flesh and whom the great man, in his hospitality, felt obliged to entertain. When Washington died, followed by his wife two years later, the stream of visitors swelled to a flood. The bulk of the property—without the slaves, who had been freed in Washington's will—passed to a series of nephews with little interest or skill in farming. By the 1850s, the crowds of travelers and pilgrims had brought the estate's owner, John Augustine Washington, to the edge of bankruptcy. The farm was a shambles and the house close to ruins. He cast about for a buyer, with an asking price of $200,000, but was turned down by both the federal government and the Commonwealth of Virginia. In a desperate moment, he considered an offer of $300,000 from an entrepreneur who proposed to turn the estate into a roadside attraction. He couldn't bring himself to say yes.

The Ladies rescued Mount Vernon. At first, though, there was only one lady—an invalid from South Carolina called Ann Pamela Cunningham. Alarmed by a letter from her mother, who on a trip north in 1852 had stopped at Mount Vernon and noted the waist-high weeds and the peeling paint, the fallen shutters and the collapsing portico, Miss Cunningham resolved that Mount Vernon would be saved, by private subscription if necessary. She composed a series of open letters "To The Ladies of the South" and posted them to newspapers from Richmond to Savannah. The thought of personal publicity horrified her, so she signed her first letters "A Southern Matron." But her ardor for the cause was unmistakable, and infectious. If the men of America would not do their duty, she wrote, then their wives and daughters would do it, for the sake of posterity:

> *While it would save American honor from a blot in the eyes of a gazing world, it would furnish a shrine where at least the mothers of the land and their innocent children might make their offering to the cause of the greatness, goodness, and prosperity of their country!* [Exclamation, of course, in the original; one of many.]

Miss Cunningham was indefatigable despite her infirmity. Over the next several years her campaign spread northward, and the coffers swelled. Shortly before the Civil War, she was able to present John Augustine with a substantial down payment on the $200,000 purchase price, the balance to be paid off in three years. He moved out on Washington's birthday, February 22, 1860, emptying the decaying house of all its furnishings except for the famous Houdon bust of Washington, a terrestrial globe the president had used in his New York office, and the key to the Bastille, presented to George Washington by Lafayette. The estate he ceded to the Ladies had dwindled to 500 acres, including the mansion where Washington had lived and died.

Originally there were twenty-two ladies, or vice regents as they are called, one from each of the twenty-two states. Today there are thirty-three, a number limited by the lodging space available on the Mount Vernon grounds, where the Ladies gather several times a year for four-day meetings to do the association's business. The association is self-perpetuating, which is to say that when one Lady retires the other Ladies choose her successor. Early on Miss Cunningham stipulated that a vice regent "should be of a family whose social position would command the confidence of the State, and enable her to enlist the aid of persons of widest influence." The tradition holds—particularly with regard to "influence." For a vice regent's duties include preeminently the raising of funds. The association's literature boasts that it has never taken a dime in government money; it has relied instead on the kindness of the very best strangers. When the Ladies decided to wire the house for electricity, for example, Thomas Edison did the job. When they thought it was time to get a fire engine, Henry Ford had one built and offered it free of charge.

Any such modernizing steps, however, have always been undertaken only after the most careful consideration. When Miss Cunningham retired as regent, in 1874, the weeds had been cut and the gardens restored, the house had been painted and partly, but painstakingly, refurnished, and the crowds continued to pass through the gates. And yet she worried for the future. Her farewell message is still read aloud when the Ladies meet for their annual Grand Council.

*Ladies, the home of Washington is in your charge—see to it that you keep it the home of Washington! Let no irreverent hand change it; let*

*no vandal hands desecrate it with the fingers of 'progress'! Those who
go to the home in which he lived and died wish to see in what he lived
and died. Let one spot, in this grand country of ours, be saved from
change!*

This was the Ladies' sacred charge—they refer to it as a sacred
charge—and it impressed itself on every aspect of their work for a century and more. With the help of the developing sciences of forensics and
archaeology, the restoration of the gardens and the house proceeded as
accurately as possible. When verisimilitude conflicted with comfort
and questions of taste, however, the Ladies favored delicacy. There were
no chamber pots in the bedrooms, no slag heaps outside the kitchen, no
pig dung littering the service roads, as there would have been, of course,
at the estate "where Washington lived and died." The Ladies created an
idealized, pristine version of Washington's home, for their intent was
not so much to instruct as to uplift. Mount Vernon was a shrine, a place
of pilgrimage. The task was appropriate to the times and to the visitors
who made the trip; their familiarity with, their reverence for, the Father
of their country was simply assumed. His virtues—of self-denial, sacrifice, patriotism, disinterestedness—were the virtues that every
American was thought to aspire to. To know the story and character of
Washington was part of what it was to be an American.

Even before he died, the popular view of Washington was elevated
far beyond anything we can imagine today. He was shrouded in religious imagery. Comparisons to Moses and Jonah were common. The
custom only intensified in the century that followed. In the 1800s his
biographers routinely capitalized the personal pronoun "Him" in
referring to their subject. The chapel at Valley Forge, built at the end
of the 19th century, dedicated one wall to stained glass tableaux from
the life of Jesus, the wall opposite to the life of Washington. In time
the religious elements fell away, but the veneration continued undi-
minished. The more secular fables of Parson Weems—inventor of the
cherry tree—passed into the *McGuffey Readers* and then into the imagination of every schoolchild. "The name of Washington," wrote Walt
Whitman, "is constantly on our lips. His portrait hangs on every wall
and he is almost canonized in the affections of our people."

Well into our own century, Washington's memory was kept alive by countless commemorations. His birthday was celebrated as a national holiday, marked by parades and fêtes and speeches and balls, and his portrait hung, if not, as Whitman observed, on every wall, at least in every classroom, staring down from above the chalkboard like a stern and unsleeping principal. His Farewell Address was read annually in special sessions to both houses of Congress. Mount Vernon was the place where this spirit of veneration could be imbibed most directly, and by the mid-1960s attendance had reached 1.3 million a year. Miss Cunningham's plan seemed to work. She had insisted that Mount Vernon never change, and it didn't. But the times did.

James Rees well remembers the moment when he knew he had a problem. Rees is the resident director of the Mount Vernon Ladies Association, the highest ranking non-Lady on the premises. Rees came to Mount Vernon in the mid-1980s. By then annual attendance had begun, for the first time in memory, to dip below one million. On this day a few years ago, he recalls, two dozen or so fourth graders had gathered on the estate grounds for a tree-planting ceremony.

"And I started making jokes," he told me recently, "you know, playing off some of the Washington myths. I said, 'Well, it's a good thing this isn't a cherry tree, or it might be in danger—you never know who might come chop it down.' And there was no reaction. Nothing. So I said, 'But I guess we could always use the wood to make some teeth.' Nothing. Blank stares.

"Now, I knew these kids' teacher—a very bright woman. In fact, she's a descendant of George Washington. And these were not dumb kids. These were kids from a privileged background. But it suddenly occurred to me: These kids don't even know the myths. We're past the debunking stage. You can't debunk misconceptions when they've got no information at all. I thought: We're all the way back at ground zero."

And so we are. Every so often the federal government's National Assessment of Educational Progress releases a report on the ignorance of American schoolchildren, provoking near-universal tut-tutting among educators and in the popular press. In 1996 NAEP found that only 17 percent of fourth graders were "proficient" in American history. The older they got, the dumber they got. Fourteen percent of eighth graders were proficient, 11 percent of twelfth graders.

But of course we knew that. What's interesting is the particular form this ignorance takes. Among fourth graders, for example, 87 percent could identify Martin Luther King's "I Have a Dream" speech; fewer than half could identify the Bill of Rights. And fewer than one in three knew that New York was one of the original thirteen colonies, while seven out of ten listed California, Texas, or Illinois. In the eighth grade, 80 percent identified the song "O Freedom" with the civil rights movement; only 41 percent associated the dropping of the atomic bomb with the end of World War II.

People like to complain that we live in an iconoclastic age. We don't. In fact, we're so overrun with icons that the word itself has become a cliché. But we've substituted a new set of icons for the old, and the strange imbalance in the historical knowledge of American students reflects the substitution. Given the multicultural enthusiasms of their teachers, it should be no surprise that children know more about, say, Harriet Tubman than Tom Paine. Black History Month—a sound idea pursued with unusual zeal in the public schools—has pretty much swallowed up the social studies curriculum for the month of February (social studies being the rubric under which "American history" is taught, when it is taught at all). Meanwhile, Washington's birthday, once universally observed in the schools and used as an excuse to dwell on the Founding Fathers, has been bundled with Lincoln's in the portmanteau "Presidents' Day." Falling within a month devoted to a celebration of African-American history, President's Day is more often than not merely an occasion to teach the young scholars that Washington owned slaves and Lincoln freed them.

Textbooks aren't much help. "I still have the history textbook I used in the fourth grade," James Rees says, "and it has ten times more pages devoted to Washington than the textbook used in the same class at the same school today." In current textbooks Washington has not been traduced so much as passed over—not ignored, exactly, but placed off to one side, like an old piece of cumbersome statuary that one can't quite bear to part with, out of some dimly felt obligation.

The recent high school textbook *United States History: In the Course of Human Events*, published by West Publishing, is a case in point.

As the historian Walter McDougall noted in a recent review, the authors of *United States History* choose to divide their subject into eleven units. Three of these recount the three hundred years from the settling of America through the colonial period and the constitutional convention to the end of the 18th century; four units cover the fifty years from the end of World War II to the present. And Washington is there, sure enough, with his very own page—as one of the book's more than 120 "People Who Made a Difference." Unlike the wholly sympathetic treatment given the other PWMD—such as Frederick Douglass, of course, and "Mother Jones," and the Japanese-American activist Gordon Hirabayashi—the view of Washington is mixed. He was a man of "ordinary talents," the students learn, "not completely successful as a military man nor as a president."

"When the Revolution succeeded," the authors write,

> *[Americans] felt justified in their choice of a leader. Praise for Washington was partly a kind of self-congratulation for their own brilliance in choosing a president who would lead them to success. In fact, it might be said that the idea of George Washington, not always the man himself, was what counted.*

Unlike other Founders, Washington has never been subjected to a successful debunking. He is undebunkable. But he is dismissible, as the work of academic historians over the past eighty years makes plain. They have sought to understand the Founding in ways that make no room for Washington's particular greatness. In *An Economic Interpretation of the Constitution of the United States* (1913), the most influential work of American history in the first half of this century, Charles Beard undertook the first mass debunking, casting the Framers as reactionary capitalists intent on insulating their riches from the grasping proles. Washington is scarcely mentioned in Beard's book. "George Washington's part in the proceedings of the convention," Beard wrote, "was almost negligible"—an odd statement about the man who was, after all, the convention's presiding officer, and without whom there might not have been any convention at all.

But Beard goes on, revealingly: "It does not appear that in public document or private letter he ever set forth any coherent theory of government. When he had occasion to dwell upon the nature of the new system he had indulged in the general language of the bench rather than that of the penetrating observer."

Richard Hofstadter's *The American Political Tradition and the Men Who Made It* was to the second half of the century what Beard's book had been to the first—the work that set the course for two subsequent generations of historians. Like Beard, Hofstadter passes over Washington in his account of the Founding. Hofstadter was a historian of ideas. And Washington was not, as Beard noted, a man of ideas, certainly not a thinker of the sort that historians nowadays favor. He lacked the incendiary brilliance of Jefferson, the sophistication of Madison and Hamilton, the rhetorical imagination of Thomas Paine. For modern historians, these qualities have proved much more attractive, and worthy of study, than the stolid, tenacious statesmanship of Washington.

The historian Paul Longmore put it well: "His gift was not the formulation of ideas, but their incarnation." Washington was indispensable to the country's Founding—became, indeed, the rock on which the country was built—because of his ability to unite his bickering countrymen, to still their passions by his very presence and resolve their disputes with utter disinterest, to embody their highest aspiration; in short, because of his character. How are intellectuals to grapple with such a man? Viewed a certain way, he seems almost uninteresting. Incorruptible, fearless, impatient with abstraction, sometimes prosaic, he falls outside the categories that professional historians have lately used to account for our past. And so he recedes, taking his place as one among the many "people who made a difference," filling the ranks somewhere between Mother Jones and Gordon Hirabayashi.

Sitting in his sunny office at Mount Vernon on a recent late-winter morning, I asked James Rees about this steady diminishment of Washington. How does he explain it?

"I suppose it has to do with lots of things," he said. "The rise of social history—filling up history with all kinds of people who'd been ignored before means there's less room for old heroes. And I suppose it has to do with the end of the great man theory of history, too. Lots of things.

"But there's something else that worries me. The qualities Washington possessed just aren't as appreciated as they were. Honesty. Good judgment. Modesty—my God, who in late-20th-century America gets credit for being modest anymore? And believe me, this is not good.

"There's this idea that leadership is changeable—that every generation redefines for itself what leadership means. Well, that's not the way we thought of it for most of our history. The qualities that made a great leader then were good for all time. But we don't think that way anymore. It's just this"—he sighed—"this whole 20th-century mindset." And with a wave of his hand he tried to dismiss the depredations of the span of a hundred years.

But he is not always so fatalistic. Rees and the Ladies understand that, as we move into the 21st century, it's the least they can do to reconcile themselves to the 20th. Earlier generations didn't demand that their heroes be "humanized." Ours, however, demands intimacy and craves familiarity. The Ladies may not be happy with the vulgarity that this sometimes requires, but they are resolute, as I say, in their determination to make Washington "fun" and "relevant."

They have added a team of media and marketing specialists to their skeletal staff. Among their innovations are two new Web sites, in which you can take a "Pioneer Farmer Quiz" and "Meet the Mount Vernon Animals." They've issued a CD-ROM about Mount Vernon with the unpleasant title "Dig Into George." A special program this spring will demonstrate that Washington was a proto-environmentalist. The mansion rooms have been refashioned to make a more "immediate experience"; in the bedroom where Washington died, the tools doctors used to bleed him rest on the bed, bloody towels are wadded on the floor, and a pan filled with theater blood sits on the nightstand. When they pass through Mount Vernon's small museum, visitors have their attention directed to a pair of Washington's oddly shaped violet sunglasses. "It's like he was a punk rocker!" a docent told me excitedly. The "multimedia mood theater" mentioned earlier, designed by a

British firm responsible for the "Vikingland" amusement park in Norway, is dazzling in its high-tech simulation of Washington's death. When I saw it the other day, with a class of fourth graders, it was a big hit. "Spooook-eeee!" one of the young historians shouted in the dark. Almost as good as *Armageddon*.

It is all intended to create at Mount Vernon an "all new Mansion experience," as the PR materials say, and it can easily be made to sound much worse than it is. The *New York Times*, for example, wrote in a front page story in February: "The directors of Mount Vernon . . . have inaugurated a $3 million public relations campaign to reposition [Washington] as a national figure with what the spinmeisters call 'heat.' Think Leonardo DiCaprio, Diana and Elvis Presley." The upshot was that Mount Vernon had at last got down with the slammin' nineties: another info-entertainment option among the dozens on offer in the Capital region—a little bit Williamsburg, a little bit Busch Gardens.

The over-hyped *Times* story was greeted with chagrin at Mount Vernon, and staffers, understandably defensive, hasten to correct its misimpressions. "We are always very conscious of going too far," Mrs. Lee told me.

But still, I said, with all the re-enactors walking around, and the special multimedia programs, don't you worry about getting trapped in the show biz?

She fixed me with a stare that could have come straight from Miss Cunningham. "We don't want to do that," she said. "We will never do that. But you have to understand, Mount Vernon can no longer be just a shrine. I don't even like that word, shrine. We have to get the children interested before they can learn about George Washington. And this is the most important thing: They must learn about this man. They need to know why he was great. If they don't—what will happen? *We need him*."

After I left Mrs. Lee, I wandered out to the piazza, the great porch with its surpassingly beautiful view of the Potomac, where Washington had entertained Jefferson and Adams and Lafayette. I sat on a Windsor chair beneath Washington's bedroom window—the bedroom where he died, two hundred years ago—and leafed through a packet the PR lady had given me. Thousands of the packets have been sent to fifth-grade teachers around the country; the Ladies have worked hard to assemble a mailing list of social studies departments, in hopes of pushing them to teach their kids about the father of their country. As you'd expect, it is loaded with gimmicks—an envelope of wheat seeds, so students can grow a crop just as "George" did; a sheet of stickers with the legend "Ask Me About George Washington!"; scratch-off quiz cards like the kind you get from McDonald's or the Lotto dealer; and an encouragement to the kids to "write a letter to George Washington and get a reply from Mount Vernon." And there's a poster, too, colorful and splashy, to replace the portrait that not so long ago hung in every classroom. To my inexpert eyes, the lesson plan looked professional, well-organized, and, for some reason, a little sad.

And I suddenly realized why. Washington has been privatized! He has been detached from the national patrimony—if we can be said to have a national patrimony any longer. And the Ladies have become a special-interest group, pleading a pet cause, just as NOW agitated for a Susan B. Anthony dollar and Indian rights groups lobbied to put Sacagawea on a postage stamp. It would take someone with more nerve than I to challenge the Ladies as they struggle, however clumsily, to return Washington to his rightful place, at the center of our historical memory. And if they are forced to use the tools of a time that finds them and their passions anachronistic, well then, they will. The Ladies do what they do because they have asked themselves an unsettling question, and because they know the answer. "If we don't do it," Mrs. Lee had said, "who will?"

# THE GREAT WAR
# AND ITS HISTORIANS

## By David Frum

*(Originally published, as "The Historians' War:
The Lessons of 1914," in the June 21, 1999, issue)*

IN THE FINAL ELEVEN YEARS of the twentieth century, time seems to have run backwards. The Red Army withdrew from central Europe, rescinding 1945. A dictatorship fell in Berlin, undoing 1933. Statues of Lenin toppled across Russia, annulling 1917. War in the Balkans was the first horror we passed on our way into the century, and it is the final horror we are passing on our way out. After nine blood-soaked decades, Europe has at last laboriously reestablished a continent-wide order nearly as enlightened, decent, and free as the one that prevailed in July 1914.

Only if we decide what lessons to learn from the First World War are we right to hope that the awfulness of the past century forms merely a detour and not an eternally recurring pattern in European history, for that war and its consequences can never really be left behind. And here—in time for the eighty-fifth anniversary of the assassination of the Archduke Franz Ferdinand and the eightieth anniversary of the signing of the Treaty of Versailles, both on June 28—are two of the most important books in many years about the war and its aftermath.

In *The First World War*, the magisterial English military historian John Keegan—author of such classics as 1976's *The Face of Battle*—writes of the war precisely as a *war*: uniquely horrible, but still intelligible in the same way that the Napoleonic Wars and the Second World

War are intelligible. In *The Pity of War*, the younger Scottish historian Niall Ferguson instead presents the war as a catastrophic caesura in world history, a calamity whose strategic and tactical aspects are perhaps the least interesting thing about it.

World War I began—as we have had cause to be frequently reminded in recent months—in a quarrel between Serbia and the Habsburgs' Austro-Hungarian Empire over Bosnia. Austria had it; Serbia wanted it. When the heir to the Habsburg throne announced he would visit Sarajevo, the capital of Bosnia, in June 1914, five young Bosnian Serbs decided to murder him.

The fatal shots were fired on the intermittently famous anniversary of the 1389 Battle of Kosovo. By torturing the plotters, the Austrians quickly discovered the involvement of the Serbian military, if not the government of Serbia, in the assassination. The Austrian government presented Serbia with a stiff set of punitive demands, while the Austrian army drew up invasion plans.

Had the Austrians struck immediately, there would probably have been no wider war: Everyone in Europe more or less agreed that the Serbs deserved what was coming to them. But the Austrians hesitated for four weeks, during which the great powers of Europe became convinced their interests were implicated in the Serbian-Austrian confrontation.

Russia supported Serbia for fear that Austria would extend its empire deeper into the Balkans. Germany backed Austria for fear that its only friend in Europe would otherwise lose a war to Russia. France joined Russia for fear that Germany and Austria would defeat its main ally. On August 1, Germany, Austria, France, and Russia all mobilized. On August 4, German troops entered Belgium, and Britain entered the war against Germany and Austria. The Ottoman Empire declared war in November 1914, and Italy in May 1915. Montenegro, Japan, Romania, Greece, and Portugal would join the Allied side; Bulgaria, the German. The United States tried for three years to preserve its neutrality, but was at last drawn in as well, declaring war on Germany in April 1917.

By the time it had come to an end, 578,000 Italian soldiers were dead, 800,000 Ottomans, 920,000 from the British Empire, 1.1 million from the Habsburg domains, 1.4 million Frenchmen, 1.8 million

Russians, and 2 million Germans. Some 15 million men were wounded, almost half of them maimed for life. At least 8 million civilians died violently or from starvation; millions more perished in the 1918–19 influenza epidemic exacerbated by the destruction of the war. In *The First World War*, John Keegan puts it starkly: More than one out of every three German boys aged nineteen to twenty-two at the outbreak of the war was killed from 1914 to 1918.

In those four and a half years of the First World War lie the causes of the Second. And that second great war, by drawing Soviet soldiers into the heart of Europe, engendered in turn the Cold War—which would not end until the fall of the Berlin Wall on November 9, 1989, seventy-one years to the day after Kaiser Wilhelm's abdication.

None of this is ever far from European minds. But for Americans, World War I looms a much smaller memory than World War II or the Civil War. Sergeant York aside, the First World War threw up few American heroes. It did not stir the profound emotions of the Civil War or the Revolution, and it lacked the moral clarity of World War II. Above all, most of the fighting was done by foreigners: America lost 114,000 men in the war, not even half as many as Romania.

As a result, the history of the First World War is a topic that Americans have tended to leave to British writers. There's nothing necessarily wrong with that. But the disproportion between America's sacrifices and Britain's (and the shabby way that the United States treated Britain after 1918) means that it takes a very fair-minded British writer to do justice to America's contribution to winning the war.

Likewise, because British wealth and power never recovered from the war, British historians are understandably tempted to wish that their country had never joined it and to write about the war in a spirit of regret over "what might have been." And from John Maynard Keynes to A. J. P. Taylor, that spirit of regret has all too often drifted into actual apologetics for Germany.

John Keegan's elegant new book, however, avoids both dangers. Keegan is a lecturer at the British military academy, Sandhurst, and the military affairs correspondent for the *Daily Telegraph*. His father, his father-in-law, and two of his uncles served in the British armies, and with *The Face of Battle*, he produced what is universally acclaimed one of the best accounts ever written of the actual experience of combat.

In *The First World War*, Keegan candidly acknowledges the amateurishness of the American soldiers (except for the Marines) and the defects of American tactics: General John Pershing refused to believe he could learn anything from the stuck-up Frogs and Limeys, and he ordered lines of doughboys to charge German machine guns in frontal assaults reminiscent of the worst slaughters of the early months of the war.

But Keegan also stresses the extent to which the United States was absolutely crucial to Allied victory. The last great German offensive, in the spring of 1918, penetrated deep into the British military zone— where the ill-fed, badly shod German troops got a look at the Allies' colossal mountains of food, clothing, boots, and ammunition. And when the offensive sputtered out, the Germans began their long retreat knowing that 250,000 Americans a month were landing in France—more reinforcements than Germany could expect in an entire year—and that four million more were in training in the United States, a force larger than the entire remaining German army.

"The consequent sense of the pointlessness of further effort rotted the resolution of the German soldier to do his duty," Keegan says. The German army in November 1918 was still intact and in occupation of foreign soil. The Allies expected at least another year of fighting. But the Germans' nerve was smashed, and had the Armistice not been sought when it was, the Kaiser's army might well have simply dissolved.

As Keegan sees it, much of the horror of the First World War was an appalling accident of timing. By 1914, the killing technologies of the twentieth century—the machine gun, the high explosive shell, the grenade—were all available. But the technologies that could coordinate them purposefully—the field radio, the tank, the airplane—were still to come. So were the lifesaving technologies that reduced casualties in the West during World War II: blood transfusions, antibiotics, and perhaps most important, trucks that could move the wounded rapidly to a field hospital.

Keegan does not make excuses for Douglas Haig and the other blood-soaked butchers of the high command; as Prime Minister Lloyd George savagely quipped: "The solicitude with which most generals in high places (there were honorable exceptions) avoided personal jeopardy is one of the debatable novelties of modern warfare." But Keegan convincingly argues that one reason World War I cast up no military men like World War II's Montgomery and Patton is that 1914 offered no possibility for them.

Keegan thus resists the temptation (succumbed to by the author of one of the early classic histories of the war, Basil H. Liddell Hart) to condemn Allied generals as idiots while leaving the impression that the German generals were a collection of Erwin Rommels. In fact, if any high command was criminally irresponsible, it was Germany's. The famous Schlieffen plan was, Keegan insists, doomed from the start. Once they reached the French border, the truckless German troops would have to get off their trains and walk. On the narrow roads of those days, a single army corps extended almost twenty miles. The great constraint on striking power in 1914 was not manpower but road surface. Keegan argues that there simply was not room on the paved roads of northern France for an army the size that von Schlieffen's plan required, and he identifies despairing hints in the text of the plan that suggest von Schlieffen himself knew it. The Germans proceeded anyway.

If "brilliance" is coming up with an idea that nobody ever thought to say before, then we pay undue honor to intellectual flash when we make brilliance the touchstone of excellence—since the most common reason that nobody has thought to say a particular thing is that the thing is wrong. In this sense, John Keegan's *The First World War* is the opposite of brilliant: It is instead lucid, impartial, and authoritative. Most impressively, it is a miracle of concision, compressing problems that have consumed entire books into two or three crystalline paragraphs. And in the same understated style as the British Imperial war cemeteries in France, it is quietly heart-rending.

"Brilliant," however, is exactly the word to describe Niall Ferguson's *The Pity of War*. Ferguson has produced a dazzlingly ambitious attempt to write not a narrative history, but a debunking of what he bills as ten "broadly held myths" about the war.

On examination, his broadly held myths usually turn out to be either not broadly held or not myths. When, for example, Ferguson attacks the notion that Germany went to war in 1914 because it felt strong, he's attacking something believed by almost nobody. The brave counterposition Ferguson claims as his own—that Germany went to war because its leaders felt weak and feared that Russian industrialization would leave them weaker still—is held by almost all modern historians.

But Ferguson's overhyping of his originality does not entirely diminish his achievement. *The Pity of War* is a fascinating volume, bristling with interesting ideas. Ferguson approaches the war from an unusual angle. He is a financial historian (his previous book was a history of the house of Rothschild), and he never loses sight of the truth that paying for the war was an absolute precondition for fighting it. The First World War was the most expensive thing the human race had ever until then done. It cost about $180 billion in the money of the day (at a time when $1 bought one-twentieth of an ounce of gold, or fifteen times as much as today), and that's not counting the reconstruction of northern France and southern Belgium or the postwar cost of caring for the crippled and the orphaned.

Cost was the crucial variable of the war. One of the enduring mysteries of the war is how Germany managed to survive as long as it did when its enemies so outnumbered it. It's true that the Germans were better fighters: "From August 1914 until June 1918," Ferguson observes, "there was *not a single month* in which the Germans failed to kill or capture" more soldiers on the Western front than they lost. But this alone would not have sufficed, given the Allies' overwhelming economic advantage. The key, Ferguson determines, was Germany's superior management of its military resources. He grimly calculates that "whereas it cost the Entente powers $33,485.48 to kill a serviceman fighting for the Central Powers, it cost the Central Powers just $11,344.77 to kill a serviceman fighting for the Entente." It was this three-to-one disparity in killing efficiency that kept Germany in the fight for four and a half years.

This same close attention to numbers leads Ferguson into some of his most ingenious but least convincing suggestions. He argues that Germany did not emerge from the war economically broken: Its internal war debt and its external reparations debt amounted to 160 percent of its gross domestic product in 1921, less than Britain's (165 percent) and substantially less than Britain's after the 1815 defeat of Napoleon (which Ferguson believes to have been close to 200 percent).

Ferguson argues that Germany could thus have afforded to honor its debt and pay as well its reparations to the Allies. The Young Plan of 1929 envisaged a payout of about 3 percent of German national income a year for sixty years—a not unimaginable stretch of time when one considers that modern Germany has been, as Ferguson notes, a net contributor to the European Union budget for forty years now. Three percent of national income is substantial, but hardly crushing: It's roughly equivalent to America's post–Cold War defense budget.

What Ferguson neglects to mention, however, is the reason that Germany's debt was as low as 160 percent. By 1921—two years before the famous German hyperinflation of the 1920s began—Germany had already inflated away most of its internal debt. Britain financed the war by dissolving its overseas investments; Germany financed the war by expropriating the savings of its middle class.

The country that Ferguson envisages paying reparations was thus not a stable, prosperous Germany, quietly accumulating trade surpluses in a free trade world; it may have been spared the terrible physical damage inflicted on France and Belgium, but it was teetering politically, locked out of world markets by discriminatory tariffs, and convulsed economically.

Ferguson is right that Germany should have been made to help rebuild France and Belgium. But it was not mere truculence that caused Germany to default. Britain and America—the winners of the war—had already defaulted on their obligation to build an international economic order in which there was room for Germany to earn the money to pay France and Belgium. In the end, the United States ended up lending Germany the money, which only made the already ramshackle financial structure of the 1920s more rickety still.

Ferguson's *The Pity of War* has attracted attention most of all for its argument that Britain ought to have stayed out of the First World War. This is the section of his work excerpted in the *Atlantic Monthly* and the argument that inspired the *New Yorker* to publish a lengthy article about him. Ferguson contends that Britain was not bound to come to the aid of France and Belgium in August 1914. True, France would have lost had Britain not. But so what?

> *Had Britain stood aside—even for a matter of weeks—continental Europe could therefore have been transformed into something not wholly unlike the European Union we know today—but without the massive contraction in British overseas power entailed by the fighting of two world wars. . . . It would have been infinitely preferable if Germany could have achieved its hegemonic position on the continent without two world wars. . . . By fighting Germany in 1914, Asquith, Grey and their colleagues helped ensure that, when Germany did finally achieve predominance on the continent, Britain was no longer strong enough to provide a check on it.*

But what grounds do we actually have to believe that the sort of rule a victorious Germany would have fastened on the continent would look like the European Union? In 1967, the leading scholar of the subject, Fritz Fischer, offered impressive documentary evidence that the German war aims of World War I bore an uncanny resemblance to the German war aims of World War II: a continent-wide system of economic exploitation.

Ferguson dismisses Fischer's work, but—in curious contrast to the painstaking care of most of the rest of the book—at this all-important juncture he substitutes unsubstantiated assertion for proof. Fischer, Ferguson says, was talking about Germany's aims in 1916. Had Germany won the war in 1915, its aims would have been less radical. Perhaps that's true, although Ferguson offers no evidence for it. But even moderate aims would have been bad enough: economic and military control of Europe from Spain to Poland by an illiberal, militarized regime.

Such a Europe would in no way resemble the modern European Union, a confederacy of democracies in which Germany happens to be the richest. A German victory would instead have ushered in a premature Cold War between two world powers, the United States and a German-ruled continent of Europe, only with the ships and armies of the illiberal great power based on the south shore of the English

Channel. Under those circumstances, Britain would have ceased to be a great power just as rapidly as it did in actual fact.

If Ferguson wanted to argue a radical position, he might have tried this one: Britain did not decline because of the First World War. Had Britain in 1919 been what it was in 1859—the world's most productive economy—all the foreign assets spent to win the war would speedily have been replaced. Ferguson ought to have thought harder about the implications of his observation that Britain spent relatively more to defeat Napoleon than it did to beat the Kaiser—but that Britain nevertheless dominated the nineteenth century economically; harder too about the observation, which he does not make, that Britain spent only slightly more to defeat the Kaiser than the United States spent to win World War II—an expenditure that did not prevent the United States from dominating the twentieth. The war weakened Britain because it deprived her of the accumulated wealth that would otherwise have cushioned her decline. But war or no war, she was declining, because of the failure of her economy to continue to lead.

The causes of this failure are much debated. Probably the best explanation is the unique strength of British trade unions, which had already by 1900 loaded onto British industry the most restrictive business practices in the developed world. But whatever the explanation, it does suggest that the right might-have-been for Britain is, "How could we have maintained our economic edge?" and not "How could we have accommodated ourselves to the Kaiser?"

There's a lesson in this for America. A country, no matter how rich, ceases to be great when it loses the heart to protect itself in a world of dangerous states. There are plenty of examples, of which the eighteenth-century Netherlands is the most familiar. The technical term for such countries is "prey." That's what Britain would have been had Germany prevailed in the First World War. That's what the United States will be on the day its readiness to defend itself falters.

# PRO WRESTLING AND
# THE END OF HISTORY

## By Paul A. Cantor

⎯⊨⎯

*(Originally published in the October 4, 1999, issue)*

WHEN THE GREAT PARISIAN HEGELIAN Alexandre Kojève searched for
an image of the end of history, he finally hit upon the Japanese tea ceremony. Coming from Brooklyn, I am a bit less sophisticated and turn
to American professional wrestling instead. For wrestling has been as
much a victim of the end of the Cold War as the military-industrial
complex. It is not just that the demise of the Soviet Union deprived
wrestling of one set of particularly despicable villains. The end of the
Cold War signaled the end of an era of nationalism that had dominated
the American psyche for most of this century. Like much else in the
United States, including the power and prestige of the federal government itself, wrestling had fed off this nationalism. It drew upon ethnic
hostilities to fuel the frenzy of its crowds and give a larger meaning to
the confrontations it staged.

The state of professional wrestling today thus provides clues as to
what living at the end of history means. It suggests how a large segment
of American society is trying to cope with the emotional letdown that
followed upon the triumph of capitalism and liberal democracy. If the
vast wrestling audience (some 35 million people tune in to cable programs each week) is a barometer of American culture, then the nation is
in trouble. Indeed, the very idea of the nation-state has become problematic. For wrestling has been denationalizing itself over the past decade,
replacing the principle of the nation with the principle of the tribe.

The erosion of national identity in wrestling reflects broader trends in American society. If one wants to see moral relativism and even nihilism at work in American culture, one need only tune in to the broadcasts of either of the two main wrestling organizations, Vince McMahon's Worldwide Wrestling Federation and Ted Turner's World Championship Wrestling. (It is no accident that one of the pillars of professional wrestling is Turner's cable TV empire, which also brings us CNN, the anti-nation-state, global news channel.) Both the WWF and the WCW offer the spectacle of an America that has lost its sense of national purpose and turned inward, becoming wrapped up in manufactured psychological crises and toying with the possibility of substituting class warfare for international conflict. And yet we should remain open to the possibility that contemporary wrestling may have some positive aspects; for one thing, the decline of the old nationalism may be linked to a new kind of creative freedom.

THE HISTORY OF PRO WRESTLING as we know it begins after World War II and is roughly contemporary—not coincidentally—with the rise of television. Wrestling provided relatively cheap and reliable programming and soon became a staple for fledgling television stations. By the 1950s—and well into the '60s and '70s—wrestling was filling the airwaves with ethnic stereotypes, playing off national hostilities that had been fired up by World War II and restoked during the Korean conflict. Wrestling villains—always the key to whatever drama the bouts have—were often defined by their national origin, which branded them as enemies of the American way of life.

Many of the villains were at first either German or Japanese, but as memories of World War II faded, pro wrestling turned increasingly to Cold War themes. I wish I had a ruble for every wrestling villain who was advertised as the "Russian Bear," but the greatest of all who bore that nickname was Ivan Koloff. Looking for all the world like Lenin pumped up on steroids, he eventually spawned a whole dynasty of villainous wrestling Koloffs. The fact that the most successful of them was named Nikita shows that it was actually Khrushchev and not Lenin or Stalin who provided the model for the Russian wrestling villain. Time and again the Russian wrestler's prefight interview was a variation on "Ve vill bury you." Nikolai Volkoff used to infuriate American opponents and fans alike by waving a Soviet flag in the center of the ring and insisting on his right to sing the Soviet national anthem before his bout began.

To supplement its Russian villains, wrestling turned to the Arab Middle East, where a long tradition of ethnic stereotyping was readily available. During the years of tension between the United States and Iran, wrestling hit paydirt with a villain known as the Iron Sheik, who made no secret of his admiration for and close personal ties to the Ayatollah Khomeini. His pitched battles with the All-American GI, Sgt. Slaughter, became the stuff of wrestling legend. Not to be left behind by the march of history, during the Gulf War the Iron Sheik reinvented himself as Colonel Mustafa, and suddenly Americans had an Iraqi wrestler to hate.

The extent to which wrestling relied on national identity to manufacture its villains should not be overstated. Some of the greatest villains were home-grown, like Nature Boy Buddy Rogers, and some of the greatest heroes were foreign-born, like Bruno Sammartino. But although ethnic stereotyping was not essential to the emotional dynamics of wrestling, it did play a crucial role. That is why the end of the Cold War threatened to deliver a serious if not mortal blow to the whole enterprise. Suddenly audiences could not be counted upon to treat a given wrestler automatically as a villain simply because he was identified as a Russian. There was a brief, almost comic era of wrestling *glasnost*, during which the promoters tried to see if they could generate drama out of the shifting political allegiances of the Russian wrestlers. The extended Koloff family was riven by internal dissent, as some sided with Gorbachev and the reformers, while others remained hardliners and stuck by the old regime. But since Kremlinology has never been a popular spectator sport outside academia, the public quickly grew bored with trying to sort out the internal politics of the Koloff family, and it began to dawn on the wrestling moguls that the end of the Cold War was a threat to their franchise.

This problem was compounded by the fact that at roughly the same time as the Cold War was ending, ethnic stereotyping began to be anathematized. By the early '90s, the WWF even seemed to be testing whether it could capitalize on the new era of political correctness. With Russia and virtually every other country ruled out as a source of villains, Vince McMahon and his brain trust searched the globe to see if any ethnic group remained an acceptable object of hatred. The result was a new villain named Colonel DeBeers—a white, South African wrestler with an attitude, who spoke in favor of apartheid during interviews. One can almost hear the wheels grinding in McMahon's head: "Russians may no longer be fair game, but no one will object to a little Boer-bashing." But wrestling fans did not take the bait. This was one of the few times the WWF misjudged its audience, proceeding as if its fans were sipping chardonnay and sampling brie instead of guzzling beer and munching on nachos. Colonel DeBeers was a flop as a villain and in some ways marked the end of a wrestling era—a last, desperate attempt to base physical conflict in the ring on political conflict outside it.

WRESTLING PROMOTERS HAVE ALWAYS been concerned that theirs is not a team sport and thus threatens to lack that extra measure of fan commitment that group solidarity can extract. Exploiting nationalist feeling had been one way of turning wrestling into something more than single combat. Instead of rooting for the home team, fans viewing a Sgt. Slaughter–Iron Sheik bout got to root for America. Or rather, America became the home team.

But there was also a germ of a team concept in wrestling's peculiar institution of the tag team—a bout in which two wrestlers pair up against a couple of opponents. And as ethnicity faded as a principle in wrestling, the WWF and the WCW began to expand tag-team partnerships into larger groupings that might best be described as extended families or tribes. The wrestlers in such tribes pool their resources to advance their careers, often illegally entering the ring to come to each other's aid, softening up each other's opponents for future matches, and generally creating trouble for any wrestler not within the tribe. These wrestling tribes adopt an outlaw pose within their larger leagues, refusing to conform to league rules and challenging the duly constituted wrestling authorities. The most famous of these groups is the New World Order (the nWo) within the WCW, which was headed by Hollywood Hulk Hogan and is constantly trying to outwit the league owners and take over the organization. It is surely one of the ironies of the end of history that in the aftermath of the Gulf War, that "vision thing" of George Bush's has left no more lasting monument than the name of a group of renegade wrestlers.

Tribal organization gives wrestling something intermediate between national identity and a purely individual identity. Fans almost have the sense of rooting for teams, since the wrestling tribes often have their own logos, uniforms, slogans, theme songs, cheerleaders, and other badges of communal or team identity. The wrestling brain trusts create ongoing storylines involving the various tribes, so that the future of the whole league, perhaps its very ownership, can seem to depend on the outcome of a given bout.

Thus the newly created tribal identities in wrestling can serve as substitutes for the old national identities. But one thing is missing—any sense of stability, the reassuring feeling of continuity that used to be provided by ethnic stereotyping in wrestling. Once a Russian, always a Russian, and, until the era of *glasnost*, that also meant always a villain as well. National identity is not a matter of choice; one is born into it and stuck with it, unless one chooses to betray one's national origins (at the height of the Koloff confusions, charges of "traitor" were routinely hurled back and forth in interviews). But in the world of wrestling today, which group a wrestler affiliates with appears to be a matter of personal choice (though in fact these "choices" are still scripted by the league). As it happens, the traditional national identities in wrestling were often made up. Both the "Manchurian" Gorilla Monsoon and the "Oklahoma Indian" Chief Jay Strongbow were in actuality Italian-Americans (Robert Marella and Joe Scarpa, respectively), and the wrestler known as Nikolai Volkoff began his career as Bepo Mongol. In the contemporary era, though, wrestling virtually acknowledges that it is manufacturing its villains, and their roles are presented as a matter of personal choice rather than national destiny.

Thus pro wrestling takes its place along with the plays of Samuel Beckett and the buildings of Michael Graves as an example of the dominant cultural mode of our age, postmodernism. The characters in Beckett's plays are not meant to represent real-live human beings, who might be said to lead an existence independent of the drama. Rather they are revealed to be fictions, consciously constructed characters who are themselves sometimes dimly aware that they are merely characters on stage. Graves's buildings are not meant to be "true" in the way the triumphs of modernist architecture were. Abandoning the modernist dogma that form follows function, Graves returns to architectural decoration, reminding us that his buildings are after all human constructions and thereby "deconstructing" them before our eyes. Pro wrestling has similarly entered its postmodern phase, in which it deliberately subverts any claims to truth and naturalness it ever had. Of course, at least since the era of television, pro wrestling has always been entertainment rather than real sport. But for decades pro wrestling at least pretended it was real. It now admits its fictionality, and indeed, like most forms of postmodernism, revels in it.

But can we confidently say that wrestling simply mirrors broader movements in our culture and politics? It is difficult to look at developments in politics and culture today and not see them as in turn mirroring developments in wrestling. Was Hulk Hogan, who dominated the 1980s, perhaps our first taste of Bill Clinton? The Hulkster—who could never talk about anything but himself, his own career, and his standing with his Hulkamaniac fans—was the model of a roguish, narcissistic, utterly unprincipled performer. While changing his stance from moment to moment, he was never held accountable by his adoring public, to the point where he seems to have gotten away with anything. If postmodern wrestling was not a forerunner of postmodern politics, why is Jesse "The Body" Ventura now the governor of Minnesota?

WHEN THE VILLAINY of wrestlers was rooted in their national identity, their evil was presented as inherent in their natures. Related to genuine political conflicts in the actual world, the evil of a Russian wrestler seemed real. But villainy has become something more fluid and elusive in the era of postmodern tribalism. Since the contemporary wrestler appears to choose his tribal affiliations, he also gets to choose whether to be a hero or a villain (again, these matters are carefully scripted by the WWF and the WCW authorities, but we are talking about how things are meant to appear to the wrestling public). The most striking characteristic of post–Cold War wrestling is the dizzying rapidity with which today's wrestlers switch from hero to villain and back again. Wrestlers used to spend their whole careers defined as either good guys or bad guys. Now they alter their natures so often that it no longer makes sense to speak of them as natural heroes or villains in the first place. The contemporary wrestler exemplifies the thoroughly postmodern idea that human identity is purely a construction, a matter of choice, not nature.

With its underpinnings in traditional notions of morality, heroism, and patriotism eroded, wrestling has turned to new sources to hold the interest of its fans. Generally these sources have been found in the dramas of private life. Televised wrestling has always had much in common with soap operas. Fans identify heroes and villains and get wrapped up in ongoing struggles between them and especially the working out of longstanding and complex feuds. Throughout its history, pro wrestling has occasionally sought to involve fans in the private lives of its warriors. Once in a while a wrestler has gotten married in the ring to his female manager or valet. (More recently—reflecting a loosening of morality—female companions of wrestlers have been at stake in matches, with the winner claiming the right to take possession of his opponent's woman.) Personal grudges have always been central to wrestling, but over the last decade they have gotten ever more personal, often involving family members who somehow get drawn into conflict inside or outside the ring.

In short, wrestling conflicts have come increasingly to resemble the appalling family feuds aired on *The Jerry Springer Show*. This is only fair, since Springer seems to have modeled his show on wrestling interviews. Wrestlers used to get angry with each other because one represented the Soviet Union and the other the United States, and the two ways of life were antithetical. Now when wrestlers scream at each other, dark domestic secrets are more likely to surface—sordid tales of adultery, sexual intrigue, and child abuse.

Here a wrestler with the evocative name of Kane is emblematic. Kane was introduced in the WWF as the counterpart of a well-established villain called the Undertaker, who often punishes his defeated opponents by stuffing them into coffins (a nasty case of adding interment to injury). Kane's aptly named manager, Paul Bearer, soon revealed that Kane is in fact the Undertaker's younger brother. Kane wears a mask to hide the frightening facial burns he suffered as a child in a fire set by his older brother, which killed their parents. Thus the stage is set for a series of epic battles between Kane and the Undertaker, as the younger brother seeks revenge against the older. Paul Bearer then reveals that Kane and the Undertaker are actually only half-brothers, and that he himself fathered the younger boy, though he neglected him for years and is only now acknowledging paternity. With its Kane storyline, the WWF crafted a myth for the '90s. All the elements are there: sibling rivalry, disputed parentage, child neglect and abuse, domestic violence, family revenge.

McMahon and his brain trust have once again proven that they have a finger on the pulse of America. In the wake of years of psychotherapy, Twinkie defenses, and the O.J. trial, they have reinvented the villain as himself a victim. No one ever felt a need to explain the evil of Russian wrestlers—they were presented as villainous by nature. But unlike his biblical counterpart, Kane is supplied with motivation for his evil, and therefore inevitably becomes a more sympathetic figure. After all, his problems started when he was just a little kid. Kane is in fact a huge man named Glen Jacobs: six-feet seven-inches tall and weighing 345 pounds. Yet when he climbs into the ring, he stands as the poster boy for the '90s—the victimized wrongdoer, the malefactor who would not be evil *if only someone had loved him as a child.*

The other victim of society now celebrated by pro wrestling is the poor, abused working man, symbolized by "Stone Cold" Steve Austin, currently enmeshed in a bitter feud with Vince McMahon and the entire power structure of the WWF. In his unceasing search for suitable villains, McMahon finally hit upon the most villainous person he could think of—himself. In the ultimate postmodern convolution, wrestling now focuses on itself as a business and makes its own corruption the central theme of its plots. McMahon has decided to build his storylines around ongoing labor-management disputes in the WWF. He is in constant public conflict with his wrestlers, trying to force them to do his bidding and above all to make his on-again, off-again champion Austin toe the corporate line.

In his quest to gain an edge on Turner's WCW, McMahon realized he could tap into the resentment the average working man feels against his boss. McMahon is always threatening to downsize the WWF wrestling staff and has surrounded himself with corporate yes-men. Austin is his perfect working class opponent—a beer-drinkin', foot-stompin', truck-drivin', hell-raisin' Texas son-of-a-gun, always prepared to tell McMahon: "You can take this job and shove it." With this storyline, wrestling has completed its turn inward, moving from the Cold War to class war. Ironically, even at the height of the Cold War, wrestling never went after Russian communism with half the fervor it now devotes to pillorying American big business. If wrestling is any indication, the United States—deprived of any mean-ingful external enemy—seems to have nothing better to do than attack itself. Why not go after a bunch of tobacco companies, for example?

The McMahon-Austin feud proved to be so successful that Turner's WCW soon began imitating it, using its chief executive, Eric Bischoff (a former wrestler himself) to play the role of corporate bad guy. Always one step ahead of his competition, McMahon went on to fuse the family soap opera aspect of wrestling with the class warfare element by involving his son, his daughter, and eventually even his wife in his corporate struggles. These storylines have become increas-ingly bizarre, with McMahon's son Shane first seeming to betray him and then revealed to have been secretly acting on his behalf all along, and his daughter Stephanie set up for a kind of wrestling dynastic marriage and then kidnapped under weird circumstances. Who would have thought a century ago when wrestling began with a simple full nelson and a step-over toehold that it would eventually culminate in a proxy fight? But that is exactly what happened when McMahon's wife and daughter shocked him by voting their shares in the WWF to make Austin CEO, thereby transforming the board meetings back in Connecticut beyond recognition. (Austin brought a case of beer to his first session as president.) No wonder McMahon is about to take his corporation public.

EVERY TIME I THINK wrestling has reached rock bottom, either the WWF or the WCW finds its way to a new moral depth. A recent plot line culminated in Austin holding a gun to McMahon's head in the center of the ring, as the nattily attired owner/operator of the WWF appeared to wet himself in terror. When one looks at wrestling's "progress" from the 1950s to the 1990s, one really has to be concerned about America's future. If wrestling tells us anything about our country—and its widespread and sustained popularity suggests that it does—for the past three decades we have been watching a steady erosion of the country's moral fiber, and America's growing incapacity to offer functional models of heroism.

On the other hand, perhaps we should cease being moralistic for a moment, recognize that wrestling is only entertainment, and try to look beyond its admittedly grotesque antics. Though it is tempting to become nostalgic for the good old days of American patriotism in wrestling, let's face it: The traditional national stereotypes did become tired, overused, and predictable. In that sense, the end of the Cold War actually proved to be liberating for wrestling, as one might hope it could be for all American society. What appeared to be a loss of ethnic stereotyping proved to be a gain in creative freedom, as wrestling was forced to scour popular culture to come up with alternatives to traditional villains. Wrestling may not be more moral these days, but it certainly is more interesting and inventive. This development suggests that maybe we all need to be thinking beyond the nation-state as our chief cultural unit.

After all, the nation-state has not always been the dominant form of cultural or even political organization. It is largely a development out of 16th-century France, and has never as fully prevailed around the world as historians would have us think. There is no reason to believe that the nation-state as we know it is the perfect or even the best unit of political organization. When Aristotle made his famous statement usually translated as "man is a political animal," what he really was saying is that man is an animal whose nature it is to live in the *polis*— the Greek city conceived as the comprehensive human community, on a scale much smaller than a modern nation-state. Thus Aristotle would have said that the nation-state is an unnaturally large and even overblown form of community.

Perhaps what appears to be the end of history is only the end of the nation-state, and humanity is now groping confusedly toward new modes of political organization, which may be at once more global and more local in their scope. Today's professional wrestling points in these two directions simultaneously. At any moment of deep historical change, it is easy to become fixated on what is being lost and fail to see what is being gained. The way wrestling has been struggling to find some kind of postnational identity reflects a deeper confusion in our culture as a whole, but one that may portend a profound and even beneficial reorganization of our lives in the coming century. Perhaps, then, when we watch—and enjoy—the WWF and the WCW, we really are wrestling with the end of history.

# THE GOD OF THE BESTSELLER LIST

## By Alan Jacobs

*(Originally published in the December 6, 1999, issue)*

I TAKE AS MY TEXT the words of a little girl to Melvin Morse, author of the bestselling *Closer to the Light: Learning from the Near Death Experience of Children*. As Dr. Morse explains it, the girl had died and gone to Heaven, only to be resuscitated and brought back to this world. And when he asked her what she had learned from her Visit to the Beyond, she considered the question carefully before answering, "It's nice to be nice."

When I was a teenager in the 1970s, magazines carried ads for posters, and the most popular of those posters offered a meditation, in what I suspect was intended to be poetic prose, called "Desiderata." Many people have come across it at some time or another, at least its more famous lines:

> *You are a child of the universe,*
> *  no less than the trees and the stars.*
> *You have a right to be here.*
> *And whether or not it is clear to you,*
> *  no doubt the universe is unfolding*
> *  as it should.*
> *Therefore be at peace with God,*
> *  whatever you conceive Him to be.*
> *And whatever your labors and aspirations,*

*in the noisy confusion of life,*
*keep peace with your soul.*
*With all its sham, drudgery,*
*and broken dreams,*
*it is still a beautiful world.*

What's particularly noteworthy about this little document is the popular conviction that it is a piece of antique wisdom, produced many centuries ago. Some of the posters identified it as "medieval" and claimed that it had been written by a monk; others dated its composition quite specifically to 1692; still others combined the two, apparently in the belief that 1692 was in the Middle Ages. (The date seems to derive from the rector of an Episcopal church in Baltimore, who typed out "Desiderata" some forty years ago on stationery that prominently featured the 1692 founding date of his church. Photocopying and careless reading did the rest.)

In fact, "Desiderata" was written in 1927 by a man from Terre Haute, Indiana, named Max Ehrmann. Ehrmann was a lawyer who worked at various times as a deputy state's attorney and a credit manager for his brother's manufacturing company—and these items from his vita may be significant. His attempt to articulate a peaceable, serene prospectus for daily life suggests that his primary concern was to maintain a sanguine and mystical temperament in a corporate and bureaucratic environment:

*Enjoy your achievements*
*as well as your plans.*
*Keep interested in your own career,*
*however humble;*
*it is a real possession*
*in the changing fortunes of time.*
*Exercise caution in your business affairs;*
*for the world is full of trickery.*

"Desiderata" is a masterpiece, of sorts, because it so perfectly completes the translation of nineteenth-century American Romanticism into the terms of modern middle-class life. You can see the process beginning to unfold back in 1836, when the Boston Transcendentalist Bronson Alcott asked a student at his Temple School about the mission of his soul—and the student replied, "I think the mission of my soul is to sell oil." Max Ehrmann is the perfect apostle of that prescient boy's gospel.

But "Desiderata" is scarcely the final word on the subject. Ehrmann's descendants now populate American bestseller lists as the stars fill the sky. The current bumper crop of books celebrating the joys of amorphous and sanguine spirituality seems to find an especially appreciative audience among people whose daily lives are spent in bureaucratized environments which, they feel, oppress their spirits. There are so many of these books that even listing them is a challenge, especially since they tend to proliferate like some uncontrollable malignancy. Clearly it wasn't enough to have the 1993 bestseller *Chicken Soup for the Soul*, for we now have reached *A Sixth Bowl of Chicken Soup for the Soul*—to say nothing of *Chicken Soup for the Woman's Soul* (the most popular of them all, with over three million copies in print), *Chicken Soup for the Soul at Work, Chicken Soup for the Golfer's Soul*, and many others.

Apparently there are a lot of people out there with no desire to vary their menu, but if they ever do drain the soup bowl of life to the dregs, they may join the millions who have thrilled to Betty Eadie's account of her "journey through death and beyond," *Embraced by the Light*. She and Melvin Morse and Raymond A. Moody Jr. (whose 1975 *Life After Life* has sold over fourteen million copies) dominate the enormous market for books that promise a sweet pastoral Beulah Land lies in store for us: No waiting, these authors all seem to say, no Day of Judgment, just immediate admission to the Place Where Everyone Is Nice. (If you wish to know more, please consult Moody's afterlife website, www.lifeafterlife.com.)

Anyone who reads these books, and the multitudes like them, will soon realize that their counsels and messages are somewhat less than earth-shakingly original and profound. But that is precisely the point. The popularity of "Desiderata" arose in large part from its power to give expression to the hopeful desires of many people: that "the world is unfolding as it should," that I am "a child of the universe," that, in short, "it's nice to be nice."

But this cannot be the whole story of the success of these books offering this vaguely spiritual message of consolation. And there is, in fact, a deeper reason for the American fascination with this kind of spirituality: It plays to the passion for having the validity of our desires confirmed by witnesses from the distant past or beyond the grave.

This phenomenon can be seen most clearly in two of the most immensely popular American spirituality books in recent years: Marianne Williamson's 1992 *A Return to Love* and Neale Donald Walsch's 1995 *Conversations with God*. Each has sold millions of copies and produced its innumerable sequels and spinoffs (Gutenberg's carcinoma striking again); Walsch's new *Friendship With God: An Uncommon Dialogue* was published on October 25, and immediately leapt into the top ten on the *New York Times* bestseller list. But what the originals mostly reveal is how deeply we Americans crave the echoing testimony of other times and places—as long as it remains merely an echo, and doesn't threaten to tell us anything unfamiliar or otherwise disagreeable. "Desiderata," in its guise as a medieval monkish meditation, brings us a confirmation of the present from the past; its recent descendants, the books of Williamson and Walsch, offer us a still louder echo: God's resounding endorsement of our every craving.

Walsch acquired what he calls "God's latest word on things" through a highly traditional method: a kind of automatic writing, in which Walsch claims to have become the Deity's amanuensis (though one with the power to scribble his own questions and responses). And what does God reveal to Neale Donald Walsch? Well, for one thing, that religious institutions, persons of religious authority, and the Bible "are not authoritative sources" for "truth about God." Instead, God says, here's what we do if we want to know about Him: "Listen to your feelings. Listen to your Highest Thoughts. Listen to your experience. Whenever any one of these differ from what you've been told by your teachers, or read in your book, forget the words."

This is certainly encouraging (not that you haven't heard it before from Timothy Leary, Abbie Hoffman, and the guy who makes those

"Question Authority" bumper stickers). But a reader with even the dimmest spark of critical reflection might be tempted to ask, "How can I tell my Highest Thoughts from my lower and presumably unworthy thoughts?"

This is a problem Walsch's God doesn't know quite how to address. He likes the sound of capitalized phrases: "Highest Thoughts" and "Who You Are" and things like that. But he is also at pains, repeatedly, to say that there is "no such thing" as right or wrong, good or bad, better or worse. "There is only what serves you, and what does not." And perhaps this is the key to identifying our Highest Thoughts: They are the ones most perfectly self-serving.

Take our thoughts about money, for instance. At one point, Walsch's God suggests that we need to "outgrow" a love of money, but when Walsch complains that he is financially strapped—"What is blocking me from realizing my full potential regarding money?"— God responds with almost gushing sympathy: "You carry around a feeling that money is bad." If only Walsch would stop feeling guilty, then he could liberate himself to make and enjoy *lots* of money. Here's a counsel Walsch is quick to warm to. "I see I have a lot of work to do," he says with evident relish. Presumably, now that *Conversations with God* has been on bestseller lists for almost four years—it's still number twenty on the *New York Times* hardcover nonfiction list—he has had ample opportunity to cultivate the requisite virtue.

The God of Marianne Williamson's *A Return to Love* bears striking similarities to the one with whom Neale Donald Walsch hangs out— which seems a confirmation of sorts. In any case, Williamson's book is also based on a revelation given through automatic writing, though in this case she was not the recipient. She draws on a hefty volume called *A Course in Miracles*, which came about in 1965 when Helen Schucman, a professor of medical psychology at Columbia University, heard a voice speaking to her that she came to believe was the voice of Jesus. Her colleague William Thetford served as amanuensis as the revelations poured forth; eventually the transcriptions made their way into print. (In the copy I saw, Jesus begins by speaking these words to Dr. Schucman: "This is A Course In Miracles®. Please take notes." One wonders who registered the trademark and where the royalties go.) But if Schucman and Thetford were the evangelists, writing this new Gospel, Marianne Williamson has turned out to be their Apostle Paul, spreading the good news far beyond its original source.

Aside from the dependence on automatic writing or "scribing," another feature shared by Walsch and Williamson is their retaining of much of the language of traditional Christianity, even down to the identification of God as a Trinity: Father, Son, and Holy Spirit. (It's Walsch who occasionally inserts references to God as "Mother," while Williamson uses "He" and "Him" throughout.) A cynical reader might see this as an attempt to borrow some external authority—especially since the resemblance to Christian doctrine is merely verbal. Walsch, for instance, reinterprets the Father as "knowing," the Son as "experiencing," and the Holy Spirit as "being." Likewise, Williamson says that the Holy Spirit "has been given by God the job of . . . outsmarting our self-hatred. The Christ does not attack our ego; He transcends it." In *A Course in Miracles*—and remember, this is Jesus Christ speaking—she quotes, "Do not make the pathetic error of 'clinging to the old rugged cross.' The only message of the crucifixion is that you can overcome the cross. Until then you are free to crucify yourself as often as you choose. This is not the Gospel I intended to offer you." (In other words, "Forget that 'Take up your cross and follow me' stuff—I was misquoted.")

"In the eyes of God," Williamson explains, "we're all perfect," and our job is merely to recognize that. Evil is an illusion. Moreover, "the word Christ is a psychological term. . . . Christ refers to the common thread of divine love that is the core and essence of any human mind." A century and a half ago, Ludwig Feuerbach brought as his gravest charge against Christianity that it is the projection of our own desires— a notion cheerfully accepted by both Walsch and Williamson, who are, when it suits them, pantheists, seeing God in all things and therefore God in us and *as* us. We like having a God who is a projection of our desires, because that God won't say anything we don't want to hear.

It never seems to have occurred to any of these authors to question the validity of what they were hearing, or to notice that when other people in the past, or in other cultures, have claimed to hear God speaking, He seems to have said very different things and to have exhibited a very different character. (The vision granted to the fourteenth-century mystic Juliana of Norwich, for example, began with an image of a crown of thorns from which blood flowed copiously; only after encountering such an image did she arrive at her famous conclusion that "all shall be well, and all manner of thing shall be well.")

I believe that I, in any case, would have been not only surprised but disappointed if I heard God speaking and He told me nothing that I couldn't have found expressed more eloquently by Ralph Waldo Emerson and Henry David Thoreau, or for that matter by Dale Carnegie and Gail Sheehy. Imagine coming down from Mount Sinai with glowing countenance, only to have to tell the assembled masses, "I have heard God, and He is Norman Vincent Peale."

How do we account for the tranquil composure, the utter lack of critical suspicion, with which Walsch and Williamson and all their kind receive their remarkably unimaginative gospels? Sad to say, the answer appears obvious: They share the universal human susceptibility to flattery, and the gods who speak to them offer nothing but flattery. "I have nothing to tell you that you don't already know," He says. "You have understood yourself, your neighbors, and your social environment with admirable clarity. Your only problem is that you don't trust your own discernment. I can neither correct nor admonish you, but merely encourage you to follow your natural inclinations, which are infallible." Or, as Walsch's God puts it, "You all think very highly of yourself, as rightly you should."

That's pretty much what these books are all about. Thus Iyanla Vanzant concludes the acknowledgments page of her popular *In the Meantime: Finding Yourself and the Love You Want* by writing, "And I would humbly like to acknowledge my Self for being willing to move through the fear, denial, confusion, and anger required to figure out why I had to write this book," and concludes the book by saying to her reader, "You, my dear, have become the light of the world—the loving light. I beseech you to do everything in your power to let your light shine." Having looked upon themselves with smitten wonderment, these authors turn and offer us the chance to indulge in the same self-celebratory gaze. (Thanks.)

All our problems, on this account, are problems of perception: We do not see things clearly. Williamson tells the story of how, when she was working as a cocktail waitress, she was unhappy until she had this realization: "This isn't a bar, and I'm not a waitress. That's just an illusion. Every business is a front for a church, and I'm here to purify the thought forms, to minister to the children of God." (But could you bring me my martini first and purify the thought forms when you're on break?) Therefore it is not moral growth, but visual or perceptual retraining that we need. And, *mirabile dictu*, what is obscured by our now-clouded sight is our own virtue. Back in the Middle Ages, people who were considered wise and discerning used to think that people are blind to their own moral failings. But now God has appeared to explain that just the opposite is true: It turns out that our moral *successes* are what we habitually disregard.

What all these books most fundamentally reject is the notion that our wills may be twisted or bent. The God of these authors never for a moment questions, or allows us to question, the validity of our desires: He merely offers superior means for realizing those desires. Thus His willingness to serve as Neale Donald Walsch's financial adviser. And Williamson's book, while it may seem at times to be more directive and to require more self-criticism—"God's plan works" and "Yours doesn't," she says at one point—in fact relies just as much as Walsch's on self-interest and self-congratulation. We should choose God's plan because it's the one that will give us what we want. "We must face our own ugliness," claims Williamson, but only to discover that it's either superficial or illusory: "The ego isn't a monster. It's just the *idea* of a monster." When we see more clearly, the bad idea disappears, to be replaced by the image of a "dashing prince." Looking back at her life, Williamson says, "there's one thing I'm very sure of: I would have done better if I had known how." We do no evil, we just make "mistakes."

Several years ago, when Woody Allen was asked to explain his affair with his wife's adopted daughter, he offered this verbal shrug: "The heart wants what it wants." This is a tautology of immense moral significance, because it indicates that there is no power capable of interrogating, much less redirecting, the heart that wants—the heart that does nothing but want. The God of these books congratulates the heart for wanting and stifles the voice of mind or conscience that would offer dissent or even query. He accomplishes this stifling by proclaiming that He merely echoes—as the entire universe merely echoes—the human heart's howl of appetite.

Am I, after all, a "child of the universe"? It's worth remembering that the phrase doesn't originate with Max Ehrmann. In Dickens's *Bleak House*, the congenitally feckless Harold Skimpole, upon seeing the orphan Esther Summerson, cries out, "She is the child of the universe," only to have the more discerning John Jarndyce reply, "The universe makes rather an indifferent parent, I am afraid."

But in one sense an indifferent parent is precisely what we want: a God who neither instructs nor disciplines, who offers neither warning nor chastisement, but who smiles wryly at our peccadilloes and laughs warmly at our charming idiosyncrasies—not a Father in Heaven but a Grandfather, as C. S. Lewis once put it.

This indifference has its dark and terrible side. Without instruction or discipline or warning or counsel, we wander witlessly in a Universe whose child we may be, but which is populated by our siblings, people just like us—which is to say, people who ardently pursue goods that are incompatible with the aspirations of their neighbors, as Thomas Hobbes pointed out way back in those Middle Ages (that is, circa 1650). The numbers of us who want to be the starting quarterback or the homecoming queen or the new executive vice president far exceed the number of desirable roles and places, and they always will. And people whose greed and lust have been certified by a celestial Parent prove—when faced with the inevitable obstacles to their aspirations—to be anything but "nice" and to be anything but concerned with "purifying the thought forms."

This is the constant threat of what Hobbes called "the war of all against all," and there's nothing "illusory" about *that* war. The irresponsibility of people like Walsch and Williamson lies in their propagating a merely verbal Deity to stroke and console our desiring hearts, reserving His condemnation only for those who would remind us, in the immortal words of the Rolling Stones, that "you can't always get what you want." If Mick Jagger can figure it out, may we not expect as much of God?

# REDISCOVERING
# SARAH ORNE JEWETT

## By Claudia Winkler

———◆———

*(Originally published, as "A Place of Her Own,"*
*in the August 26–September 2, 2002, issue)*

THE NAME SARAH ORNE JEWETT, for those to whom it means anything at all, evokes principally the landscape of southern Maine and
the particular serenity of her 1896 novel *The Country of the Pointed Firs*.
Because she captured there the harmonies of undramatic lives lived out
in their native place, Jewett deserves the attention of modern readers
too prone to overlook so pallid a thing as contentment. And she
remains worth reading for another reason: her role as mentor to a bet-
ter-remembered and greater artist, Willa Cather.

Early classified (and nowadays mostly dismissed) as a "local col-
orist"——doing for Maine what the likes of John Fox Jr. had done for
Kentucky, Thomas Nelson Page for Virginia, and Edward Eggleston for
Indiana——Jewett was *rooted* in a way almost no American is anymore.
She was born in 1849 in the inland port of South Berwick, upriver from
Portland, the daughter of a prosperous and cultivated doctor. As a girl,
she accompanied her father on his visits to patients, taking in the ways
and speech of the local people. Her first story was published when she
was nineteen, and soon her work was appearing regularly in the *Atlantic
Monthly*, edited by the young William Dean Howells. With his encouragement, she produced three novels: *Deephaven* in 1877, *A Country
Doctor* in 1884, and her masterpiece, *The Country of the Pointed Firs*. She
died a few years after suffering injuries in a carriage accident, in 1909.

Jewett's writing enjoyed immediate success. Before she was thirty, she was "a fully arrived celebrity," as an early biographer put it, and she was swept into the literary circles of nearby Boston. Upon reading *Deephaven*—a youthful precursor to *The Country of the Pointed Firs*—for the third time, the poet John Greenleaf Whittier wrote her a fan letter. Two years later, she was a guest at the seventieth birthday party of the literary lion Oliver Wendell Holmes.

Along the way, Jewett became friends with the publisher of the *Atlantic*, James T. Fields, and his wife, Annie, and after Fields's death, Annie and Sarah were companions. For some years, they kept a Boston salon at 146 Charles Street, where they hosted the literati of the day— meeting, there and on trips to Europe, such luminaries as Henry James, Kipling, Tennyson, Matthew Arnold, and the Dickens family.

Through all this exposure to high culture, Jewett never deviated from her own vocation as a chronicler of simple country life. Her characters live close to nature, in isolated homesteads and small seaports. Above her desk she kept a line from Flaubert: "To write about ordinary life as one would write history."

Her most famous story, "A White Heron," is emblematic. In the story, a little girl walking her cow home through the woods encounters a young man with a gun. He is an ornithologist, come in search of a white heron. He spends the night at the girl's house and offers the dazzling sum of $10 to any who will lead him to the great bird's nest. He is kind and attractive. Wanting to please him, she slips away to climb the tallest tree at dawn, to see the white heron's first flight and so discover its nest. Her plan works perfectly—until the moment comes to tell. Remembering how she had seen the great bird "flying through the golden air and how they [had] watched the sea and the morning together," the child realizes "she cannot tell the heron's secret and give its life away."

Her decision involves sacrifice, for the stranger has awakened intimations of adventure in a wider world. But her loyalty to the woods and its creatures is decisive. The story ends: "Whatever treasures were lost to her, woodlands and summer-time remember! Bring your gifts and graces and tell your secrets to this lonely country child!"

Fidelity, this time not to nature but to vocation, is also the theme of *A Country Doctor*. Anna Prince, an orphan, is raised in the town of Oldfields by a kind widower, Dr. Leslie, who recognizes her aptitude for his profession. Eventually, Anna herself comes to see medicine as her God-given calling. Like the girl in "A White Heron," she is fleetingly tempted by romance but hews to her chosen path and finishes medical school. She quietly disregards the "fettering conventionalities" upheld by some disapproving townsfolk and relatives, and earns their respect for her healing art. She finds joy in serving the people of Oldfields and environs, not only by relieving their bodily pains, but also by acting as their comforter, confessor, and "interpreter of the outside world."

This coherence of work and surroundings, and the selfless devotion to the good of others, are reprised on a higher literary plane in *The Country of the Pointed Firs*. Where "A White Heron" is crudely symbolic (the woodland child is named Sylvia; the man is never without his gun) and *A Country Doctor* intermittently reads like a tract ("our heroine" is actually likened to Christ), *The Country of the Pointed Firs* has the individuality of fully realized art.

The central characters are three single women. Two are widows—Almira Todd, a sixty-seven-year-old herbal healer, and her elderly mother, Mrs. Blackett—while the third, never named, is a writer who rents a room from Mrs. Todd for a few months' summer stay in the coastal village of Dunnet Landing. This third woman is the narrator, the outsider through whose eyes we discover this place.

The plot is nearly nonexistent: a succession of scenes, many consisting merely of conversations. A Milton-quoting retired sea captain pines for the wider horizons of bygone whaling days. A grief-stricken widowed fisherman knits as he remembers his beloved wife, whom he honors by striving to maintain her standards of housekeeping. Mrs. Todd recounts the saga of "poor Joanna," the daughter of a good family who, crossed in love, retreated to Shell-heap Island and lived there as a hermit till she died, whereupon the whole town turned out to bury her on the island in accordance with her wishes.

At the heart of the book is an account of a day trip to Green Island by Mrs. Todd and the narrator. This farthest offshore island is where Mrs. Blackett lives and farms, with an "odd" aging son named William who never left home.

Mrs. Blackett is Jewett's finest creation. At eighty-six, she has seen "every trouble" short of her own death, yet she is light-hearted and light-footed. She is discerning, too—Almira Todd speaks of "mother's snap and power o' seein' things just as they be"—and, above all, generous. Her hospitality is "something exquisite," and of tact, which is "after all a kind of mind-reading," she has the "golden gift."

After the visitors have eaten a meal of fish chowder and explored the island while Almira gathers pennyroyal and other herbs for her syrups and elixirs, the visitors come into the farmhouse for a last cup of tea. William, conquering his shyness, sings for them, and his mother joins in the old Scottish and English tunes and Civil War ballads.

Then, just before their farewells, while Almira is bundling up her herbs, Mrs. Blackett invites the summer guest into her bedroom to sit in her rocker and see the finest view in the house. The room is plain. There is a Bible on the lightstand, and a pair of glasses and a thimble. A striped cotton shirt Mrs. Blackett is making for William is neatly folded on the table. "I sat in the rocking-chair," records the narrator, "and felt that it was a place of peace, the little brown bedroom and the quiet outlook upon field and sea and sky. I looked up, and we understood each other without speaking. 'I shall like to think o' you settin' here today,' said Mrs. Blackett. 'I want you to come again. It has been so pleasant for William.' "

As drama, it barely registers on the Richter scale. Yet perhaps the serene climax of *The Country of the Pointed Firs* conveys why Willa Cather could quote Henry James as saying of Jewett, "She had a sort of elegance of humility, or fine flame of modesty. She was content to be slight if she could be true."

Willa Cather knew Sara Orne Jewett briefly, during the sixteen months before Jewett's death. It was Louis Brandeis's wife Alice who took Cather—by this time no longer a refugee fresh from Nebraska, but an accomplished New York journalist and story writer in her early thirties—to the house on Charles Street. Cather showed Jewett her stories and took to heart the older woman's advice: to work at writing fiction full time, and write what she knew.

"Write it as it is, don't try to make it like this or that," Cather summed up the injunction. After a false start with her first novel, the pseudo-Jamesian *Alexander's Bridge* (1912), she turned seriously to her "home" material, and by 1918 had published all three of her prairie novels, *O Pioneers!, The Song of the Lark,* and *My Ántonia.*

Cather remained deeply grateful to Jewett, her only female mentor, and in 1925 wrote the introduction to a new edition of *The Country of the Pointed Firs* and other stories, lauding them as "almost flawless examples of literary art." She even likened *The Country of the Pointed Firs* to *Huckleberry Finn* and *The Scarlet Letter.* In an expanded essay on Jewett published in 1936, Cather was more restrained, saying only that Jewett, like Twain and Hawthorne, possessed that "very personal quality of perception, a vivid and intensely personal experience of life, which make a 'style.' "

It was a style that was rapidly becoming passé. By the 1930s, literary fashion was running to Fitzgerald and Hemingway, Eliot and Joyce. Wrote Cather: "Imagine a young man or woman, born in New York City, educated at a New York university, violently inoculated by Freud, hurried into journalism, knowing no more about New England country people (or country folk anywhere) than he has caught from motor trips or observed from summer hotels: what is there for him in *The Country of the Pointed Firs?*"

But the kinship between Cather and Jewett transcends fashion. Indeed, it consists partly of an indifference to fashion. Both are very American artists, responsive to nature, to landscape, and to people who live close to the land. Neither bothers much with politics or high society; both write about religion. Neither woman married or successfully portrays romantic love in fiction (Jewett doesn't try). Both are most at home writing about, as critic Joan Acocella says of Cather, "noble-minded people living in small towns."

It was a subject embedded in their life histories. Growing up in out-of-the-way places—Jewett in South Berwick, Cather in Red Cloud, Nebraska—they had some similar experiences. Each received her early education mainly through her friendships with adults. Just as Jewett accompanied her father on his medical rounds, so Cather attached herself to a German piano teacher, devoured the library of a Jewish couple, and rode out in the buggies of both of Red Cloud's doctors, peppering them with questions about science. Like Jewett, she reprised all this in fiction. In *The Song of the Lark* Thea Kronborg grows up to become not a doctor but a singer, yet Dr. Archie remains her life-long friend.

In other ways, however, Jewett and Cather's biographies—and their writing—sharply diverged. Jewett, whose fiction evokes a single, integrated culture, never really left home. South Berwick is only seventy miles from Boston. As a young woman, she could move into a cosmopolitan adult world without cutting her New England roots. She always spent summers in the Maine house where she grew up. She died in the house where she was born.

The contrast could hardly be greater with Cather, who early lost any chance for such stable belonging. When she was nine, her family made the wrenching move from their farm near Winchester, Virginia, to Nebraska, where they lived first on the prairie, then after a year in a town of about 1,200. Going to college meant the University of Nebraska, in raw Lincoln, scarcely gouged from the frontier. Work as a journalist and teacher took her to Pittsburgh, then New York. She traveled in the southwest and in Europe, and ultimately settled in Greenwich Village, summering in New Brunswick, Canada, and spending the fall months in Jaffrey, New Hampshire, where she is buried. Not surprisingly, Cather gave her stories widely varied settings: Nebraska, eastern Colorado, Chicago, Pittsburgh, Virginia, a French battlefield, New Mexico, seventeenth-century Quebec. Her last, unfinished novel was set in medieval Avignon.

Similarly, where Jewett wrote about people in their indigenous surroundings, Cather studied exiles: Bohemian immigrants and Scandinavian pioneers on the plains, farm girls in town and small-town girls in the big city, a Colorado pastor's kid on the stage of the Dresden Opera, and French priests in the lonely far reaches of the New World.

In 1925, when Willa Cather prepared her new edition of Jewett and wrote that introduction so lavishly praising *The Country of the Pointed Firs*, she herself was incubating what would prove to be her own most nearly perfect book, *Death Comes for the Archbishop*, published in 1927. There are enough affinities between *Death Comes for the Archbishop* and *The Country of the Pointed Firs* to suggest that Cather's great New Mexico novel was nourished by her reflections on Jewett's masterpiece.

Cather's essay praises Jewett's book for its structure—"so tightly built and significant in design"—and its inherent beauty. In both respects, her own book resembles it. Like *The Country of the Pointed Firs, Death Comes for the Archbishop* is a succession of episodes virtually without plot. A young French missionary working in Ohio is named the first bishop of New Mexico. He goes there, explores his immense diocese on horseback, encounters some singular personalities, has certain adventures, plants a garden, builds a cathedral, grows old, dies.

True, Bishop Latour's relationship with a second central character, his boyhood friend and vicar, Father Vaillant, appears throughout the book—rather as do the relationships among the three women in *The Country of the Pointed Firs*. And Cather's novel is held together by two other constants: the omnipresent scenery of desert and canyon, mesa and arroyo, stone and adobe; and the central thread, the bishop's everyday, faithful performance of his life's work. Nevertheless, *Death Comes for the Archbishop* is mostly discontinuous close-ups and freestanding scenes, strung together like beads on a string.

Cather could have been talking about her own New Mexico novel-in-the-making when she wrote, "The *Pointed Fir* sketches are living things caught in the open, with light and freedom and air-spaces about them. They melt into the land and the life of the land until they are not stories at all, but life itself." This capturing of life itself is what the artist strives for, and Cather begins her introduction with an observation of Jewett's from their correspondence about how it is achieved: "The thing that teases the mind over and over for years," Jewett wrote, "and at last gets itself put down rightly on paper—whether little or great, it belongs to Literature."

In *Death Comes for the Archbishop*, Willa Cather gave fullest expression to two themes that had teased her mind persistently for years: the Southwest and Christianity. (A third such theme—the French domestic arts as carriers of civilization—is present here but reaches full flower only in her next book, *Shadows on the Rock*.)

Cather first visited the Southwest in 1912. She returned again and again to explore the old towns and Spanish missions and cliff dwellings; and she steeped herself in the memoirs of early explorers and missionaries (including the originals of her Bishop Latour and Father Vaillant). She used the Southwest as the backdrop for a somewhat contrived passage of *The Song of the Lark*; then again for the middle section of *The Professor's House*, a book otherwise set in a midwestern college town—and published the very year of her essay on Jewett. Perhaps as she contemplated Jewett, who embraced her Maine material so unreservedly, Cather glimpsed what it would mean to devote an entire novel to the Southwest. "If [the writer] achieves anything noble, anything enduring," Cather wrote in her introduction to Jewett,

> it must be by giving himself absolutely to his material. And this gift of sympathy is his great gift; is the fine thing in him that alone can make his work fine. He fades away into the land and people of his heart, he dies of love only to be born again. The artist spends a lifetime in loving the things that haunt him, in having his mind "teased" by them, in trying to get these conceptions down on paper exactly as they are to him and not in conventional poses supposed to reveal their character.

So, too, Christianity "haunted" Cather—another link with Jewett. Reared in a Baptist home, Cather attended Episcopal services as a young woman, but it was only in 1922, when she was nearly fifty, that she was confirmed in the Episcopal Church. Apparently something had ripened in her own religious life in the years just before she undertook her reconsideration of Jewett and went on to write a whole novel about Catholic priests. Fresh from meditating on *The Country of the Pointed Firs*, with its deft interweaving of place, ethos, and personality, Cather produced a book saturated with a sense of place, about two men living out lives consecrated to God.

That it could influence, so profoundly, a book as good as *Death Comes for the Archbishop* is sufficient reason to take another look at *The Country of the Pointed Firs*, an American classic whose memory seems to have faded even among the well-read. It is a book whose power and beauty are difficult to sum up. It leaves in the mind of the reader, as Cather wrote, "an intangible residuum of pleasure; a cadence, a quality of voice that is exclusively the writer's own, individual, unique. A quality that one can remember without the volume at hand, can experience over and over again in the mind but can never absolutely define."

At the beginning of that essay on Jewett, Cather placed an epigraph from the poetess Louise Imogen Guiney: *But give to thine own story / Simplicity, with glory*. That word "glory"—while apt for Cather, never one "content to be slight"—doesn't ring quite true for the self-effacing Jewett. Closer to the mark are the words of an early commentator who praised Jewett's "sweet, sane knowledge of life." The chronicler of a world where conversation is a kindness—where "fitness" is an ultimate tribute and self-forgetfulness is "the highest gift of heaven"—Jewett gave a great deal to the more restless and ambitious literary heir who so warmly acknowledged the debt. To the overstimulated, worldly wise reader of today she has at least as much to offer.

# THE LAST PUBLIC POET: REREADING ROBERT LOWELL

## By Joseph Bottum

*(Originally published in the August 4–11, 2003, issue)*

ROBERT LOWELL BEGAN HIS poetic career by espousing religion—with all the marital fidelity of a gigolo on the make. A convert at age twenty-four, he quickly lifted from his Catholic moment a complex metaphysics, a large system of artistic imagery, and a Pulitzer Prize for his first full collection of poems, *Lord Weary's Castle*, in 1946.

What he couldn't seem to get from Catholicism, however, was enough of what he went looking there to find: content, mostly—something to write about that seemed worthy of the astonishing power his poetic voice had from the very beginning of his career. Sanity, too, wasn't waiting patiently for him in the pews. And so, in the early 1950s, he abandoned the Church to chivvy, over the next twenty-five years, both his poetry and his mental health through a long series of alternatives—the diseased memories of his childhood, the corpus of world poetry, the rage of 1960s politics, the sum total of human history, and finally even his ex-wife's letters—all in the attempt to find a topic sufficient to match the ability he had to express it.

Nothing except the recreation of the world itself—nothing except being God, in fact—could have satisfied the cosmic ambitions of his poetry, and it's tempting to say that he was never really serious about any of the subjects he took up in his writing. But even to begin thinking this way is to sound ridiculous, for Robert Lowell was perhaps the most *serious* poet America has ever known—our last poet of high seriousness, as it happens, and also our last public poet.

There's much to dislike about the man. To read Ian Hamilton's 1982 biography *Robert Lowell*, or Paul Mariani's 1994 *Lost Puritan*, is to see that Lowell's family life was a godforsaken mess (even discounting the large portions that can be blamed on his frequent bouts of madness). To read his exchanges with Diana Trilling about the 1968 student riots at Columbia—or the brouhaha surrounding his withdrawal from Lyndon Johnson's 1965 "White House Festival of the Arts"—is to see that his politics were routinely silly (again, leaving aside the merely crazy parts). And to read even such fond portraits as Eileen Simpson's delightful 1982 memoir *Poets in Their Youth* is to see that his interventions in the literary world—over Ezra Pound's 1949 Bollingen prize, for instance, or the management of the Yaddo writers' retreat—were invariably peculiar and occasionally vicious (ignoring, one last time, the insane bits).

But to read the thousand pages of his *Collected Poems*—finally published this summer, a quarter century after Lowell's death in 1977 at the age of sixty—is also to see how little his failings matter to his poetry. Right or wrong, he had the voice of public authority, and he did his work in the public eye. *Time* magazine put Lowell on its cover in 1967, anointing him America's national poet. The newspaper gossip columns noted his marriages and divorces. He won book awards as though they were discount coupons: the National Book Award for 1959's *Life Studies*, a second Pulitzer for 1973's *The Dolphin*, and nearly every other literary prize imaginable (except the Nobel, for which he probably died too soon; curious to think that if Lowell had lived, he would be only eighty-five today).

From the first poems he published, he seemed to belong to the great tradition of poetry; whether great or not himself, he appeared the very incarnation of literature at the time. "The age burns in me," he wrote, and he was right. A new poem from Lowell was an *event*, something to be talked about, in a way that we haven't seen since. No general reader under age forty knows what it means to have a public poet in America, and hardly any general reader over forty has followed a poet since. Poetry remains popular these days, in its fashion. But poems no longer seem things of public importance. Something went out of poetry when Robert Lowell died.

Something went out of America, as well.

In *The Armies of the Night*—an account of the 1967 anti-Vietnam march on the Pentagon that contains an astonishing amount of Lowell worship—Norman Mailer described the poet as having "the unwilling haunted saintliness of a man who was repaying the moral debts of ten generations of ancestors." The poet Elizabeth Bishop once jokingly complained that Lowell had a certain authority just because he was a *Lowell* and not, say, her Uncle Artie: Simply to recite the names in his family, from James Russell Lowell to Amy Lowell, was to talk about America itself. Robert Traill Spence Lowell IV was born in 1917 in Boston—where the Lowells talk to the Cabots, and the Cabots talk only to God—the child of three hundred years of *Mayflower* ancestors. Every other president in the history of Harvard was a relative. Boston society consisted entirely of his cousins.

Not that it made much difference while he was a child. He was banned from the Boston Public Garden for fighting, and his school-mates nicknamed him "Cal": after the Roman emperor Caligula, according to one version of the story, or Shakespeare's Caliban, according to another version—either way, not exactly the image one wants for a boy. After his freshman year at Harvard and a knockdown argu-ment with his weak father and overbearing mother, he fled south to spend the summer camped on the lawn of the poet Allen Tate. The Southern Fugitives quickly drew him in, and he transferred to Kenyon College, where he studied with John Crowe Ransom and became friends with fellow student Randall Jarrell.

After graduation, he converted to Catholicism, worked on his first chapbook of poems, *Land of Unlikeness*, and married the young Catholic novelist Jean Stafford. It was a curious home. Describing her life after Lowell, Stafford once explained how nice it was to live in a place where it was all right not yet to have won the Nobel prize. Lowell insisted on daily Mass and limited the family reading: "no newspapers, no novels except Dostoevsky, Proust, James, and Tolstoy." He also became what he called a "fire-breathing" conscientious objec-tor, and in 1943 he sent an open letter to President Roosevelt and various newspapers denouncing the war and refusing to serve. It was, he later admitted, a "manic statement," but the government's hand was forced by publicity he received, and he was convicted of draft evasion, serving several months of a one-year sentence in federal prison.

The book *Lord Weary's Castle* followed in 1946. There was a little of Allen Tate in it and a lot of T. S. Eliot. But mostly it seemed a completely original stew of American Puritanism, European Catholicism, and New England history—all in service of a serious modernism, written in the traditional rhyme and meter that had seemed anathema to modernism. "The Quaker Graveyard in Nantucket" is the longest poem in the collection, and it reads almost as though John Milton had decided to rewrite *Moby-Dick*. An elegy for one of Lowell's Winslow cousins, lost at sea, it begins with hard enjambments and thick, loud lines—as though the language itself were compelled to match the topic:

> *A brackish reach of shoal off Madaket,—*
> *The sea was still breaking violently and night*
> *Had steamed into our North Atlantic Fleet,*
> *When the drowned sailor clutched the*
>     *drag-net.*

"The Drunken Fisherman," "At the Indian Killer's Grave," and many other poems in the collection turned American Puritanism into the great tradition of Catholicism—in lines that read as though they had been carved from granite. Jonathan Edwards's theology was a perpetual fascination of Lowell's, and "After the Surprising Conversions" renders as poetry one of Edwards's famous letters about New England's Great Awakening:

> *September twenty-second, Sir, the bough*
> *Cracks with the unpicked apples, and at dawn*
> *The small-mouth bass breaks water, gorged*
>     *with spawn.*

What was America to make of all this? The nation decided to sink to its knees in awe. A Guggenheim fellowship followed the Pulitzer, and Lowell at age thirty was free to do anything.

What he did was divorce Stafford, leave the Church, and go insane for the first of what would be many times. After his recovery, he married the writer Elizabeth Hardwick, and, in 1951, published his second collection of poems, *The Mills of the Kavanaughs*. Though it contained poetry as good as "Falling Asleep Over the Aeneid," which followed the mood of *Lord Weary's Castle*, the book seemed in many ways to flounder. The title poem is a long dramatic monologue spoken by a young woman in Maine whose voice, Randall Jarrell suggested, sounded just the way a girl would sound if she were Robert Lowell— but who ever met a girl like Robert Lowell?

No one could mistake the voice in his third book, *Life Studies*, published in 1959. Suffering more breakdowns and undergoing a cycle of institutionalizations and psychiatric therapy, he rolled together autobiographical prose and poems, still rhythmical though mostly unrhymed—creating the foundational document of an entire school of American poetry, the juggernaut of the "Confessional Poets." The collection ends with its best poem, "Skunk Hour," which owes something to Elizabeth Bishop's descriptions of animals but moves in what seemed at the time like nothing else in American verse:

*A car radio bleats,*
*"Love, O careless Love, . . ." I hear*
*my ill-spirit sob in each blood cell,*
*as if my hand were at its throat. . . .*
*I myself am hell;*
*nobody's here—*

*only skunks, that search*
*in the moonlight for a bite to eat.*

Helen Vendler has noted how mild some of these poems in *Life Studies* actually are: *blue threads as thin / as pen-writing on the bedspread.* But—as Lowell described his childhood, his parents, his time in prison, and his insanity—the reviewers latched on to the shocking poems exposing his family's failures and sins. "Commander Lowell," a son's blast at the weakness and ineffectuality of his naval-officer father, is a brutal poem from beginning to end, but the cruelest moment comes in the final lines, where Lowell adds to the description of his father's meaningless old age the lines: *And once / nineteen, the youngest ensign in his class, / he was "the old man" of a gunboat on the Yangtze.*

The volume *Imitations* followed in 1961, Lowell's eccentrically chosen but brilliantly rendered set of translations of everything from ancient Greek to modern German. And then, with 1964's *For the Union Dead*, Lowell turned the confessional voice from himself to the outside world of politics and social ruin. The title poem moves from childhood memories of the "old South Boston Aquarium" to a meditation on Augustus Saint-Gaudens's bronze memorial for Robert Gould Shaw, the young white Bostonian abolitionist who was killed leading a black regiment in the Civil War: *Two months after marching through Boston, / half the regiment was dead; . . . Their monument sticks like a fishbone / in the city's throat.* The poem ends:

> *When I crouch to my television set,*
> *the drained faces of Negro school-children rise*
>     *like balloons. . . .*

> *The Aquarium is gone. Everywhere,*
> *giant finned cars nose forward like fish;*
> *a savage servility*
> *slides by on grease.*

The mood continued through 1967's *Near the Ocean*, with its strange amalgam of politics and Lowell's attempt to explain why he no longer believed in God in the volume's most famous poem, "Waking Early Sunday Morning": *Pity the planet, all joy gone / from this sweet volcanic cone.* But the bouts of insanity grew more frequent, and though he was, if anything, *more* political in his actions—marching on the Pentagon, campaigning for Eugene McCarthy—his poetry began to seek madly in history some connection between his jumbled brain, his jumbled politics, and the jumble of the human condition.

His private life got messy again, as well. In 1970, he left his wife Elizabeth Hardwick to take up with a married Irish woman, Lady

Caroline Blackwood, whom he married in 1972 after their divorces came through.

Perhaps it's not surprising that at this point his publishing also grew confusing. He'd begun to keep a poetic journal, a sort of sonnet sequence of daily events in the newspaper and his personal life. He published it in 1969, as *Notebook 1967–68*, and then revised it to publish it again the next year as *Notebook*, and then revised it yet one more time when he published three books in 1973: *History*, which contains the political and public sonnets from *Notebook; For Lizzie and Harriet,* which contains the personal poems; and *The Dolphin*, which relates his abandoning of the same wife Lizzie and daughter Harriet that he celebrated in *For Lizzie and Harriet.* Just to make matters worse, *The Dolphin* uses quotations from Hardwick's private letters, splashing across the literary reviews her attempts to cope with his madness and keep their marriage of twenty years together.

Lowell published his *Selected Poems* in 1976 and his last collection, *Day by Day*, in 1977. That fall, deciding to leave his new wife for his old, Lowell flew from London to New York and died of a heart attack in the taxi on his way into Manhattan from the airport—a sixty-year-old man, carrying the painter Lucien Freud's portrait of Caroline Blackwood back to Elizabeth Hardwick's house.

The general reader of literature can now walk many of the poetic battlefields of the twentieth century with little more emotion than the tourist's usual wonder at how much blood was spilt to gain so little ground. Along that low wall, the Georgians made their last, doomed stand. That hilltop over there is where contemporary modernism was decided, the high mandarins easily crushing the populist, lowbrow rebellion from the likes of Vachel Lindsay, Robinson Jeffers, and Carl Sandburg. Across that nearby field the Beat berserkers once howled their way close to victory before their charge was at last turned back.

Perhaps as a result, recent school anthologies have begun to agree on something like a canon of twentieth-century American poetry. William Carlos Williams has won, and Stephen Vincent Benét has lost. Hart Crane has surprisingly faded, and Wallace Stevens has unsurprisingly shone. Delmore Schwartz has been washed under by the great wave of the world, while Sylvia Plath has made it safe to shore. Amy Lowell is out, and Robert Lowell is . . . well, what is he these days? Time will revisit some of these judgments. Time *ought* to revisit some of these judgments. But what will time make of Lowell?

This is a moment of decision about Lowell—not the final judgment of hundreds of years' reading, but a real moment, nonetheless, at which we must decide where he belongs in the pantheon, thanks to Frank Bidart, Lowell's longtime "amanuensis and sounding board," who has finally finished editing the *Collected Poems*.

The book has its peculiarities. Bidart writes in the introduction about the importance of Lowell's first chapbook, *Land of Unlikeness*, but then prints it only as an appendix. He doesn't print at all the 1970 version of *Notebook*, though he insists it is an independent work, not to be conflated with the volumes Lowell mined from it in 1973. Still, with a thousand pages of poetry and a hundred and fifty pages of judicious and informative notes, Bidart's edition of Robert Lowell's *Collected Poems* is enough to be going on with.

What makes judgment difficult is the fact that we have so few public poets with whom to compare Lowell. Poetry done in private, even by famous poets, is distinct from poetry done in public. Public poetry aims at different targets, it speaks to different purposes, and it is judged by different standards—primarily by the standard of responsibility, for it has a claim to speak, with the special insight of its unique gift of language, on what are or ought to be the public issues. And, more to the point, it has an audience that agrees to listen while it makes its claim.

As it happens, not all poetry on public events succeeds at being public poetry, in this sense. Though they'd settle for being the unacknowledged legislators of the world, all poets want really to be the *acknowledged* legislators: They want to pronounce, and they want us to listen. But, as demonstrated by the recent tempest over Mrs. Bush's attempt to invite poets to a White House tea just before the Iraq war, contemporary poetry is missing both the voice of public responsibility and the ear of the responsible public—a kind of high, morally serious agreement between poets and their readers.

Whatever that was exactly, Lowell had it. There's no denying that a great deal of his work draws in the reader brilliantly. The person who isn't mesmerized by "The Quaker Graveyard in Nantucket," "Skunk Hour," and "For the Union Dead" has forgotten what poetry is. In the middle of his career, he could finish the dramatic monologue "To Speak of Woe That Is in Marriage" with a woman's lament: *Each night now I tie / ten dollars and his car key to my thigh. . . . / Gored by the climacteric of his want, / he stalls above me like an elephant.* Later in life, he could begin "Waking Early Sunday Morning" with the lines:

> *O to break loose, like the chinook*
> *salmon jumping and falling back,*
> *nosing up to the impossible*
> *stone and bone-crushing waterfall—*
> *raw-jawed, weak-fleshed there, stopped by ten*
> *steps of the roaring ladder, and then*
> *to clear the top on the last try,*
> *alive enough to spawn and die.*

And yet, good as he was, the simple truth is that Richard Wilbur, Elizabeth Bishop, and Anthony Hecht were all better at some of the things he attempted. In retrospect, the Pulitzer committee in 1959— that *annus mirabilis* for the creation of "Confessional Poetry"—may have been right to have given that year's prize to W. D. Snodgrass's confessional *Heart's Needle* rather than Lowell's confessional *Life Studies*. When Lowell tried craft, J. V. Cunningham proved the better craftsman. When Lowell tried learned drunkenness, John Berryman's first volume of *The Dream Songs* had already captured the field. When Lowell tried high-voltage effusions, Allen Ginsberg and Gregory Corso made him look lethargic. Randall Jarrell was a better critic, Delmore Schwartz was a better literary operator, and Lowell's students Sylvia Plath and Anne Sexton were more successful at self-dramatization.

But Robert Lowell was not just respected, or famous, or infamous, as his contemporaries aspired to be. He insisted—and succeeded in his insistence—that his work be judged by the standard of public responsibility. Of course, judged that way, Lowell was also a massive failure, as irresponsible a public poet as English literature has known since Percy Shelley.

The examples are endless. This is a man who, at nineteen, could become engaged, leave Harvard, knock his father to the ground for daring to say something less than complimentary about his fiancée, and then promptly abandon the fiancée to go study poetry with Allen Tate and John Crowe Ransom. (When Tate, trying to explain that Lowell was not invited to stay for the summer, joked that the house was so full he'd have to pitch a tent on the lawn, Lowell promptly went out and bought a tent.)

This is a man who could first make failed attempts to enlist in both the Navy and the Army, and then write an open letter calling the Second World War "a betrayal of my country." This is a man who could break his wife Jean's nose twice, once in a car accident and once with his fist. And what, besides unforgivable, are we to call his capping his career as a confessional poet by publishing extracts of his ex-wife Elizabeth's letters?

And yet, through it all, he somehow kept the public's ear. He had *gravitas*, we all agreed, and nothing could take it away from him. His conversion from Tate and Eliot's high modernism in *Lord Weary's Castle* to the confessional poetry of *Life Studies* didn't actually make his public thoughts private; it made his private life public.

The fact that he was a Boston Lowell helped, of course. But there was more to it than the last gasp of the Back Bay social world. Lowell took himself as seriously as America took him, and through it all, he wanted poetry to matter in a way that hardly anyone these days appreciates once seemed possible; his work aimed at the sum total of creation. For that purpose, his later fascination with the self and politics was a poor substitute for his early fascination with religion, but all his fascinations sought to provide his poems with the seriousness and public standing he felt they deserved.

Did they in fact deserve it? Now that the dust has settled, we can look back and decide. Lowell was always better at suggesting that connections exist than he was at explicating those connections; he was always better at showing us a mind that believes in a conjunction than he was at convincing us the conjunction is real. But so what? Much of the poetic mind is taken on faith, an agreement we make that poets' skill at language gives them some insight worth our time to pursue. "The Quaker Graveyard in Nantucket" ends:

*You could cut the brackish winds with a*
  *knife*
*Here in Nantucket, and cast up the time*
*When the Lord God formed man from the*
  *sea's slime*
*And breathed into his face the breath of life,*
*And blue-lung'd combers lumbered to the kill.*
*The Lord survives the rainbow of His will.*

Robert Lowell survives the rainbow of his own will—a little tattered, a little less important than we once thought him, but still alive, still the genuine thing.

# VIRGIL'S *AENEID*—AND OURS

## By Robert Royal

———

*(Originally published, as "Virgil Lives!," in the September 29, 2003, issue)*

IN STANDARD HISTORIES OF LITERATURE these days, Virgil tends to be characterized as a fairly gifted versifier and coiner of a few memorable phrases: "Arms and the man I sing," "Love conquers all," "I fear the Greeks, even bearing gifts." The *Aeneid*—his epic poem about the founding of ancient Rome, in ten thousand dactylic hexameter lines— was once the dominant classical epic in the West, and Dante justly made Virgil his first guide in the *Divine Comedy*. But from the nine- teenth century on, Virgil has faded somehow—until he has reached near dismissal, in our own age, as the poor man's Homer: Caesar Augustus needed a heroic poem to justify his rule over the Roman Empire, we have been told, and Virgil obligingly wrote one for him. That's apparently all we need to know about the *Aeneid*—and all we need to know about Virgil, too.

Every schoolboy once knew a fuller story. Born in 70 BC, Publius Vergilius Maro had a long and close history with the future emperor—in some legends, going all the way back to Virgil's youth, in which he is supposed, as a farm boy from the northern Italian city of Mantua, to have cured some of Augustus' horses. His literary tal- ents surfaced early. The *Eclogues*, ten pastoral poems, were so obviously superb that Cicero called him Rome's second greatest hope (reserving first place to himself). And the fourth *Eclogue* had a curious career: Written in the last few decades before Christ, it predicted the birth of a miraculous boy who would restore the mythical Golden Age. Later Christian readers applied this to Jesus and regarded Virgil as a prophet and magician. His four books of the *Georgics*—a seven-year effort on agricultural subjects—won him further praise.

But the twelve books of the *Aeneid*, on which Virgil spent his last decade, were quickly judged a masterpiece of Latin literature. We owe the poem's survival to Augustus. Virgil fell ill on his way to Greece, where he intended to spend three years polishing his poem, and died in 19 BC in the eastern Italian port city known today as Brindisi. A perfectionist, on his deathbed he asked friends to burn the manuscript. Fortunately, Augustus overruled this dying wish and had a pair of literary scholars bring out the text *summatim emendata*, with only slight editing.

Part of the explanation for Virgil's modern decline is classicists' general preference for Greek sources over Latin—which began slowly in the Renaissance and gathered irresistible momentum through the sheer power of eighteenth- and nineteenth-century German scholarship. The catastrophic decline of reading knowledge in Latin among the generally educated in twentieth-century England and America also contributed to the shrinking of Virgil's natural audience.

But something political seems to be at work in the dismissal of Virgil, as well. His closeness to Augustus (and the emperor's well-known desire to maintain a façade of classical tradition while covertly recasting it in Roman imperial form) has deeply shaped approaches to the epic, for good and later for bad, over the centuries. The first half of the *Aeneid*, in this reading, is Virgil's *Odyssey*; it tells of a Trojan warrior named Aeneas, who wanders the Mediterranean after the fall of Troy and eventually founds the city of Rome. Similarly, the second half of the *Aeneid* is Virgil's *Iliad*, recounting battles in Italy and connecting Roman history with the heroic age of the Trojan War.

After World War II, T. S. Eliot tried to resuscitate the Latin poet, declaring that, "Our classic, the classic of all Europe, is Virgil." Translations of the *Aeneid* continue to appear. Between the imperial reading of Virgil and the general anti-imperial feeling of the twentieth century, however, the *Aeneid*'s high place in the Western literary canon was doomed. A few recent scholars, partly reflecting contemporary sensibilities, have detected ambiguities in the poem that raise the question of whether Virgil was, in his heart of hearts, a true believer in empire. The damage, however, was done: Except for this mop-up operation on a few critical questions, Virgil's position in literature seemed fixed—at a moderate height—forever.

Eve Adler, a classicist at Middlebury College in Vermont, may have just changed all that with her new book, *Vergil's Empire: Political Thought in the Aeneid*. (The traditional English spelling of the poet's name is "Virgil." Adler follows the trend in some recent classical studies of spelling the name "Vergil," more closely reflecting the Latin.) After this analysis, it will be difficult to think of Virgil merely as a gifted imitator of Homer. If Adler is right, Virgil had ambitions at least as grand as his Greek predecessor—and with good reason.

*Vergil's Empire* draws heavily on Leo Strauss for the political analysis of the *Aeneid*. Something of a secret teaching may be glimpsed behind the imperial screen, she argues, which emerges most clearly near the center of the text, where Aeneas' descent into the underworld signals the shift from wandering to battles. But her sensitive and penetrating reading of many passages in the *Aeneid* does not reduce Virgil to a Procrustean bed of Straussian proportions. This book is stunningly original. Indeed, Adler's account of Virgil's views on universal empire has urgency not only for literary studies but for our reflections on empire in the current global situation.

Adler believes that Virgil is powerfully grappling not only with Homer, but with Lucretius, his Latin predecessor in the first century BC. The first great poet after Rome's clear emergence as the classical superpower, Lucretius presents a problem—what we might call the anxiety of influence—for all the Augustan Golden Age poets: Horace, Virgil, Sextus Propertius, Ovid. But Virgil, in Adler's reading, is much more of a philosopher than he is often thought, and Lucretius offered a particular challenge for Virgil—because of the Epicurean philosophy Lucretius laid out in his book-length poem *De Rerum Natura* (On the Nature of Things).

The followers of Epicurus were materialists who denied the existence of the gods and sought as tranquil a life as this world can offer in private enjoyments and material comforts. Politics, in particular, was strictly avoided as leading to pointless troubles. Thus, Lucretius begins the second book of *De Rerum Natura*:

> *Pleasant it is, when over the great sea the*
> *winds shake the waters,*
> *To gaze down from shore on the trials of others;*
> *Not because seeing other people struggle is*
> *sweet to us,*
> *But because the fact that we ourselves are free*
> *from such ills strikes us as pleasant.*
> *Pleasant it is also to behold great armies*
> *battling on a plain,*
> *When we ourselves have no part in their peril.*
> *But nothing is sweeter than to occupy a lofty*
> *sanctuary of the mind,*
> *Well fortified with the teachings of the wise,*
> *Where we may look down on others as they*
> *stumble along,*
> *Vainly searching for the true path of life.*

There are many signs that the young Virgil was an Epicurean and that he never wholly repudiated that philosophy in adulthood. But Adler believes Virgil detected a fatal flaw in the Epicurean system, which he presents most memorably in the contrast between Aeneas and Queen Dido, and between Rome and Carthage. It may be true that the radically rational philosopher is freed from fear of both the gods and death—while limiting himself to rationally moderate pleasures. But such philosophers are so rare as to be of almost no social effect. Almost always, those who free themselves from traditional religion find themselves, like poor Dido, subject to *furor*: anger and lust. Epicurus was far too optimistic about our ability to tame these demons, and in his desire to spread this philosophy to the entire populace, Lucretius threatens the civic order. Indeed, he invites his own destruction, for the retired life of the Epicurean philosopher depends upon the existence of a peaceful city, which the passions unleashed by disbelief in the gods will not produce.

If Virgil had been a Straussian *avant la lettre*, he might have contented himself with suggesting that for the sake of private tranquillity the philosopher should connive at public religiosity, even though false, as a means of restraining and educating the masses. That would enable the philosopher to achieve his proper happiness—and the masses to enjoy as much good fortune as they are capable of.

But in Adler's reading, Virgil goes a step further; he has been affected by Epicurean materialism, but is not wholly certain of the ultimate truth about nature. The shortcomings of Epicureanism, however, convince him beyond all doubt that arms and religion are needed to remedy evil tendencies in human nature.

Virgil's powerful mind actually leads him to recast almost all the usual elements of this debate. For example, the arms he sings are not simply a continuation of the old heroic ethos of Homer. That, *Vergil's Empire* notes, is certainly one way to confront the fear of death, but it is ultimately as rare as the way of the Epicurean philosophy. Nor does the turn to domestic pleasures satisfy Virgil. In Homer, Odysseus refuses Circe's offer of immortality because of his loyalty to Penelope and Ithaca. But that too is only a half-measure against death. In Virgil, Aeneas both braves death in battle and seeks a new city for the Trojan gods. His main virtue is *piety*—something without precedent in the Greek stories of Achilles and Odysseus.

Thus, Adler argues, Virgil is consciously seeking to surpass Homer as well as Lucretius. So new and radical was this shift in the ancient world, Adler claims, that it raised the question of whether Virgilian piety—a mixture of duty, religiosity, and loyalty—is compatible with manliness (the root meaning of the Latin word "virtue"), as the ancients understood it. Epicureans could claim heroic virtue in rejecting the consolations of religion, even if they lived relatively unstrenuous lives. But if we also reject the Homeric combination of martial valor and human domesticity—a combination traditionally embraced by the Romans—what's left?

Not much of the dominant classical systems, but part of Virgil's genius is to have discovered another ethos, one that acknowledges something like divine providence in history, especially in the fated nature of Rome. Aeneas and his men will suffer along the way (*lacrimae rerum*, or the sorrows attending all human affairs, is another Virgilian idea that used to be a cultural commonplace). But Jupiter, Rome's greatest god, promises early in the poem (in Robert Fitzgerald's translation):

> *Young Romulus*
> *Will take the leadership, build walls of Mars,*
> *And call by his own name his people Romans.*
> *For these I set no limits, world or time,*
> *But make the gift of empire without end.*

This was already an unusual claim within the classical understanding of the world and of time. Aeneas' visit to the underworld in Book Six of the *Aeneid* even led people to believe that Virgil was a "naturally Christian soul," *anima naturaliter christiana*, in the Middle Ages. His new vision of piety and Aeneas' fateful journey have echoes (which Adler does not mention) of Abraham setting out for the Promised Land at the divine command, another point of contact with the Biblical tradition. But Aeneas returns from the underworld to our life through the gates of ivory, which Virgil goes out of his way to explain is the portal of false dreams and prophecies. As daring and inventive as Virgil was, he knows that pure reason comes up against a limit—although a human limit that Virgil was occasionally tempted to cross.

One reason for his hesitation was his worry about what the city he envisioned, even if it was a kind of holy city, might lead to. It would not be enough for Aeneas to found another city like Troy; that city would be subject to perpetual danger from neighbors, as were all the squabbling city-states of classical Greece. The city that Aeneas had to found would be universal, as Jupiter promised, without limit in space or time. That was the only way it could fulfill its divine mission. As Aeneas' dead father tells him in Hades:

> *Others will cast more tenderly in bronze*
> *Their breathing figures, I can well believe,*
> *And bring more lifelike portraits out of marble;*
> *Argue more eloquently, use the pointer*

*To trace the paths of heaven accurately*
*And accurately foretell the rising stars.*
*Roman, remember by your strength to rule*
*Earth's peoples—for your arts are to be these:*
*To pacify, impose the rule of law,*
*To spare the conquered, battle down the proud.*

In this, perhaps the most famous passage in the poem, the classical arts and sciences are not rejected, but are left to other peoples, subordinated in Virgil's vision to the god's specific demands of the sacred city. Anything less would be radically deficient in establishing a stable peace, given the nature of the world and human nature.

Yet Virgil does not stop even here. Contrast this, as *Vergil's Empire* does, with Dante. Trying to solve the problem of clashing city-states in his own time, Dante argues in his essay on monarchy, implausibly, that possession of universal empire would quell the emperor's temptations to tyranny. Nonetheless, Dante's vision of the Christian God does provide "a term to men's desire or love in a fully adequate object," says Adler. Virgil, still linked in many ways to the old pagan mythology, lacks any such notion of absolute love. Even though he announces in the first few lines of the *Aeneid* that Aeneas has to bring gods into Italy, force and religion run up against a limit in the classical cosmos because the neediness of all living beings finds no final remedy among men or gods: "It would be folly," Adler writes, "to hope for the disarming of the erotic passions by reason in any but the rarest philosopher, and certainly in any ruler: Dido in spite of her philosophic tutor, certainly pious Aeneas, and ultimately even Jupiter himself are subject to the furor of these passions." In a world where even the highest god is characterized as committing rapes and abductions, what hope is there for a perpetual peace under a human emperor?

It is one measure of Adler's achievement in *Vergil's Empire* that even though she—along with Virgil—cannot answer that question, she is worth reading very carefully, not only for what we can learn about a step in the development of the West, but what we can learn about our time as well. It is no accident that the modern equivalent of Epicureans—materialistic, disdainful of religion—tend to be overly optimistic about human nature and to resist the idea that we need war or other forms of coercion to restrain vice. It is equally no accident that the modern equivalent of Virgilians—with a religious vision about the need for the right kind of piety in the human city—are more likely to view both arms and religion as essential to the good of the United States and the restraint of evil in the world.

An empire, even a benevolent one, may overreach, of course. And Virgil hints that there are, humanly speaking, perhaps even seeds of self-dissolution in the most providential and perfect of empires. And so we oscillate between force and restraint, unmindful of their deeper meaning—still caught in the dynamic perceived by Virgil and brilliantly revived for us by Eve Adler in *Vergil's Empire*. The time has come to restore Virgil's epic poem to its place at the center of Western literature—both for its poetic qualities and because we have not surpassed the *Aeneid* or the world it portrays. In many ways, we are still living in it.

# BOB DYLAN'S LYRIC VERSE

## By Christopher Hitchens

*(Originally published, as "America's Poet?," in the July 5–12, 2004, issue)*

"NOT ALL GREAT POETS—like Wallace Stevens—are great singers," Bob Dylan once suggested. "But a great singer—like Billie Holiday— is always a great poet."

It would be an enterprise in itself to disentangle the many ways in which this brief statement is dead wrong. The antithesis, if it is meant as an antithesis, between poet and singer, is false to begin with. The "not all" is based on a nonexpectation: How many poets have been singers at all? Certainly not Dylan Thomas, the Welsh boozer and bawler from whom Bob Dylan—a Jewish loner from Hibbing, Minnesota, who was born as Robert Zimmerman—annexed his *nom de chanteur.*

Other cryptic or pretentious observations, made by Bob Dylan down the years, have licensed the suspicion that he's been putting people on and starting wild-goose chases for arcane or esoteric readings that aren't there. There are also those who maintain that Dylan can't really sing. (This latter group has recently been reluctantly increasing.) Of his ability as a poet, however, there can be no reasonable doubt. I used to play two subliterary games with Salman Rushdie. The first, not that you asked, was to retitle Shakespeare plays as if they had been written by Robert Ludlum. (Rushdie, who invented the game, came up with *The Elsinore Vacillation, The Dunsinane Reforestation, The Kerchief Implication,* and *The Rialto Sanction.*) The second was to recite Bob Dylan songs in a deadpan voice as though they were blank verse.

In addition to the risk of the ridiculous, it can become quite hypnotic. Try it yourself with "Mr. Tambourine Man": It works so well, you hardly care that a tambourine man can't really be playing a song. "Lily, Rosemary and The Jack of Hearts," "Chimes of Freedom," and "Desolation Row" all have the same feeling.

But as a guide to Dylan's poetic moments, do we really need help from Christopher Ricks, author of *Keats and Embarrassment*, editor of T. S. Eliot's juvenilia, instructor on the funny side of *Tristram Shandy*, and all-around literary mandarin? Need him or not, we now have Ricks—who, in *Dylan's Visions of Sin*, performs over five-hundred pages of literary criticism on the lyrics. Reading Dylan as the bard of guilt and redemption, Ricks takes his stand on the recurrence in the songs of the seven deadly sins, only just balanced as they are by the four cardinal virtues and the three theological virtues (or heavenly graces: faith, hope, and charity).

It's Ricks's own potentially deadly virtues that bother me. What temptation should one avoid above all, if one is a former professor of English at Cambridge? The temptation to be matey, or hip, or cool—especially if one is essaying the medium of popular music. But Ricks begins his book like this: "All I really want to do is—what, exactly? Be friends with you? Assuredly I don't want to do you in, or select you or dissect you or inspect you or reject you."

The toe-curling embarrassment of this is intensified when one appreciates that Ricks is addressing his subject, not his reader. Why did he leave out other verbs Dylan had in that song: *simplify you, classify you, deny, defy, or crucify you?* And surely, he's already at least "selected" him?

Then, accused by one of his usually admiring rivals in Dylanology, Alex Ross of the *New Yorker*, of "fetishizing the details of a recording," the prof resorts to unbearable archness. ("What me? All the world knows that it is women's shoes that I am into.") Some of Ricks's jokey attempts at making puns work ("cut to the chaste"), but "interluckitor" is a representative failure. This last is coined to deal with a claim by Dylan, made in 1965, that every song of his "tails off with—'Good Luck—I hope you make it.' " Such a claim, if taken seriously, would in any case vitiate most of Dylan's claims to profundity.

Having said that distinguished academics ought not to try and be ingratiating with the young, I pull myself up a bit and realize that true Dylan fans are probably well into their fifties by now. It must have been in 1965 that I first heard what Philip Larkin called, in a quasi-respectful review of *Highway 61 Revisited*, his "cawing, derisive voice." And it will be with me until my last hour. Some of this is context. The "sixties" didn't really begin until after the Kennedy assassination (or "Nineteen Sixty-Three," as Larkin had it in another reference), and Bob Dylan was as good a handbook for what was supposedly happening as Joseph Heller. Much of it of course also had to do with the sappiness, in both "sap" senses, of adolescence. Yet even at the time, I was somehow aware that Dylan wasn't all that young, and didn't take "youth" at its face value. A good number of his best songs were actually urging you to grow up, or at any rate to get real. Dylan respected his elders, most notably Woody Guthrie. And he was braced for disillusionment. *How does it feel? Don't think twice, it's all right. It's all over now, baby blue. I was so much older then, I'm younger than that now.*

Ricks essentially wants to argue that Dylan has always been swayed by the elders and that his verses consistently defer to the authorities. How else to explain, for example, the many latent affinities between "Sad-Eyed Lady of the Lowlands" and the Book of Ezekiel? The kings of Tyre, the dying music, the futility of earthly possessions. . . . That's Covetousness taken care of, with Pride (or at any rate hubris) given a passing whack into the bargain. Six sins to go.

Ricks has no success with Greed (as he admits) and not much with Sloth, either. There is a good deal of anomie and fatalism in Dylan; a fair amount of shrugging and dismissal and an abiding sense of waste and, equally often, of loss. It's pervasive but nonspecific in "Time Passes Slowly," which Ricks interrogates without any great profit. So I pushed on to Lust, and was taken aback.

"Lay, Lady, Lay" is one of the great sexual entreaties, and it has in common with "I Want You" and "If You Gotta Go, Go Now" a highly ethical reliance on the force of gentle persuasion. There is no blackmail, moral or otherwise, and no hint of a threat or even a scene in the event of nonconsummation. But nor is there any doubt of what the minstrel wants: *His clothes are dirty but his hands are clean. / And you're the best thing that he's ever seen.* Of this false modesty and abject flattery,

Ricks astonishingly says that "his hands are clean because he is innocent, free of sin: no lust, for all the honest desire, and no guile." Had
Dylan written "his clothes are dirty but his mind is clean," this might
have been believable. And is there no guile in the succeeding stanza?

*Stay, lady, stay, stay with your man awhile*
*Why wait any longer for the world to begin?*
*You can have your cake and eat it too.*
*Why wait any longer for the one you love*
*When he's standing in front of you?*

Ricks then moves to a laborious comparison with Donne's "On
His Mistress Going To Bed," at which point I thought, well, as soon as
I turn the page he'll stop clearing his throat and make the obvious
metaphysical connection to Andrew Marvell and "To His Coy
Mistress." But no. And here's the clue to Ricks's method. The words
"bed," "show," "see," "man," "hands," "world" he says all appear in
both Donne and Dylan, while the words "unclothed" and "lighteth"
appear in Donne, balanced by "clothes" and "light" in Dylan.

Shall we agree that all the words just specified are in somewhat
common use today, and were in equally ordinary employment in the
seventeenth century? Whereas, if you care to glance again at the Dylan
lines I just cited, not only do you think at once of Marvell's *Had we but
world enough and time / This coyness, Lady, were no crime* (which gets
"lady" in there, right enough, and in delicious apposition to "world" at
that), but you also find yourself grappling with Marvell's gentle but
urgent sense of delay and frustration. Dylan further beseeches the lady
to stay *while the night is still ahead* and to *have [her] cake and eat it too*:
Metaphysically speaking this is not so remote from Marvell's reminder
that the darkness of death will last an awfully long time, while in the
grave the worms may dine long and well. This is something different
from Donne's poem, which swiftly becomes a near-raunchy celebration
of achieved carnal knowledge of someone familiar to him. Finally,
Marvell speaks beautifully and seductively about keeping the sun in
motion since there's no chance of making it stand still, and Dylan
longs to see his beloved "in the morning light," having banished the

night in the only way that lies open to him. I hope I don't boast about my own poor exegesis, but Ricks's procedure is more like that of the people who pore over Bible codes or kabbalistic crossword puzzles.

Dylan's version of anger is sardonic and bitter: an exemplary match for the "cawing, derisive" tones noted by Larkin. In "Masters of War," "Only a Pawn In Their Game," and "The Lonesome Death of Hattie Carroll," he said to the military-industrial complex and the racists, in effect, "You win. For now. But for now you also have to live with your shame. And judgment will follow, and is coming." (I have always hoped, for this reason, that Joan Baez was wrong in claiming that Dylan wrote "When The Ship Comes In"—his most Jeremiad and vengeful poem—in response to bad service at some hotel.)

"The Lonesome Death of Hattie Carroll" was based on a real event in 1963: the lethal beating of Hattie Carroll by William Zanzinger in a Baltimore hotel. Zanzinger's lenient treatment by the courts fired Dylan into a hot rage, yet producing his most glacial and most measured poem of outrage and contempt. He simply relates the story, with deadly counterpoint as between the rich and careless white man and the dispensable black servitor. The song never uses the words "black" or "white," as Ricks points out, but just: *He owns a tobacco farm of six hundred acres*, while she *emptied the ashtrays on a whole other level*. Thus is the plantation relationship recast and, as Ricks rightly says, "it's a terrible thing that you know this [their respective colors] from the story." But then again, as Ricks also emphasizes, Dylan's affecting line *And she never done nothing to William Zanzinger* is a sort of clue. I have always thought that this was Dylan ventriloquizing, without condescension, the "Black English" demotic comment on the affair. Ricks improves on my intuition by giving the example of James Baldwin in *The Amen Corner*: "He hadn't never done nothing to nobody."

*Doomed and determined to destroy all the gentle*, in Dylan's haunting phrase, Zanzinger slew Hattie Carroll *with a cane that he twirled around his diamondring finger*, and who would pass up the chance to recall the first murderer, Cain, in this context? Not Ricks, who also calls attention to the words *lay slain by a cane* and to the triple repetition of the word "table," which closes three consecutive lines. "Does this *-able*" he inquires, "prepare for the word that soon follows, 'cane'? Cain and

Abel, masculine and feminine endings?" Well, no, I shouldn't think so. Whatever the song is about, it most decidedly isn't about fratricide. And Cain and Abel—scarcely unique metaphors where murder is concerned—appear in other Dylan songs under their own names. Ricksian hermeneutics has its limits.

I could, nonetheless, have used some more counsel from Ricks about the title. In what way was Hattie Carroll's death "lonesome"? There is an unmistakable sentimentality in this word; a tear-jerking note that is wondrously absent from the song itself. Insufficient guidance is forthcoming: Ricks proposes without much brio that Dylan "perhaps" wanted the word to evoke a contrast between Hattie's death and the crowded hotel. But with or without that "perhaps," ultimately, everybody dies alone.

Ricks's closing thought is superior. He argues that T. S. Eliot understood the difference between writing religious poetry and writing poetry religiously, and that Dylan with "The Lonesome Death of Hattie Carroll" has written politically rather than merely writing a political song. That seems to be a distinction well worth observing, most especially at a time like the present with its ephemeral garbage of pseudoprotest. ("We've suffered for our music—now it's your turn.") The finest fury is the most controlled. One still feels a generous anger when listening to the song—incidentally, William Zanzinger turned up again a few years ago in the Baltimore courts, for leasing black people squalid, waterless cabins that he didn't even own—and the pairing of generosity with anger (annexed from Orwell out of Dickens) might license some interpenetration of sin and virtue, or even sin with grace.

It's back to hermeneutics in Ricks's study of "Love Minus Zero / No Limit," which occurs in the chapter on "Temperance." As you will recall, the song begins *My love she speaks like silence / Without ideals or violence,* while in a succeeding verse:

> *In the dime stores and bus stations*
> *People talk of situations*
> *Read books, repeat quotations*
> *Draw conclusions on the wall.*

For Ricks, this is Belshazzar's feast in the fifth chapter of Daniel: "In the same hour came forth fingers of a man's hand, and wrote over against the candlestick upon the plaster of the wall of the king's palace. And this is the writing that was written: MENE, MENE, TEKEL, UPHARSIN. This is the interpretation of the thing: MENE; God hath numbered thy kingdom and finished it."

Building upon this, Ricks insists that the biblical "candlestick" furnishes Dylan not only with his song's reference to candles and matchsticks, but the biblical word "numbered" may have a relation to the "Minus Zero" in Dylan's title. This same chapter of Daniel has the words "people," "tremble," "wise men," and "gifts"—and also "spake," "said," and "that night." What more could one want as proof of the direct influence of the prophet Daniel upon the song?

Something more, as it happens. *The words of the prophets are written on the subway wall*, as Paul Simon and Art Garfunkel were to say in "The Sounds of Silence," and it was as obvious to me the first time I heard "Love Minus Zero / No Limit" as it is today that Dylan was alluding to graffiti: a special emphasis in that time and place. If you really want to connect Babylon to Dylan, you might have better luck with "Lily, Rosemary and the Jack of Hearts": *The cabaret was silent— except for the drilling in the wall.*

At the same time I was digesting all this in *Dylan's Visions of Sin*, I noticed that Ricks deals with an obvious contradiction in his account (the king being "reduced" to the pawn) in the following evasive manner: "'Even the pawn must hold a grudge.' Even the king? Even Dylan, whom I ungrudgingly admire?" This is ingratiation raised to the level of unction. I remember the first time that I ever felt a qualm about Dylan's claims. It was early on as well: He said that he had written "A Hard Rain's Gonna Fall" at the time of the Cuban missile crisis—and he had been in such an apocalyptic hurry that every line could be the first line of another song. Even in my early teens, I knew that that was bravado.

Oddly, perhaps, Ricks spends almost no time on the influences that Dylan actually does affirm or the influences that we know about. "Blowin' In the Wind" borrows from an old slave spiritual called "No More Auction Block," with its haunting words about "many thousands gone." Dylan was actually sued by Dominic Behan, brother of Brendan, for plagiarizing not only the tune but the concept of "The Patriot Game" for his "With God on Our Side." More recently, his song about a Japanese yakuza was tracked down to an obscure but identifiable source, while the deft Daniel Radosh has blogged a near-perfect match between Dylan's "Cross the Green Mountain" (written for Ron Maxwell's movie *Gods and Generals*) and Walt Whitman's "Come up from the Fields, Father." If I had to surmise another influence, it would be William Blake, not just for the speculative reasons given by Ricks but because, as Blake phrased it: "A Last Judgment is Necessary because Fools flourish."

Even secularists often find themselves thinking things like that, and there is a store of words in the Bible that springs ready-made, as it were. Thus, Ricks could well be correct in thinking that Dylan's "how many times" is an echo both of "How long, oh Lord, how long?" and of Christ's injunction in Matthew on the number of times that it might be needful to turn the other cheek. (He may also be right, though coming down-market more than he likes, in discerning a vague sacred/profane overlap between "I Believe in You" and "Smoke Gets In Your Eyes.")

But Christianity as a religion of peace and tolerance and forgiveness is not, superficially at least, compatible with ringing phrases about judgment and the sword: In order to believe in the apparently kindly and reassuring verses about taking no thought for the morrow, one had better have a lively sense of the second coming. This was the line that Dylan actually did take in his born-again period, where he spoke of "spiritual warfare" as well as his "precious angel," and warned that there would be no hiding place on the day. But this, which produced some of his most beautiful writing (and singing) would appear to have been as lightly affected as the gritty dustbowl socialism which the Old Left was already denouncing him for abandoning as far back as 1964. Dylan dropped it and kept moving on.

Indeed, I am sure I remember Ricks welcoming him "back," as it were, when he came up with "Most of the Time" about fifteen years ago. But here, and in his discussion of this superbly apt and lovely and troubling song, I began to write heavy notes in the book's margin:

"Most of the time," Ricks writes, " 'Most of the Time' consists of repeating the words, 'most of the time.' " [Marginal note: *Oh no it doesn't.*] Unbelievably, Ricks manages to go on for a half-dozen pages about this song, without ever achieving the realization that it is one of the most vertiginous, knife-edge accounts of a post-love trauma ever penned. You should only listen to the song if you are not currently trying to persuade yourself that "it" is all over and that you are all over "it."

Ricks wraps up blandly: "It is only most of the time that the man in this long black song succeeds in being *not disturbed*. But he is halfways there. On the other hand, 'She's that far behind.' One too many mornings and a thousand miles behind, to be exact." [In the margin: *To be inexact, you mean, you fool. She's right behind him and in front of him and all around him, all of the time. His attempted banishment of her is a hopeless failure! What have you got in your veins—tap water?*]

There follows a lengthy Ricksian contrast between the words of Dylan's song "Not Dark Yet" and Keats's "Ode to a Nightingale." Not, you understand, that our author wants to be taken too seriously. "I don't believe that Keats's poem is *alluded* to in Dylan's song. That is, called into play, so that you'd be failing to respond to something crucial to the song unless you were familiar with, and could call up, Keats's poem." [In the margin: *Oh no, of course, not that.*] After all, the deep connection between Keats's *My heart aches, and a drowsy numbness pains* and Dylan's *Well, my sense of humanity has gone down the drain* is transparent neither in sense nor rhythm.

It is true that the words "dark," "shadow," and "day"—together with "sleep" and "time," or their cognates—are to be found in both sets of verses. I am quite ready to believe that Dylan had a subliminal memory of being taught the poem in school. But Renata Adler did much better than this, during the 1968 Republican convention that nominated Nixon in Miami. Surveying the sea of placards with their jaunty slogan "Now More Than Ever," she suddenly recognized that it came from verse six of the "Nightingale" ode: *Now more than ever seems it rich to die, / To cease upon the midnight with no pain.*

I think that might have afforded Dylan a smile, and possibly Ricks too. But only one of them has an attitude to sin that is in any sense original.

# HAWTHORNE AND THE AMERICAN PAST

## By Wilfred M. McClay

⬥

*(Originally published, as "Land of Hope and Fear,"*
*in the August 16–23, 2004, issue)*

OF ALL THE COMPLAINTS LEVELED at the canon of nineteenth-century American books, the hardest to credit is the charge that they are conventional and comfortable—like picturesque little pleasure boats plying the sunny surface of American life.

How then does one account for the unsettling preoccupations of those authors: the desperate God-grappling of Herman Melville, the macabre fixations of Edgar Allan Poe, the fevered omnisexuality of Walt Whitman, the nature-intoxicated anarchism of Henry David Thoreau? This doesn't sound like the stuff of which genteel outings at the lake are made. In fact, such a list makes one wonder whether there has ever been a great national literature more full of craziness and inflationary excess, more indifferent to measure and proportion, more riddled with anxiety and self-doubt.

Americans seem generally unaware of their literature's disquieting features. Take, for example, the exalted status accorded *The Scarlet Letter*, Nathaniel Hawthorne's 1850 masterpiece, the first indisputably great work of American literature. For much of the twentieth century, an acquaintance with *The Scarlet Letter* was considered an essential part of American education. But it's hard to imagine a more bizarre candidate for a literary rite of passage—or one better calculated to establish a permanent aversion to classic literature.

This seems especially true for students who've grown up in the age of Bill and Monica. What they find in *The Scarlet Letter* is the story of a minister and a married woman who had a love affair and feel bad about it afterward—especially the man, a sensitive fellow who also turns out to be a hypocrite and a bit of a coward. The woman, an impressively resilient spirit who bore a love-child out of that furtive encounter, is publicly humiliated. The minister chooses to conceal his part in the matter, although profound feelings of guilt gnaw away at him. The cuckolded husband schemes to get even, while degenerating into an ever-more loathsome monster in the process. In the end, everyone lives (or dies) unhappily ever after. Pretty depressing stuff, when you consider how much better off everyone would have been, if they could just have . . . well, *gotten over it* and *moved on.*

It's hard to improve on what one of my students said during a class discussion of Whittaker Chambers's passionate and gloomy autobiography, *Witness.* "The dude just needed to chill," he murmured, gazing down at his fingertips, a tiny smile playing upon his lips—affectless contempt expressed in perfect twenty-first-century pitch. The other students nodded agreement.

They say the same, and then some, about Nathaniel Hawthorne and his characters. Leaving aside the spidery intricacies of the prose in *The Scarlet Letter*, and the lack of action in its plot, what really dooms the novel for present-day readers is the alien intensity of its moral universe. Part of Hawthorne's message makes sense to them, the part they've been trained to hear—that the Puritan religious and social code (as he understood it) was excessive, cruel, sexist, and inhuman, that it wrung all beauty and joy from life, and that the actions of the avenging husband, Roger Chillingworth, though he was technically the wronged party, were ultimately far more sinister than those of the unconfessed adulterer, the Reverend Arthur Dimmesdale, and his near-blameless lover, Hester Prynne.

But what they can't comprehend is what all the fuss is about—why Dimmesdale felt so guilty, why he couldn't confess, why what he and Hester did was in fact a grievous sin, why our sins and the sins of our forebears are inseparable from who we are, why those sins must be paid for, why it is almost impossible to pay for them fully, and yet why sins that remain unacknowledged and unconfessed and unpaid will surely destroy our souls. The central premise in Hawthorne's imaginative world—his insistence that the weight of the sinful human past, in one's own life, in the life of one's family, and in the life of one's city and country, can never be denied or wished away—is completely lost on a generation raised on smug therapeutic platitudes.

Given such difficulties, one might have hoped that the academy at least would keep Hawthorne's reputation alive. But Hawthorne has had a rough time of it in recent years. The problem, of course, is politics. Much of the Hawthorne scholarship emanating from academic English departments during the past two decades has been dominated by "New Historicism," which has tended to reduce Hawthorne to little more than the sum of his unacceptably skeptical or reactionary positions on the burning issues of his day: slavery, abolitionism, women's rights, the conditions of the laboring classes, movements of radical social reform. To make matters worse, he was an ardent American nationalist and expansionist. Surely an author so politically benighted must have produced works that "inscribed" all the worst features of American life.

Accordingly, in an influential 1991 book, Sacvan Bercovitch disparaged *The Scarlet Letter* as an ideologically conservative work of "thick propaganda," a "vehicle of continuity" that opposed radical change and celebrated the tawdry American icons of "gradualism and consensus." Other critics, such as Jane Tompkins, concluded that Hawthorne's high literary reputation has been undeserved, having been propped up artificially by patriarchal networks of critical opinion. It increasingly seems that the only point of keeping Hawthorne around is to have him handy as a whipping boy.

Thankfully, though, an important counterbalance to these influences has come from the biographical literature on Hawthorne. Generally produced by writers operating on the fringe of the academy, that literature presents Hawthorne in a richer and more multidimensional way. It's too much to hope that the recent appearance of Brenda Wineapple's *Hawthorne: A Life* and Philip McFarland's *Hawthorne in Concord* signal a turning of the scholarly tide. But both books are well written and sensibly argued, with only a modicum of interpretive excess or psychoanalytic license. They may help keep alive the possibility of a more respectful audience for Hawthorne, during a dry season that dismisses him too easily, and may need him more than it suspects.

Born in Salem on the Fourth of July in 1804, Hawthorne was a paradox from start to finish: The isolated and brooding child of an old and rooted family, he became the first great literary voice of a boisterous, restless new nation. He found endless ways of embodying this tension, in a life that was both cautiously provincial and perpetually unsettled. He was both deeply proud of his Puritan family pedigree and deeply troubled by it, not least by the fact that his great-grandfather John Hathorne had been one of the judges in the infamous Salem witchcraft trials. It was part of the family tradition, one that formed the basis for his novel *The House of the Seven Gables*, that the family house retained a curse brought down upon it by that forebear's deeds. Hawthorne may have changed the spelling of his own surname partly to avail himself of the American promise of a fresh beginning. But at the same time he never ceased to acknowledge and even wallow in that heritage—in ways that profoundly affected his view not only of his own past but also of America.

His father was a sea captain who died in Dutch Surinam of yellow fever when Nathaniel was four. So he grew up with his eccentric, reclusive mother and sisters in an entirely female house. His own tendencies toward introversion and bookishness were only accentuated by a youthful foot injury, which kept him indoors a great deal of the time. By the time he went off to college at Bowdoin in 1821, he was already fairly certain that he would not aspire to any of the conventional masculine careers of business, the clergy, the law, or medicine. Instead, he was already setting his sights upon becoming "an Author, and relying for support upon my pen." But those ambitions also had a nationalistic tinge to them, for he hoped, as he told his mother, to produce works that would be regarded as equals to the "proudest productions of the scribbling sons of John Bull."

The years at Bowdoin were important for a variety of reasons. He came out of his shell a bit and initiated some of the most lasting relationships of his life, notably his friendship with Franklin Pierce, a future president of the United States. In the company of Pierce and other Bowdoin friends, he discovered a passion for partisan politics, settling easily into the political sympathies of a Jacksonian Democrat, an outlook that would stay with him for the rest of his life and help immunize him against the Whiggery and evangelical reformism that dominated his literary circles.

Such a rough-and-tumble practical mindedness in politics might seem out of character with his authorial ambitions, but the two were united by a strong sense of American cultural destiny. The commencement address at Hawthorne's 1825 graduation—delivered by fellow graduate Henry Wadsworth Longfellow and entitled "Our Native Writers"—offered a passionate plea for a new American literature, "springing up in the shadow of our free institutions." Such words spoke directly to Hawthorne: The desire to have a hand in creating such a distinctive American literature was, as Brenda Wineapple says, "the secret ambition lodged like a thorn in his own heart."

That ambition would be a long time in the realization. After college, he returned to Salem, and spent a mysterious twelve years living in his mother's home, incubating his talent, publishing stories here and there (usually in near-complete anonymity), and struggling with the fears and loathings that such a self-imposed isolation must have imposed upon him. Others found it incomprehensible that Hawthorne, who was an extraordinarily handsome man, with captivating eyes that were, in the admiring words of Elizabeth Peabody, "like mountain lakes seeking to reflect the heavens," chose to withdraw into the blue chamber of his soul.

But his exalted sense of authorial calling was accompanied by an equally powerful apprehension that the work of a writer could not qualify as "man's work." Writing became, as Wineapple neatly puts it, "a source of shame as much as pleasure, and a necessity he could neither forgo nor entirely approve." Yet out of this tortured state would finally come, slowly but surely, a body of short fiction that would eventually make up his *Twice-Told Tales* (1837), the work with which he finally emerged in the public eye. This collection of tales, all of them previously published, included some of his best-known short stories, "The Fountain of Youth" (later retitled "Dr. Heidegger's Experiment"), "The Minister's Black Veil," and "The May-Pole of Merry Mount." He had already written such classic stories as "Roger Malvin's Burial," "My Kinsman, Major Molineux," and "Young Goodman Brown," but chose, for his usual mysterious reasons, not to include them.

One could plausibly argue that Hawthorne was at his most inspired in his early short fiction. Certainly the reader can see the characteristic lines of his thought, in ways that would not be much altered or improved upon in the later work. The penchant for symbolism and allegory is there—together with the spooky echoes of past sins and the creepy defamiliarization of ordinary life, which is seen to hide strangeness and horror beneath its thin veneer. These early stories all show the typically static Hawthornian characters frozen compulsively in moral dilemmas, often self-chosen, and they all run on a prose style that conveys gauzy, dreamlike distance rather than novelistic clarity and specificity.

Take for example the baffling story "Wakefield," which had first appeared two years earlier in *New-England Magazine*. The story is presented as an imaginative reconstruction derived from scraps in an old periodical, just the kind of framing device Hawthorne loved to use. An ordinary Londoner named Wakefield—"a man of habits"—leaves his wife, allegedly on a short business trip to the country, of no more than a few days. But, for reasons that are never explained, involving some deep and inscrutable psychological compulsion, he decides not to return. He does not go to another woman, or to some faraway place to begin a new life. Instead, he rents rooms on a street near his home, and stays in them, living there incognito for twenty years. He was "spellbound," an illustration of the principle that "an influence, beyond our control, lays its strong hand on every deed which we do, and weaves its consequences into an iron tissue of necessity."

During those twenty years, Wakefield feels compelled occasionally to spy on his wife, although always with an electrifying feeling of terror at the thought of being found out. He sees her grow old and portly, adjusting resignedly to her widowhood, a woman whose "regrets have either died away, or have become so essential to her heart, that they would be poorly exchanged for joy." Then, just as suddenly as his decision to depart, Wakefield decides to return and put an end to "the little joke" that he has played "at his wife's expense." We are left in the dark about how he was received. But Hawthorne means the story to show how easily each of us, if diverted from the comfortable and familiar, can find himself "the Outcast of the Universe."

It is vintage Hawthorne, a weird and troubling little story, filled with misogyny and bottomless despair. Yet writing such a tale, far from being an act of symbolic transgression, was surely an act of self-disclosure, and self-mortification—for Wakefield was, in part, Hawthorne himself, or Hawthorne as he feared he was becoming. "He always puts himself in his books," opined his sister-in-law Mary, wife of the great educator Horace Mann, "he cannot help it." And as Hawthorne confessed poignantly to Mary's sister, Sophia Peabody, who would soon be his wife: "Thou only has taught me that I have a heart. . . . [W]ithout thy aid, my best knowledge of myself would have been merely to know my own shadow—to watch it flickering on the wall, and mistake its fantasies for my own real actions."

He knew that the literary artist could easily become a heartless man like Wakefield, "dissevered from the world," reduced to being an observer, a wraith cut off from the world he claims to understand. Brenda Wineapple's biography adroitly traces the twists and turns in this relentless struggle of authorship, showing it as a continuously formative theme in Hawthorne's entire career.

Of course, there is more to Hawthorne's life than his years of painful and anonymous alienation. The great virtue of Philip McFarland's charming and immensely readable *Hawthorne in Concord* is to show us the writer not as the radically isolated man he imagined himself, but as a member of a lively community of writers and thinkers: Emerson, Thoreau, Fuller, the Alcotts, the Manns, and the Peabodys.

But Concord, although a place of unusual happiness for Hawthorne in the early years of his marriage, was far less important to him than Salem, and the fact that he lived in Concord on three separate occasions (and is buried there) does not appear to have translated into the town's having any particular significance for him. The more enduring reality about Hawthorne seems to have been his restlessness, his inability to be content in any setting—whether Salem, Maine, Boston, Concord, West Roxbury, Lenox, West Newton, or Liverpool, where his friend President Pierce had appointed him U.S. consul. Even Rome, a place where the past was never dead, reminded him in the end of a rotting corpse. The sense of place was, for Hawthorne, a haunted and confining thing at best.

If, however, one thinks less about place than about milieu, then McFarland's angle of vision becomes very useful, for it reminds us of how enmeshed Hawthorne was in many of the most characteristic enthusiasms of his day. It is customary to see him as the soberly pessimistic countervoice to Emerson's wild optimism, the cautionary voice of the repressed past, the unredeemed present, and the unreformable future. That is true, but not true enough. Hawthorne was a Jacksonian Democrat, not a Burkean neo-Calvinist, let alone a neomedievalist crypto-Catholic.

He shared with his longtime friend John O'Sullivan a belief in "the essential equality of all humanity," in which "all ranks of men would begin life on a fair field"—and believed that it was America's destiny to spread this doctrine across the continent. And it was he, and not Emerson or Thoreau, who was willing to go live in George Ripley's utopian experimental community Brook Farm for seven months (and then, like any red-blooded American opportunist, turn his experience into publishable prose with *The Blithedale Romance*).

"I should like to sail on and on forever," Hawthorne mused when returning from Rome, "and never touch the shore again." But the alienated artist, with his joys and fears, is by now an exhausted, even tiresome, theme. We hardly need the assistance of Hawthorne to understand it, and if his reputation were to hang on that alone he would not really deserve the high status he is granted.

In fact, however, there are deeper themes in Hawthorne that may never before have been as salient as they are in our own times. Consider a story such as "The Birthmark," in which a scientist insists on removing from his beautiful wife's left cheek a crimson birthmark, her sole imperfection, and inadvertently kills her in the process. Or "Earth's Holocaust," in which a fire begun to rid the world of its "accumulation of worn-out trumpery" ends up consuming everything and leaving the world no better. Or "The Celestial Railroad," in which the hard path of Bunyan's *Pilgrim's Progress* is replaced by an easy and convenient railway, which carries its comfortable passengers straight to Hell. Or

"Rappaccini's Daughter," a complex allegory in which a beautiful young woman, as an experiment in the control of nature by her scientist father, has been raised on a diet of poisonous plants to make her self-sufficient—and ends up being killed by her lover when he administers an antidote to the poison.

Unfortunately, the interpretations of such stories offered by both Wineapple and McFarland fall short. Wineapple sees "The Birthmark" strictly as a tale of sexual anxiety, in which "a man confronts marriage, and hence sexuality, with horror." And she reads "Rappaccini's Daughter" as a "biographical palimpsest" in which the evil doctor is his father-in-law, or perhaps Emerson, and the young woman "represents a woman's struggle to free herself " from confinement and "irresolute men." McFarland says that "Earth's Holocaust" is less a story than a sketch "reflecting on contemporary issues." These readings may or may not be accurate. But they are essentially trivial in comparison to the profounder meanings that leap off of these pages today.

These irony-filled allegorical tales, with their constant reversals and inversions, are also warnings about the moral perils of human efforts to gain mastery over the terms of human existence. What a bitter sadness it would be, Hawthorne reflected at the end of "Earth's Holocaust," if "Man's agelong endeavor for perfection" served only to "render him the mockery of the Evil Principle, from the fatal circumstance of an error at the very root of the matter." And what was that error? It was lodged in the heart, "the little, yet boundless sphere"; all the misery of the world derives from that "original wrong." The human heart is where the problem is, and where the only solution can be found. "If we go no deeper than the Intellect," he warned, striving, "with merely that feeble instrument, to discern and rectify what is wrong," then the result will be no more substantial than a dream.

Hawthorne's appeal to the heart over the intellect aligns him, once again, with the romanticism that was sweeping through the salons of Boston and Concord in his day. So too do the dreamy and gothic elements in his fiction. But if Hawthorne was partly a romantic, he was even more of a Hebrew prophet, a throwback to the Isaiah who reviled the hardened and self-satisfied hearts of his contemporaries—and prophesied that the hidden things would come to the light and the pitiful wisdom of the wise would be destroyed. If he was not quite able to reembrace the Christian theology of his forebears in all its details, his invocation of an "original wrong" was a long and respectful bow to the explanatory power of their most fundamental assertion.

At its best, Hawthorne's prose achieves an uncanny quality in which the fiber of familiar reality gives way. McFarland casually compares this effect to the "magical realism" of Gabriel García Márquez, but that utterly fails to capture the terrifying moral energy swirling through Hawthorne's writing. Especially in a handful of his most powerful short stories, Hawthorne's work forces us to observe the essential moral value of *fear*. This hasn't been a popular thing to notice since the Enlightenment's disenchantment of the world, and it is completely at odds with the therapeutic ethos that now reigns. If the fear of God is the beginning of wisdom, however, it is so partly because such fear protects us against the fatal presumption of mastery—a fearlessness much more to be feared than fear itself.

# THE TRILLING IMAGINATION

## By Gertrude Himmelfarb

*(Originally published in the February 14–21, 2005, issue)*

A RECENT CASUAL, DISMISSIVE reference to Lionel Trilling recalled to me the man who was the most eminent intellectual figure of his time—certainly in New York intellectual circles, but also beyond that, in the country as a whole.

So, at any rate, he appeared to me many years ago. And so he appeared to his contemporaries, who thought it entirely fitting he should have received (a few years before his death) the first of the Thomas Jefferson Awards bestowed by the National Endowment for the Humanities. It is typical of the man that the lecture he delivered on the occasion, "Mind in the Modern World," was exhortatory rather than celebratory, cautioning us about tendencies in our culture that diminished the force and legitimacy of the mind, tendencies that were to become obvious to others only many years later. Here in 2005—on the centenary of his birth in 1905—it is interesting to reflect upon the quality of Lionel Trilling's mind, a quality rare in his time and rarer, I suspect, in ours.

I was a budding Trotskyite in college when I came across Trilling's 1940 essay on T. S. Eliot in *Partisan Review*. I had read only a few of Eliot's poems; "Prufrock" was a particular favorite among my friends at the time, much quoted and affectionately parodied. (In a Yiddish parody Saul Bellow liked to recite, *In the room the women come and go / Talking of Michelangelo* became *Talking of Marx and Lenin*. I forget how it rhymed.) But I had never read Eliot's essays or the journal he edited, the *Criterion*, which had ceased publication before the out-break of World War II. I was, however, a faithful reader of *Partisan Review*, which was, in effect, the intellectual and cultural organ of Trotskyites (or crypto-Trotskyites, or ex-Trotskyites, or more broadly, the anti-Stalinist Left). Many years later I remembered little about Trilling's essay except its memorable title, "Elements That Are Wanted," and the enormous excitement it generated in me and my friends. Rereading it recently, I experienced once again that sense of excitement.

Trilling opened by quoting an essay by John Stuart Mill on Coleridge, written a century earlier. That essay had angered Mill's rad-ical friends, Trilling said, because it told them they could learn more from a "religious and conservative philosopher" like Coleridge, who saw "further into the complexities of the human feelings and intel-lect," than from the "short and easy" political discourses of their own mentor, Jeremy Bentham.

In this spirit, Trilling introduced his own radical friends, the read-ers of *Partisan Review*, to T. S. Eliot, another "religious and conserva-tive" thinker. Trilling did not, he hastened to say, mean to recommend Eliot's "religious politics" to their "allegiance"—only to their "atten-tion." He reminded them of their own precarious situation a year after the outbreak of war in Europe: "Here we are, a very small group and quite obscure; our possibility of action is suspended by events; perhaps we have never been more than vocal and perhaps soon we can hope to be no more than thoughtful; our relations with the future are dark and dubious." Of only one thing about the future could we be certain: our "pledge to the critical intellect."

That pledge recalled to Trilling not only Mill's invocation of Coleridge but also Eliot's "long if recalcitrant discipleship" to Matthew Arnold. (Trilling did not have to remind his readers of his own, less-recalcitrant discipleship to Arnold; his book on Arnold, a revision of his doctoral dissertation, had been published the previous year.) Trilling quoted Arnold on the function of criticism: "It must be apt to study and praise elements that for the fullness of spiritual per-fection are wanted, even though they belong to a power which in the practical sphere may be maleficent."

The audacity of this essay on T. S. Eliot is hard to recapture now. Trilling suggested that there were, in the philosophy of his own circle, elements that were wanted. More, that these elements were wanted "for the fullness of spiritual perfection." And, finally, most provocatively, that these elements were found in a thinker whose ideas could well be maleficent in the practical sphere. Yet it was precisely in the practical sphere—not as a poet but as a political thinker—that Trilling commended Eliot to readers of *Partisan Review*.

Although the journal had broken with the Communist party three years earlier, *Partisan Review* was still committed to a radical politics and to an only-somewhat reformist Marxist philosophy (of the kind espoused by, say, Sidney Hook). But *Partisan Review* was also committed to a modernist literary vanguard—Eliot, Joyce, Yeats, Pound, Proust, and all the rest—who all too often had conservative, if not reactionary, views on political and social affairs. The editors were entirely candid about the disjunction between literature and politics; indeed, it was the Communists' insistence upon a party line in literature as in politics that contributed to the break from the party. And this disjunction made for a tension in the journal that was a constant challenge to the ideological pieties of its readers.

What Trilling was now proposing, however, went well beyond the reassertion of that disjunction between literature and politics. Where others found Eliot interesting in spite of his politics, Trilling found him interesting *because* of his politics: a politics not only conservative but religious, and not only religious but identifiably Christian. And this, to readers who were, as Trilling said in his usual understated manner, "probably hostile to religion" (and many of whom, he might have added, like himself, were Jewish). Where John Stuart Mill had cited Coleridge's *On the Constitution of Church and State* as a corrective to Benthamism, Lionel Trilling recommended Eliot's *The Idea of a Christian Society* as a corrective to Marxism. The Left had simplistically assumed that Eliot "escaped" from "The Waste Land" into the embrace of Anglo-Catholicism.

But even if this were true, it would not be "the worst that could be told of a man in our time." Surely, Trilling observed, Marxist intellectuals, who had witnessed the flourishing and the decay of Marxism, should appreciate the intellectual honorableness of Eliot's conversion. They need not follow Eliot's path to theology, but they could emulate him in questioning their own faith.

Marxism was not the only thing that Trilling (by way of Eliot) called into question. He challenged liberalism as well. Totalitarianism, Eliot had said, was inherently "pagan," for it recognized no authority or principle but that of the state. And liberalism, far from providing an alternative to paganism, actually contained within itself the seeds of paganism, in its materialism and relativism. Only Christianity, Eliot argued—the "Idea" of Christianity, not its pietistic or revivalist expressions—could resist totalitarianism, because only Christianity offered a view of man and society that promoted the ideal of "moral perfection" and "the good life." "I am inclined," Trilling quoted Eliot, "to approach public affairs from the point of view of the moralist."

Trilling hastened to qualify his endorsement of Eliot in "Elements That Are Wanted"; he did not believe morality was absolute or a "religious politics" desirable. But Eliot's vision of morality and politics was superior to the vision of liberals and radicals, who had contempt for the past and worshiped the future. Liberals, in the name of progress, put off the realization of the good life to some indefinite future; radicals put off the good life in the expectation of a revolution that would usher in not only a new society but also a new man, a man who would be "wholly changed by socialism."

Marxism was especially dangerous, Trilling found, because it combined "a kind of disgust with humanity as it is and a perfect faith in humanity as it is to be." Eliot's philosophy, on the other hand, whatever its defects and dangers, had the virtue of teaching men to value "the humanity of the present equally with that of the future," thus serving as a restraint upon the tragic ambition to transcend reality. It was in this sense, Trilling concluded, that Eliot bore out the wisdom of Arnold's dictum. Eliot's religious politics, while maleficent in the practical sphere, contained elements wanting in liberalism—"elements which a rational and naturalistic philosophy, to be adequate, must encompass."

I do not know how Trilling's radical friends reacted to this essay, but I do remember the effect it had on some radicals of my own, younger generation. For us, it was a revelation, the beginning of a dis-affection not only with our anti-Stalinist radicalism but, ultimately, with liberalism itself. Trilling has been accused (the point is almost always made in criticism) of being, not himself a neoconservative, to be sure, but a progenitor of neoconservatism. There is much truth in this. Although he never said or wrote anything notable about the "practical sphere" of real politics (he was not a "public intellectual" in our present sense, commenting on whatever made the headlines), he did provide a mode of thought, a moral and cultural sensibility, that was inherently subversive of liberalism and thus an invitation to neo-conservatism.

The volume of essays he published ten years later, *The Liberal Imagination*, is often cited as evidence of Trilling's conservative (or neo-conservative) disposition. Oddly enough, when I looked, some years ago, for "Elements That Are Wanted" in that volume, I could not find it, although the preface clearly alluded to it. (Nor was it included in *The Partisan Reader*, a collection of essays from *Partisan Review* published in 1946.) When I remarked upon this omission to Diana Trilling, she could not account for it, and the essay, under the title "T. S. Eliot's Politics," reached book form only in 1980 in *Speaking of Literature and Society*, the final volume of her edition of Trilling's *Collected Works*.

One much-quoted passage in the preface to *The Liberal Imagination* suggests Trilling deliberately dissociated himself from conservatism: "For it is the plain fact that nowadays there are no conservative or reac-tionary ideas in general circulation." What is not generally quoted are the qualifications that followed, which almost belie the assertion. If there are no conservative ideas, there are nonetheless conservative "impulses," which are "certainly very strong, perhaps even stronger than most of us know."

For that matter, liberalism itself is more a "tendency" than a set of ideas. Goethe had said there are no "liberal ideas," only "liberal senti-ments." But sentiments, Trilling observed, naturally and impercepti-bly become ideas, and those ideas find their way into the practical world. *"Tout commence en mystique,"* he quoted Charles Péguy, *"et finit en politique"*—everything begins in mysticism and ends in politics. Thus liberalism is no more a party of ideas than conservatism; indeed, without the corrective of conservatism, liberal ideas become "stale, habit-ual, and inert." But the best corrective to liberalism is literature, because this is "the human activity that takes the fullest and most pre-cise account of variousness, possibility, complexity, and difficulty."

These words and variants upon them—"contingency," "complica-tion," "ambiguity," "ambivalence—appear as a refrain throughout all of Trilling's work; they are his unmistakable signature. In one writer after another—in Eliot, Wordsworth, Keats, Austen, Dickens, James, Hawthorne, Tolstoy, Joyce, Flaubert, Babel—Trilling found "elements that are wanted" in the liberal imagination: elements that testify to the essential humanity of man and defy political machination or social engineering.

The last book he saw into print before his death in 1975, *Sincerity and Authenticity*, seemed a departure in tone and substance from his earlier work, for in it, philosophers—Rousseau, Diderot, Hegel, Nietzsche, Burke, Sartre, Marx, Freud—played a starring role, min-gling with his usual cast of literary figures. This juxtaposition of phi-losophy and literature is exhilarating and sometimes startling. Trilling sums up his discussion of Rousseau, for example, with the line: "Oratory and the novel: which is to say, Robespierre and Jane Austen."

He justified that odd coupling: "This, I fancy, is the first time the two personages have ever been brought together in a single sentence, separated from each other by nothing more than the conjunction that links them." But they were not, he insisted, "factitiously conjoined: They are consanguineous, each is in lineal descent from Rousseau, cousins-german through their commitment to the 'honest soul' and its appropriate sincerity." ("Honest soul" refers back to yet another figure in this linkage, Hegel.)

The final chapter of *Sincerity and Authenticity* returned to the sub-ject that had long occupied its author, personally as well as intellectu-ally. Trilling, his wife, and his son had been in psychoanalysis for many years, and he had given much thought to it, as theory and as therapy. Now, in the context of authenticity, Freudianism presented a special challenge. Where most psychoanalysts regarded the therapeutic prac-tice as an effort to identify and overcome the inauthentic nature of man, to make conscious what was unconscious, Trilling insisted the unconscious had its own authenticity—the "Authentic Unconscious," as the title of his chapter put it. And where others were troubled by Freud's *Civilization and Its Discontents* because it seemed to take a bleak view of human beings and their potentialities, Trilling saw it as evidence, once again, of "the essential immitigability of the human condition, . . . its hardness, intractability, and irrationality."

The initial—and enduring—inspiration for much of Trilling's work was perhaps Freud, even more than Eliot, Arnold, or his other literary heroes. In the first of his essays on Freud, published shortly before "Elements That Are Wanted," Trilling described the peculiar synthesis of romantic antirationalism and positivistic rationalism that gave Freudianism the distinctive characteristics of "classic tragic realism." Another piece, written two years later, dealt with Karen Horney's revisionist mode of psychoanalysis. Horney, Trilling explained, was popular in liberal circles because she posited "a progressive psyche, a kind of New Deal agency which truly intends to do good but cannot always cope with certain reactionary forces." Freud's view of the psyche was less "flattering," Trilling admitted, but more in accord with "the savage difficulties of life."

An even more provocative version of this theme, "Freud: Within and Beyond Culture," was delivered as a lecture in 1955 and reprinted ten years later in *Beyond Culture*. Here Trilling posed the issue as biology versus culture: biology representing the "given," the immutability of man's nature; culture, the forces of society ("civilization," as Freud put it) that strove to alter and overcome biology. Again Trilling challenged the dominant liberal, progressive orthodoxy—and his audience for that lecture was the New York Psychoanalytic Society, which epitomized such orthodoxy. Unlike most of his listeners, who regarded any idea of a "given" as "reactionary," Trilling insisted that the givenness of our biological condition was, in fact, "liberating"—liberating man from a culture that would otherwise be absolute and omnipotent. "Somewhere in the child, somewhere in the adult, there is a hard, irreducible, stubborn core of biological urgency, and biological necessity, and biological *reason* that culture cannot reach and that reserves the right, which sooner or later it will exercise, to judge the culture and resist and revise it."

. Trilling was responding to the problem George Orwell had posed so dramatically in *Nineteen Eighty-Four.* Reviewing that book when it appeared in 1949, Trilling made clear that Orwell was not, as liberals liked to think, merely attacking Soviet communism. "He is saying, indeed, something no less comprehensive than this: that Russia, with its idealistic social revolution now developed into a police state, is but the image of the impending future and that the ultimate threat to human freedom may well come from a similar and even more massive development of the social idealism of our democratic culture." A few years later, reviewing another book by Orwell, Trilling repeated this theme: "Social idealism" is not the only thing that can be perverted into tyranny; so can any idea "unconditioned" by reality. "The essential point of *Nineteen Eighty-Four* is just this, the danger of the ultimate and absolute power which the mind can develop when it frees itself from conditions, from the bondage of things and history."

Trilling could not have anticipated the ultimate perversion of this tendency half a century later: the mutation of social engineering into genetic engineering. Today the imperious "mind" is even more bent upon that "ultimate and absolute power," as it attempts to free itself from the bondage of all conditions, things, and history—indeed from the bondage of biology itself. One can imagine a volume of essays by Trilling entitled *Beyond Biology.*

In "Elements That Are Wanting," Trilling quoted Eliot as writing from "the point of view of the moralist." He might well have made that statement of himself, understanding "moralist" in the largest sense of that term. He did not reflect much upon the kinds of moral questions, or "moral values," that occupy us today: marriage, family, sex, abortion. What interested him was the relation of morality to reality—the abiding sense of morality that defines humanity, and at the same time the imperatives of a reality that necessarily, and properly, circumscribes morality. He called this "moral realism." Even before the appearance of Orwell's novel, Trilling wrote about "the dangers of the moral life itself," of a "moral righteousness" that preens itself upon being "progressive."

*Some paradox of our natures leads us, when once we have made our
fellow men the objects of our enlightened interest, to go on to make them
the objects of our pity, then of our wisdom, ultimately of our coercion.
It is to prevent this corruption, the most ironic and tragic that man
knows, that we stand in need of the moral realism which is the product of the free play of the moral imagination.*

"Moral realism" is Trilling's legacy for us today—for conservatives
as well as liberals. Conservatives are well disposed to such realism,
being naturally suspicious of a moral righteousness that has been often
misconceived and misdirected. And their suspicions are confirmed by
the disciplines upon which they have habitually drawn: philosophy,
economics, political theory, and, most recently, the social sciences,
which are so valuable in disputing much of the conventional (that is to
say, liberal) wisdom about social problems and public policies.

The element that is still wanting, however, is the sense of variety,
complexity, and difficulty—which comes, Trilling reminds us, primarily from the "experience of literature," and which at its best informs
the political imagination as well as the moral imagination.

V

MINIATURES

# LIBERAL SPORTS / CONSERVATIVE SPORTS

## By Fred Barnes

*(Originally published, as "From Bradley to Barkley,"
in the February 5, 1996, issue)*

MAYBE I'M SLOW. But it wasn't until a conversation with my friend
Bob that I realized how ideological American sports have become. Bob
asked if I'd been to the Washington Redskins football game the day
before. Nope, I answered, I gave away my tickets and went to the
University of Virginia soccer game instead. Bob was thunderstruck.
How could I pass up football for that wimpy, boring sport? It took a
few minutes before it dawned on me what Bob was really getting at:
I'd passed up a chance to watch a *conservative* sport, football, for a *liberal*
one, soccer. Bob was on to something.

Nearly all sports, I've concluded, are either conservative or liberal.
Really. The conservative ones are rough, individualistic, obsessed with
winning, just as Newt Gingrich is in politics. Liberal sports are nonviolent (mostly), collective, and less than triumphal—in a word,
McGovernesque. It's obvious boxing, wrestling, football, and basketball (1990s-style), which involve lots of physical contact and one-on-
one confrontation, are conservative. But baseball, soccer, and basketball (1960s-variety), where violence is supposed to be kept to a
minimum and intricate teamwork matters, are liberal. And not as fun
to watch.

There are a couple more things that make a sport liberal. If it's one
in which women are as thrilling to watch (and almost as good) as

men—you know, tennis and swimming and soccer—it's liberal. Or if liberals love to participate in the sport or profess to enjoy watching it, it's also liberal.

Baseball is liberal because liberals have idealized the game. Yes, George Will has contributed to this, but he actually likes baseball. I suspect many liberals who extol baseball don't. After all, the game is often slow and boring. But liberals lurched to baseball's defense when football threatened in the 1960s to become America's number one sport. Liberals hate football. Not only because it's violent, but also because its biggest enthusiasts are southerners and Catholics, two disproportionately conservative groups.

Worse, football players often pray after scoring. Baseball, at the major-league level anyway, is heavily represented in the Northeast and Rust Belt, liberal stomping grounds. To stop football's advance, liberals concocted myths, like the one about a surge of domestic violence during the Super Bowl. Liberals detest the Super Bowl. It's the summit of conservative sports.

There's another telltale sign of a liberal sport: Winning isn't paramount. So marathon running, where the important thing is not finishing first but just finishing, is a liberal's delight. And soccer, in which many games end in ties, is too. The fundamental liberal vision of sports was stated two decades ago by a pro basketball player, Neil Walk. He said no score should be kept. Rather, basketball should be judged like ballet, for its artistry. Walk wasn't kidding. The conservative view was expressed by longtime Redskins coach George Allen (father of the Virginia governor): "Winning is life. Losing is death." Vince Lombardi made the same point: "Winning isn't everything. It's the only thing."

Certain sports are both liberal and conservative—call them Clintonsports. Hockey is the best example. Remember when the Soviets first sent their hockey squad to North America? The players were a model of socialist teamwork. The played hockey the way liberals like. Now contrast them with the star-oriented U.S. and Canadian players, who are frequently selfish, hogging the puck and taking outlandish shots. They play the game the conservative way. Occasionally, there's a happy medium in Clintonsports, a mixture of brilliant teamwork and individual entrepreneurship—the Brazilian soccer team, for instance.

That brings us to basketball, the sport that went from liberal to conservative. The classic liberal team was the New York Knicks of Bill Bradley's day. Bradley had few one-on-one skills. He was "the open man," reliant on others to screen his man so he could get off an unimpeded shot. The Knicks won several championships with that style of play. Basketball isn't like that anymore. The closest thing to the old Knicks is today's Cleveland Cavaliers, and they're barely over .500. Now one-on-one matchups are everything; even the short white guys in the NBA dunk, and basketball is far more exciting. Charles Barkley has replaced Bill Bradley. Barkley, by the way, is a conservative Republican. You know what Bradley is.

But if you're a Gingrichian Knicks fan, don't fret. It's fine to like liberal sports. I enjoyed the UVA soccer game far more than recent Redskins games. Under coach Bruce Arena, who'll coach the U.S. Olympic soccer team this summer, Virginia isn't methodical and low-scoring. The Virginia way combines clever teamwork with individual flair, fast-paced play, high scoring, and lots of victories. A moderately conservative style, I'd say.

# AT THE RED LOBSTER
# WITH TAMMY FAYE

## By Matt Labash

*(Originally published in the November 11, 1996, issue)*

"OH! CAN WE GO TO RED LOBSTER? Can we? It's my favorite!" she implores, batting those pig-bristle eyelashes that make me want to buff my shoes and/or get my car washed. "She" in this case is the former Tammy Faye Bakker (now married to one Roe Messner), and I have followed her to a book signing for her autobiography, *Telling It My Way*, in her natural habitat of pool-supply stores and outlet malls. "Surely, darlin'," I say, "I will fulfill your culinary fantasies," wanting her to feel comfortable in this white-trash Virginia milieu so I can execute my elegantly simple plan: to impale her over lunch.

Though I loathe "civic journalism," I am not opposed to occasionally slipping into the less sententious role of God's Avenger. Sure, He's already chastised Tammy Faye plenty: her ex-husband's imprisonment, her Lucille Ball voice, and the brief stint on a daily talk show costarring the insufferable Jm (yes, Jm) J. Bullock. But I have come to finish the job, because not since Emperor Nero has anyone dissuaded so many people from Christianity—people whose only religious dalliance might have come through the murky prism of fallible televangelists.

"Typical Baptist," huffs her defensive husband Roe Messner (a former PTL Club contractor who stole her from an incarcerated Jim), as I ask prickly questions over Red Lobster's patented cheese bread. He's referring to the historic rivalry between the Baptists (me) and Assemblies of God (them)—King-James-punching Tub-Thumpers and Glossolalial Pew-Jumpers respectively. But Tammy remains unflustered, suckling her beer-battered shrimp. "Oooh, this is awesome! Now what were you saying, hon?"

I was saying, Doesn't she feel any remorse for cheapening the faith with over-the-top extravagances: getting gullible biddies to raid their savings to fund grand hotels and waterslides for the-Holy-Ghost-comes-to-Branson-style variety shows, while she and Jim were collecting million-dollar salaries and air-conditioned doghouses (she claims it was only heated) and paying hush money to paramours and taking 25-city "Farewell For Now" tours complete with full orchestra and inspirational dancing waters?

"All we wanted was to minister," she says in between slurps of white-cheddar mashed potatoes, "and give Christians a place where they could come and feel they had something as good as Las Vegas." But what about the his-and-hers Rolls Royces?

She removes a lemon-pepper-shrimp shell from her teeth. "Mine was a Mercedes," she corrects me, "and I drove one simply for protection. It was a heavy car."

"Did you ever maybe consider . . . a Buick?" I rejoin. But I'm losing my edge, partly out of pity (when I questioned whether their ministry was completely altruistic, she asked, "What does that word mean?") and partly because she's still such a peach in those T. J. Maxx zebra prints and see-through plastic chukkas with that pancake-batter foundation and ice-cream-dipper rouge. "The makeup only takes five minutes to put on," she says, and I'm inclined to believe her, since it doesn't appear to be applied with much precision.

Of the stacks of books she signed at the empty Super Crown, she says, "Nobody will buy them. People are sick of me." And now I start feeling guilty for judging somebody so gloriously obtuse. After 10 years on the losing side of punchlines and resigned to her fate as an iconic cliché, all she can do now is construct counterfeit realities (she blames "Jerry Foulwell" for most of the PTL sordidness) and anesthetize herself with discount pleasures: like hawking her new wig line, or Fashion Bug shopping sprees, or eating at Red Lobster on a surly reporter's expense account. Perhaps if redemption is unattainable, one must take solace in the $7.99 Mix & Match Shrimp Combo.

She can feel me going maudlin on her when she tries to pick me up. "You're a smartass, Matt—I like you," she says. "But I love all things chocolate." So it was one Fudge Overboard and two spoons as Tammy let me tug on her wig while she catalogued where she purchased all her fake jewelry. She tells me she never talks to Jim anymore. "I feel sad that he's been treated so wrong. He's a very lonely man, and I feel sorry for him."

She's seceded from the Bakker legacy on paper, if not in the public's consciousness. But deep down she knows that they share an insoluble bond through their love of the ministry, their love of duplicity, and now, through the seafood-lover within.

As Jim writes of his first post-slammer meal in his new hardcover mea culpa, *I Was Wrong*: They "took me to a Red Lobster restaurant. . . . I knew I was in trouble. The menu was so big! . . . There were just too many choices. Tears welled up in my eyes. . . . I ordered a seafood platter with a little bit of everything on it. Wow!"

# MY FATHER'S MANSION HAS MANY HOLES

## By David Brooks

=====

*(Originally published in the January 20, 1997, issue)*

IT WAS DUSK. Brooks flicked a speck of lint from his velvet smoking jacket, poured himself a finger of Chivas, and held out his glass so that it could capture a few ounces of water leaking down from the bathroom upstairs. Through the hole in the ceiling that had been cut by the plumber who had desperately tried to stem the leak, Brooks admired his wife as she applied the last of her emeralds. "Shall we dine eight-ish?" he asked. They were soon to host another of their glamorous Washington dinner parties. Her reply was drowned out by a sudden crash. It was the handiwork of the laborers outside, who were trying to rebind the wall of the house to its foundation.

It was dusk. Brooks retired to his sitting room and admired the way his two-year-old daughter had mastered the power saw a carpenter had left on the floor between visits. Contractors had begun making such frequent visits to the house that many were leaving tools and toothbrushes overnight. Many were having their mail forwarded. He sipped meditatively on his cocktail, trying to remember whether this was the night Oscar de la Renta was bringing Princess Stephanie. Regardless, it would—as it is every night at the Brooks salon—be an evening of glittering conversation, witty bons mots, and sensual elegance, as each guest would outdo the last in trying to explain why mud was mysteriously seeping up through the tiles of Brooks's downstairs bathroom. Ensconced in a wing chair, Brooks drifted off into a reverie: A bank official hands him a foreclosure notice and the Brooks family is forced to move out of this house (bought in September, just three months before) and into a dingy apartment. A smile swept across Brooks's face. If only!

It was dusk. The cool gray of the winter's evening permeated the abode; one of the painters had earlier in the day detected water seeping through the wall around the circuit-breaker box, so the power had been shut off. The air was frosted with anticipation of the coming candlelit dinner. The bracing winter's wind whispered a dulcet tune through the holes around the window frames. Brooks watched his breath cloud and drift upward, until crystalline drops of condensation mimicked the water stains on the ceiling.

It was dusk. The sounds of children's play resounded from the kitchen, where Brooks's six-year-old son was trapped under some of the cabinetry that had been shaken loose when the dishwasher door again fell unaccountably to the floor. Halfway extricated, the boy was able to motion to his mother upstairs through the other hole in the ceiling, the one caused by the leaking pipes in the children's bathroom.

It was dusk. A sense of calm enveloped the Brooks household, disturbed only by the twinkling of jackhammers as the landscapers regraded the ground around the back room, where water had been flooding into the basement. Martha Stewart had stopped by to pick up a few tips on gracious living, and Brooks put aside his collection of E. B. White essays. For a few minutes the halls resounded with gasps of admiration as Miss Stewart gushed over the piquant French furnishings, the graceful rococo settees, and the witty yet elegant gash in the dining-room wall (so admired by the curatorial staff of the National Gallery at the recent fund-raiser hosted by the Brookses). The wall had been breached accidentally by a carpenter who was trying to install bookshelves in the adjacent living room.

It was dusk. Brooks wandered into the kitchen and savored the rich aroma of homemade bread—the flour ground from the bones of the realtor who had made a 6 percent commission selling him the house. The kitchen wall, last painted during the Coolidge administration, was itself an edifying canvas of twentieth-century food stains.

Twirling an elegant '73 Merlot in preparation for its uncorking, Brooks could only admire the energy with which the home's previous owners had practiced their drilling techniques on the kitchen walls.

It was dusk, and in the growing darkness the house was transformed into a symphony of melodious murmurings. From the basement came the gentle lapping of rain puddles brushing against the family book collection, still waiting for the shelves to be completed. The sump pump was doing its work much as the legendary Sisyphus did his, and with as much effect. Brooks could hear the sound of shuffling feet outside the front door. The guests had arrived five minutes earlier, but Brooks had not heard them because the doorbell didn't work. Knocks followed. As Brooks entered the foyer, he heard the creaking of the ceiling above just before it came down on his head. As the paramedics lifted him onto the gurney he looked up where his roof had once been and beheld the infinite vastness of the sky. It was dusk.

# MY DETESTED FELLOW PILGRIMS

## By Joseph Epstein

*(Originally published in the April 28, 1997, issue)*

"CHRIST," THINKS THE WIFE of Harry Morgan, the hero of Hemingway's *To Have and Have Not*, "I could do that all night if a man was built that way." But, of course, a man isn't. Men aren't built other ways as well. "Men don't like complicated food," says one spinsterish character to another in a Barbara Pym novel.

I would like to add another male deficiency. With the exception of those who make their living in and around the places, men don't have much museum stamina—the ability to spend hours contemplating works of art, even the greatest works of art, with anything like the same concentration women seem able to bring to the job.

I base my opinion on a by-no-means random opinion sample: my wife and I. My wife can, in the museological equivalent of Mrs. Harry Morgan's sentiment, go all night. And I? I have just returned from a week in England, where I visited only two museums: the Courtauld in London and the Fitzwilliam in Cambridge. Both have what I think it fair to call small but select collections of painting and sculpture and art objects. Yet in both places I felt my attention wandering. I longed for fresh air. Surrounded by grand works of art, I nonetheless wished to be—elsewhere. I used to say that my museum stamina extended to roughly 90 minutes. I fear it is now under an hour, and shrinking.

The energy for the acquisition of culture seems to be diminishing in me. I used to want to read—and, truth be known, own—all the world's excellent books. This desire has departed, sent packing by the realization that it can't be done. I don't care enough about opera to want to see all the world's operas, though I continue to want to hear as much serious music as possible. I once thought I wanted to see all the world's—or at least all the Western world's—great paintings and sculpture, and I still do, but I shall evidently have to do so in half-hour sessions.

The slightly alarming thought occurs to me that I may already have seen too much art, and thus have become, without my quite knowing it, jaded. Owing to the ease of contemporary travel as well as to the ingenuity of contemporary curators in putting together "super" shows, I have doubtless seen ten times the art that a man of my equivalent level of culture was able to see a century ago.

A few months ago I was in a Park Avenue penthouse once owned by Helena Rubinstein, whose walls were all but papered with Renoirs and other paintings, so little space was there between works. I found myself deeply unmoved and greatly unimpressed. If you have seen one Renoir, as the late and not-too-soon forgotten Spiro Agnew said about slums, you have seen them all. Or so I concluded, as I plowed into my dessert, oblivious to the art all around me. Let the Renoirs go hang, I said to myself, which was what they were already doing. As I say, jaded.

This past summer I was in Philadelphia and went to the Barnes Foundation, a peculiar museum out on the Main Line. The brilliant accumulation of a most eccentric man, a physician who made his fortune selling an antiseptic called Argyrol, the Barnes Foundation contains 60 Matisses, 69 Cézannes, and 180 Renoirs, and much else that seemed to me dazzling, all mounted in the most higgledy-piggledy fashion. Taken together, it was as pleasing a museum experience as I have had in recent years, though the rooms were awash with art gobblers such as myself.

A break for lunch, then on to the super Cézanne show at the Philadelphia Museum of Art. I had written away for tickets months before. When I arrived, vast lines had already formed, and I joined what Henry James, in a not dissimilar situation, once called "my detested fellow pilgrims." (Since I first encountered it, I have found the phrase immensely useful for dealing with the problem of tourism and snobbery, or the dislike for people who are all too identical in their interest to you: that they are fellow pilgrims doesn't mean you can't detest them.) Although the Cézanne paintings were splendid, the crowd wasn't, and my stamina, after my session at Dr. Barnes's joint in the morning, was at low ebb. I had, clearly, over-arted myself.

In viewing art, it may be that less is more. It may be, too, that I have to put myself on an art diet. No more super shows; no attempts to do large museums in one fell, or even a triple fell, swoop. Abstinence may be required.

Perhaps a year's layoff would be helpful. After that I might be able once again to view a Gauguin or a Chagall or a Picasso as something more than very costly wallpaper, which is what these artists' paintings have pretty much become for me. A year off—who can say?—might remove the pink from the cheeks of all those Renoir ladies and put it back in my own.

# THE NO-LUCK CLUB

## by Christopher Caldwell

*(Originally published in the December 1, 1997, issue)*

SOME PEOPLE MAKE FUN of my old Honda. Not much longer, baby. My wife was in a car dealership the other day when she was invited to enter a contest to win a four-wheel-drive Mercedes. As I see it, the thing is practically in the driveway.

I've always been drawn to contests and games, a trait that surely comes from my grandmother. Nana's weakness was Bingo, known to her and other Massachusetts Irish as Beano. "Beano" because Protestant legislators had banned Bingo nights once it became clear they were the source of half the revenue of the Catholic church. That the very same game became legal when renamed "Beano" was an early lesson in legalistic sophistry for the Boston Irish, not that they were ever a people desperately in need of such lessons.

If you were willing to leave your parish, you could play Beano every night of the week, which is what Nana did. But her luck was bad. She never won anything in her entire life.

Sorry, she did win one thing. One weekday, my mother took her to some gargantuan ladies' luncheon. There was a raffle, and in a packed room, Nana's number came up. The prize was "Got My Mojo Workin'," a tape by a jazz/fusion crooner called "The Incredible Jimmy Smith." When Nana got home, brandishing the cassette with considerable pride, I offered to play it on my tape recorder. So there we sat at the kitchen table in Lynn, Massachusetts, my 62-year-old grandmother and me, while Jimmy Smith, as Incredible as they said he would be, went through a series of suggestive grunts:

> *Oh! Ohhhhhhhhhh! Mo-jo!*
> *Ugh! Ugh! Mohhhhhhhh-jo!*
> *Work it! Ugh!*
> *Work it, baby! Ugh!*

At that age, I liked all music, from "All Along the Watchtower" to "Julie, Julie, Julie, Do Ya Love Me?," but here I drew the line. Nana didn't like it much, either. We shut it off before Jimmy had even got his mojo halfway worked.

I had been similarly deceived by the *Saturday Evening News*, for which I had a paper route. One winter afternoon a flyer arrived announcing that any paperboy who got ten new subscriptions could enter a drawing to meet Don Awrey of the Boston Bruins. Awrey was my favorite hockey player. God knows why—he was a scrub defenseman on the great Bruins teams of Orr, Esposito, and Bucyk. But that only made me the more desperate to seize this rare chance to pay him homage.

My neighborhood was *Lynn Item* territory. Potential subscribers for the *Salem News* were thin on the ground, and to get them meant transforming my paper route from a tight little 12-minute circuit into an afternoon-consuming odyssey through neighborhoods I'd never seen before. But, after weeks of bicycling through sleet, with my fingers frozen to the handlebars, I got them. My name got drawn. And one night in January, my father drove me up to a gymnasium in Beverly to sit at the hero's knee.

The gym was stifling; there were at least 200 kids there, sitting on folding chairs with their chain-smoking dads in tow. It was clear that I'd been had. There hadn't been any subscription contest. Anyone who'd wanted to come could have come. I probably wasn't even going to get to shake Don Awrey's hand.

Awrey started off with a contest of his own, waving an official Boston Bruins winter hat in front of the throng. "I'll give this hat," he said, "to anyone who can tell me why, when someone scores three goals, it's called a hat trick."

It was a no-brainer. Two hundred *Salem News* paperboys were wildly waving, screaming, "I know! I know!" Awrey lifted his head, looked down his nose, and pointed directly at me. Then he said something I'll never forget. He said:

*"Uh . . . the little girl in the back."*

I plead long hair (it was 1973) and the smoke in the gym. I said, more pipingly than I would have liked, "I'm a boy." Then I explained—at least I'd get a hat out of it—that whenever a player scores three goals, fans throw their hats on the ice.

Awrey said only one word: "Wrong." He turned to the (now-chuckling) crowd and explained that the right answer was that in 1890-something, a Montreal hat merchant promised that anyone on the Canadiens who—

I told my father we had better go. What a humiliation. No handshake, no prize, no food, my eyes watering (from all the smoke, of course). And the realization that for this I had saddled myself with a paper route that would have prostrated Lewis and Clark.

Awrey left the Bruins soon after. He was ignominiously benched for the 1976 Stanley Cup finals, a coaching decision that still tops my list of Great Moments in Sports History. But I'm more magnanimous now. After all, this stuff is small beer for someone who'll be tooling around in a four-wheel-drive Mercedes in another week or so. Then I'll be saying, "Hey, Don Awrey, who's the little girl now?"

# A CHILD'S CHRISTMAS
# IN PIERRE

## By Joseph Bottum

*(Originally published in the December 28, 1998, issue)*

MY FATHER ALWAYS INSISTED on an early Christmas breakfast—a huge feast of eggs poached in milk, and bacon and hashbrowns and pancakes and marmalade and grapefruit and a sort of sweetened toast whose name I can't remember, but it tasted like corrugated cardboard with cinnamon and sugar sprinkled on top.

And then, after that groaning meal, nothing. No lunch, no snack, no Christmas gingerbread, no nuts, no fruit. None of the fancy chocolate a cousin had sent from England, none of the *bûche de Noël* my sister taught us to make when she came back from her junior year in France with her bangs cut at a Parisian angle and her diary filled with recipes. Nothing until two o'clock, or three or four or, one year, even five, when the ravenous aunts had begun to snip at each other in hunger, and the starved uncles were arguing angrily in the living room about how many terms Sigurd Anderson had been governor of South Dakota, and the children—past the wheedling stage, past the whining stage, past the stage of sitting on the kitchen floor and weeping for food— were crouched together on the sofa, dumb with misery.

But then at last the kitchen door would swing open in a blast of steam and smoke and relief. And the dining-room table would fill with a turkey or a goose, rolls and salad and green beans, little glass bowls of watermelon pickles with tiny three-pronged forks beside them, and cranberries plopped whole in sugared water, boiled until they started to burst, then set aside to cool. "You see," my father explained every year as we sat down to eat, "this is the way to do it: A big breakfast to stretch your stomach, then no lunch, so by dinner time you're really ready for a full Christmas meal."

They can't have all been there the same year, but my memory puts together on the table sweet potatoes and yams, butternut squash and the white potatoes mashed with milk and butter that—in one of those sneaky family traditions by which chores get divvied up—we were told only Uncle Harlow could make well. But there was always the onion and breadcrumb dressing into which my father dumped two, three, four little white tins of sage, sneaking back into the kitchen to add more when he thought no one was looking.

And then there were pies: made from pieces of cooked pumpkin kept in the freezer since October, apples up from the cellar, Mason jars of mincemeat, a chocolate-pudding pie from a packaged mix that one of my aunts brought. My father would look at the pie, then at us children, and then at my mother. And my mother would shake her head, but my happy, roly-poly aunt never noticed and gave us big pieces anyway.

The food was all more enormous than it was complicated. The only elaborate thing I remember my mother making for Christmas was an aspic, a sort of clarified gelatin made from a consommé of veal bones and flavored with tomatoes. I have no idea where she got the idea—South Dakota didn't run much to French cooking—but she would spend hours working on it, and the result always looked to me and my sisters like Jell-O made with tomato juice.

Every year, the relatives would ooh and ah as the aspic was triumphantly brought on a platter to the table, and the crisis of the children's refusal to eat it would escalate from parental glares to harsh whispers to my father banging the table and forbidding us in a loud voice to have dessert or play with our new toys until we finished our portions. At last, while our parents snuck out to the porch to recover their nerves, our favorite uncle would pick up our plates, along with his own, and head off to the kitchen, whistling. He didn't like the stuff either.

It's almost impossible to separate those Christmas dinners from one another. There was the year Great-aunt Fern fell asleep by the fire, the year the cousins from Milwaukee came, the year the car was snowed in and we had to walk to my grandmother's house bearing tomato aspic through the streets of Pierre. But the menus never changed. And though the Christmas season was church and presents and staying up till midnight and carols, memories of Christmas Day itself are mostly memories of food.

But that's what almost everyone who writes about Christmas knows. In Washington Irving's *Bracebridge Hall*, in Charles Dickens's account of Bob Cratchit's feast, even in Dylan Thomas's childhood Wales, at the center of recollection stands the contrast between the cold weather outside and the steaming dishes inside.

The sudden burst of abundance in the midst of winter: That's what Christmas is about. That—and the other thing, the other sudden burst of abundance in the midst of winter, almost two thousand years before.

# MY CAREER IN FORGERY

## By Tucker Carlson

*(Originally published, as "Doctored Letters,"
in the March 15, 1999, issue)*

I GOT A CALL FROM A FRIEND of mine the other day asking if I'd write a letter of recommendation on her behalf to a medical school. No problem, I said. I write a lot. I can handle it.

A week later, I still hadn't finished the letter. It seemed simple at first. Ivy League schools have liberal admissions boards. Why not tailor my recommendation accordingly? My friend, I explained in my first draft, is a promising scientist with a heart, a cross between Dr. Spock and Al Sharpton. More than merely an aspiring doctor, I wrote, she is a "dedicated community organizer," a "longtime activist," a—and this was my personal favorite—"passionate advocate for the rights of the underserved." Whatever that means.

They'll love this, I thought, and got ready to lick the envelope. Then I stopped. Should I really do this? I wondered. I'm not even sure my friend is a Democrat. Is it really a good idea to bet on the reaction of college administrators? As I learned long ago, probably not.

Senior year in college, one of my roommates wound up in trouble with an economics professor. As was his custom, he hadn't bothered to show up to a single economics class all semester, assuming that he could learn everything he needed to know in 48 coffee-soaked hours and do decently on the exam. Days before the midterm, though, he got a call from the professor. "Just thought I'd let you know," she said, "that I've counted each one of your absences against you. No matter how well you do on the test, there is no way you can pass my course."

My roommate was stunned, and once I heard about it, so was I. Failing a class simply for doing no work? Outrageous. We decided to find a solution. But what could excuse a semester's worth of missed economics classes? A death in the family seemed a bit dramatic. Conscription struck us both as implausible, at least in peacetime. Then it came to me: How about mental illness? Perfect, we agreed.

He located a textbook on abnormal psychology, I sat down at the keyboard, and we got to work. It wasn't easy to settle on a malady. We scanned the index. Agoraphobia? Too obscure. Schizophrenia? Too scary. Chronic fatigue syndrome? Too hard to explain. We settled for something that sounded grave but not dangerous, acute neurotic depression.

"Dear Professor," began the letter, which was signed by a nonexistent psychologist from Maryland, "I am writing to you on behalf of a patient of mine who is also a student of yours." The psychologist had very official-looking stationery, and he seemed to take an almost avuncular interest in my roommate, Bill. Bill, the shrink explained, had loads of problems. In addition to being a suicidal alcoholic, Bill still bore scars from growing up amid "maladaptive family patterns." Bill's behavioral symptoms, the psychologist wrote, constituted "a textbook example of acute neurotic depression." And indeed they did, since we copied them verbatim from the textbook. "Bill," the letter said, "has difficulty concentrating, exhibits a high level of anxiety and apprehensiveness, together with diminished activity, lowered self-confidence, constricted interests and a general loss of initiative."

I remember chuckling as I typed the letter, sure it would reduce the professor to weepy sympathy. But just to be sure, we ended on a note of hope: "Fortunately, Bill has responded well to a combination of antidepressant and antianxiety drugs. He is also attending Alcoholics Anonymous. I am in regular contact with Bill and think that his chances for recovery are good."

Good, but not a sure thing. The last paragraph was pretty explicit about college professors' role in Bill's recovery process: "The road to wellness is often a long one, but I believe that, if given the chance, Bill can rise above his recent past. Of paramount importance are instances in which Bill can meet success in tangible ways. He is an impressive and likable young man who needs opportunities to redress his mistakes. I hope that this letter has made Bill's situation clear and you will show him the sensitivity that his full recovery requires."

That ought to do it, we thought. We were right. Within about two hours, the dean of students summoned Bill to his office and kicked him out of school.

What had seemed to us like a clever excuse looked to the administration like cause for immediate hospitalization, not to mention a potential liability nightmare. The dean asked Bill to pack his bags and be off campus by nightfall. "We just don't have the facilities to meet your needs," he said with what seemed like sadness.

It was another three years before Bill got his undergraduate degree, though I'm happy to report that everything worked out well enough in the end. He graduated at the top of his class from a well-regarded law school. I was in the front row when he got his diploma, the most relieved person in the room.

# THE HIGH SCHOOL ALUMNI FOOTBALL TRADITION

## By Victorino Matus

*(Originally published, as "Team McDonald's,"
in the November 29, 1999, issue)*

FOR MOST PEOPLE, New Year's and birthdays are the annual events that remind us we're getting older. Another year, another birthday. But for me, that prompting comes in the form of football. Not the kind you watch, the kind you play.

Every Saturday-after-Thanksgiving for the past 10 years, my high-school friends and I have joined in a classic game of football. And every year it is played the same way: full-body contact, with no padding or protection whatsoever. The teams are usually seven on seven, and everyone, at some point in the game, gets the ball—and gets crushed. The first years after high-school graduation, we played from mid-morning until mid-afternoon. We'd take a short half-time break, guzzle down some Gatorade, and then head back to the field to inflict further damage.

But college life, free of required phys ed, had an impact on our play. With each Thanksgiving that passed, our guys were looking heavier and running slower. Receivers were starting to slip and fall, and the quarterback's beergut made him less agile and more prone to getting sacked. And injuries began to mount. Bloodied knees, gashed lips, sprains, and pulled muscles.

One player, who had previously suffered a dislocated shoulder, aggravated it when he was monster-tackled. Another time, a friend with a size 15 shoe stepped on a guy's neck. One of us still plays despite pins in his upper arm from an arm-wrestling contest gone awry. And once, a player brought his buddy from college who was hit so badly we had to take him to the hospital for an ultrasound. That was the last time we saw him.

Still, we kept on playing, knowing that pain and punishment awaited us.

Then two years ago, my team faced a motley crew who were strangers to all of us but one—and even he scarcely knew these distant in-laws and assorted hangers-on. Obviously, though, they were out for our blood—especially their wiry quarterback, who had a shaved head and tattooed arms. It was an unpleasant experience, with a few personal scuffles, and in the end they ripped us to shreds. Those guys, for reasons apparent, weren't invited back last year.

My friends and I took comfort in the thought that that game was an aberration. But last year, we came up against our younger selves: my classmates' younger brothers. Most of them had played varsity football in high school, and one was playing at the college level. We lost badly. Some of us had taken up smoking and were just plain out of breath. This time, there was no escaping the knowledge that we weren't the athletes we had been in 1990.

Now, none of us is quite 30 years old, but we're getting close. In the back of our minds, there's long been a nagging question we have chosen to ignore. But a few weeks ago, two players brought it up: *What if we used flags?*

The very notion sent shivers up my spine. And most of my teammates thought the same: To take down an opponent not by tackling him to the ground but rather by pulling off tiny flags attached to his waist with Velcro? It was a slap in the face. To end a decade of tackle football with humiliating flags would be to admit we were all washed up. That our bodies have had it. One of us pleaded that if we play again this year under tackle rules without protection, we'll lose George, and "he's got to support a wife and kids!" True. So maybe George doesn't *have* to play.

There probably isn't a better example of male stubbornness (and possibly stupidity) than this. That we will someday go from tackling to touching is inevitable; we're all more or less resigned to this by the time we're 30. Just not now.

As the years fly by, that post-Thanksgiving game is a jealously guarded constant in our lives. Same time, same place, even same weather—in 10 years, the Saturday after Thanksgiving has never failed to produce a crisp autumn morning beneath a cold blue sky. Some of us are indeed married, some have children; some still live in Jersey, while the rest have gone to big cities. But when we step onto that field, it's as if we were still in high school.

After the big game, both teams head over to McDonald's for a second Thanksgiving feast. And for the time being, we continue to eat as if we had the metabolism of 18-year-olds. It's awfully hard to consume $10 worth of McDonald's, but after the game, some of us come pretty close. A favorite postgame meal is the Surf 'n' Turf (Big Mac and Filet-O-Fish). And everyone supersizes.

Come to think of it, the eventual shift from tackle to flag football might not even be the ultimate proof of our getting older. Perhaps an even crueler blow will come the day one of us forgoes the Double Big Mac for a Grilled Chicken Salad. Now that will be the final insult.

# THE IDIOCY OF RURAL TELEPHONY

## By Richard Starr

*(Originally published, as "Toeing the Party Line,"
in the November 20, 2000, issue)*

I'M NO RED-DIAPER BABY, but I grew up hearing lots of talk about the party line. This had nothing to do with politics. The party line was the phone line we shared with the neighbors—a rapidly dying practice, according to an article in *USA Today*. There are apparently only 5,000 of these multihousehold phone lines left in the country, and they won't be around much longer. But party lines were once a way of life for millions of (mostly rural) Americans. And before this venerable institution finally disappears, it's worth pausing to remember . . . just how hideously awful it was.

The *USA Today* reporter paints a somewhat romantic picture: "Though the lines lacked privacy," he writes, "they helped build a sense of community." No, I'm sorry. This is like saying, Though the abandoned pickup trucks in my neighbor's front yard were unsightly, they helped create valuable habitat for wild rodents. If your idea of "community" includes eavesdropping, prying, and unusually authoritative gossip, then you should mourn the passing of the party line.

For the uninitiated, this is how it worked. The party line was a ploy by Ma Bell to make customers unhappy and thus willing to shell out for more expensive private service—to up-sell, as the marketers now say. The more neighbors you shared the line with, the cheaper your rates. As a Bell engineer wrote in 1899, the up-selling strategy "cannot be accomplished unless the service is unsatisfactory. It therefore requires that enough subscribers be placed on a line to make them dissatisfied and desirous of a better service."

But Bell and its smaller rivals underestimated the willingness of people to put up with lousy service in order to save a buck. When I was a kid in backwater Indiana in the early 1960s, there were probably eight or ten families on our line. That went down to four by the time I was in high school, which was still three too many as far as I was concerned. Anyone on the party line could listen to anyone else's calls, and often did.

Nowadays, if you call, say, your cable company to scream that your TV has flickered out just when Regis was about to ask the $250,000 question, you get put on hold and receive a recorded warning: "All our calls are monitored to ensure proper service." Same thing if you happen to call a military office. Instead of hello, you are advised by Private Bailey: "This is not a secure line." Try living this way on a daily basis. Such warnings—how to put this?—can inhibit a frank exchange of views. (I've always assumed, by the way, that no company actually monitors its calls to ensure proper service; they just want to intimidate their entry-level employees into a semblance of politeness. But that's another rant.)

Of course, you never actually got a warning from your neighbors. At best, you might be put on notice by Mrs. Smith's heavy breathing or Mr. Jones's tobacco chewing. But you had to assume someone was listening all the time. You think you were nervous asking someone out on a first date? Try doing it with a heavy breather listening in. This is a recipe for a lifetime of phone paranoia and self-consciousness.

For anyone who has ever lived with a party line, the first private line is one of those milestones of modernity, like indoor plumbing or central heating. Which is why we're probably more astonished than most people to hear someone with a cell phone freely sharing her intimate conversation with a sidewalk full of strangers. The breach of manners is not what jars us; when we overhear a private phone conversation it seems not a novelty but a throwback—almost a willful rejection of progress. Like choosing to beat the dust out of your winter coat with a stick instead of sending it to the cleaners.

People basically had two approaches for coping with the party line. The well-behaved majority kept their calls brief and to the point. Assuming that anything they said might very well end up in the public domain, they led phone lives of impressive decorum. The others, like people today who heedlessly forward dirty e-mail jokes to everyone in their company (you know who you are), seemingly cared little for their public reputations. Indeed, there were more than a few provocateurs who enjoyed saying something outrageous whenever they heard the rattle of Miss Brown's dentures. They knew she couldn't utter a reproach without exposing herself as a snoop.

It's a little-known fact that the highest rates of violence in America have traditionally been not in the city but in rural areas. Sociologists claim not to understand this, but I suspect they overlook the role of the party line.

# SELF-DISTRACTION

## By Andrew Ferguson

*(Originally published in the January 1-8, 2001, issue)*

I COULD HARDLY WAIT to sit down and finally get to work writing this little essay—the deadline is fast approaching and I have a delightful subject this week that I think you'll enjoy—in fact it's the kind of small, delicate subject that a skilled writer likes to hold up to the light as he would a jewel, turning it first this way then that, playfully allowing each unexpected facet to disclose itself in its own fashion, at its own pace, to the wry amusement of the reader, who in response feels those quiet little bursts of recognition and who, when he finishes the essay, senses that the world is somehow fresher than it was just a moment ago, somehow more alive, as though lit from within—and I was just about to do this when the mailman dropped the new issue of *Backpacker* magazine through the mail slot and it splayed on the floor in the entryway.

I seldom read *Backpacker*, having no interest in the outdoors. A friend gave us a subscription, though, as a present. And my kids seem to enjoy it. I went to pick it up—just to keep things tidy, because I can't work if a room is messy—and before dropping it into the basket where we keep our magazines, I happened to flip through it and what do you know: an article on tickborne diseases. In an amazing coincidence, tickborne diseases is one topic I had never, ever had the slightest interest in reading an article about. I threw myself onto the sofa and instantly began reading. As it happens the article was predictably repellent, so after about an hour and a half I set the magazine aside and happily returned to my desk and to the subject that excites me at the moment and which, as I say, I think you'll enjoy as well.

But first, I think I mentioned my feelings about work and tidiness. I simply cannot work amid clutter, sometimes. Across my desk were several—no, many more than several—colored pencils scattered higgledy-piggledy, courtesy no doubt of my children, who I referred to earlier. I carefully returned the pencils to the pencil holder and then pulled them out again and lay them in a row according to length, sorting them by color also, when it became clear that a number of them needed to be sharpened. I can't stand dull pencils. The pencil sharpener was downstairs, which is where I was headed when the radio started playing Schubert's "Unfinished" Symphony. Schubert: an amazingly talented guy, right? A guy very dedicated to his craft, correct? Then why didn't he finish the symphony? What gives with that? As luck would have it, several years ago I bought a book called *The Lives of the Great Composers.* I went back upstairs and fished around for it, finally laying hands on it in a dusty old box tucked beneath the crawl space. I thumbed my way through to the Schubert entry, and found the answer to my question. I've filed it away mentally, for an essay on Schubert I'm hoping to write pretty soon, probably after I finish this one.

When I returned to my desk I realized I'd forgotten to sharpen the darn pencils! Going back downstairs I was rethinking the subject of this essay when my mind offered up a really clever turn of phrase that I mean to use: Discussing pundits who always say "On the one hand, on the other hand," I'm going to say, "Thank god there are no three-handed pundits!" Isn't that good? I rushed back upstairs to jot it down. It's difficult to convey the feeling of satisfaction a writer gets when these little jokes come to him unbidden—almost like finding that perfect subject for an essay, where the piece "just writes itself," as we say. But there was still the problem of the dull pencils. Back downstairs I saw that my wife had been cleaning house. Next to the pencil sharpener she had stacked at least a dozen cookbooks. I arranged them first by size, then by cuisine, then alphabetically by author, and dusted off the shelf she'd removed them from. When I went to the basement to find more lemon Pledge, I couldn't find it, although I did find several pairs of my kids' old shoes. Such tiny feet they had! I admit I fell into a kind of reverie. Then, at last, after noticing my son's basketball was deflated, searching for the air-pump, and reinflating the ball, as well as an old football and a volleyball that lay nearby, I eagerly headed back to work.

At my desk I put nose to grindstone. There are many magical moments in the writing life. You're just sitting there and the words flow, and the writing and the subject matter become one, and it's as though you the writer were a mere onlooker, a privileged witness to the act of creation. The pleasure is almost sensually intense. I can't get enough. I mean it.

# AN OPEN LETTER TO THE MARYLAND OFFICE OF UNEMPLOYMENT INSURANCE

## By David Tell

*(Originally published in the January 28, 2002, issue)*

Dear Sirs:

Thank you for your recent "Notification of Assessment and Pending Civil Action" wherein I am informed that unless I make good a $3.53 tax debt by January 25, the State of Maryland will send "the sheriff" to seize my house and sell it "at [my] expense." That seems a reasonable plan to me, though I must say I wonder about the January 25 deadline, which is almost a week away. Shouldn't you come seize and sell my house immediately, while you're still confident of its location? I mean, by January 25 a man could easily box up a house like mine, truck it to Baltimore, and reassemble it right in your office. Where, *entre nous*, nobody can ever find anything. My advice: Contact the sheriff today.

Fact is, you probably should have contacted him a long time ago. Several years back, I conducted a major phone and letter campaign to alert the State of Maryland that there was this nice lady who'd begun helping us clean the house, and here was her name and Social Security number, and this was how much money we were giving her, and would you please tell me what taxes I'm supposed to pay, and so on. Boy, were you guys nice about this. First you let months and months go by, totally ignoring the matter. Then, after this grace period had ended, you were kind enough to send me a personalized form letter explaining that, because months and months had gone by during which I'd never once made a contribution to the "Maryland Unemployment Insurance Fund," the sheriff would soon be coming to seize and sell my house.

State law requires that a persuasive rationale for this program be kept permanently on file in your office, where . . . well, we've been over that already, haven't we? So I can only guess, but I figure it this way: If I pay approximately $3.50 to the unemployment insurance fund four times a year, your office, should I ever fire the cleaning lady, will step in and replace the wages she's lost. And since, as the document you've sent me points out, I'm currently paying this woman an annual salary of $133,333.20 for six hours of work each week, the stakes are obviously very high: You need all the $3.50 contributions you can get.

Better yet, you need to keep this tax a secret, even as I'm begging you to tell me about it, thereby guaranteeing that I will fail to pay, thereby giving the sheriff cause to seize and sell my house. That way, the lady who helps clean said house won't have to work there without an unemployment safety net—because there won't be a "there" for her to work in the first place. Best of all, because I won't have "fired" her, your office won't have to pay her a dime, am I right?

As I say, this seems a reasonable plan to me. You should have carried through on it when you first had a chance. Instead, you let me slide. And we both know I've been taking shameless advantage of you ever since. I've completed the two hours of paperwork necessary to pay every last one of those $3.50 quarterly tax bills. Occasionally—such is the mind of the habitual law-abider—I've even waited until after the due date, just to get you thinking it was finally time to seize and sell my house, only to dash your hopes by sending in the full $3.50, plus a $35.00 penalty, plus interest. Top of the world, Ma!

At least, that always used to be my attitude. But lately—I don't know, maybe I'm getting old—the thrill is gone. So let's talk turkey, shall we? Says here in your Notification that I am delinquent with my unemployment-fund contribution for the third quarter of last year. Once upon a time I would have thought it relevant to mention that the check I sent you for that quarter actually cleared my bank some time ago. But such are the petty concerns of yesterday, and for now I am concentrating on the deeper truth suggested by your otherwise laughable complaint. To wit: Any man, like me, who pays his Sunday-afternoon housekeeper $133,333.20 a year belongs in a lunatic asylum, not left free to roam the streets of Bethesda. Informal research I've conducted in the neighborhood confirms this hunch. Hundreds of families around here have cleaning ladies like ours. But nobody else pays even a fraction what we do. Also, I've noticed that whenever I bring up the Maryland Unemployment Insurance Fund, people invariably get all nervous and change the subject, as if I'd exposed myself or something.

Clearly I am a danger to the community. You'd best send the sheriff, pronto.

Incidentally, will he be seizing the house *and* its contents, or just the house? I'd really like to keep the piano, for example. And speaking of the piano, would now be a good time to ask you about unemployment insurance for the music teacher who comes over on Saturdays? Or would you prefer to wait until we find a new place to live? Please advise.

Sincerely,
David Tell

# ASHES TO ASHES

## By Christopher Caldwell

———

*(Originally published in the February 24, 2003, issue)*

WHEN I WAS GROWING UP, every adult in my family smoked: both parents, all four grandparents, and every single uncle and aunt. It was Camels (and Dutch Masters "President" cigars) for the men, Viceroys for the women. There was an ashtray on every surface flat enough to accommodate one—coffee table, dinner table, end table, guéridon, countertop, newel, television, washing machine, workbench. Near the winter solstice in New England the sun comes sideways into the house all day long, so that, at holidays, any room where the adults were gathered would have a head-high, gunmetal-colored blanket of smoke so opaque that it would cast a line of shadow along the wall. We repainted the inside of the house every year, and the first fresh coat of white my father dabbed on would glow against the year's accretion of airborne tar like white icing on carrot cake.

Since smoking today is cast as an extremely bad choice that a certain rotten or pitiable type of person makes, it may be necessary to add that the adults in my family struck me growing up—and strike me still—as well above the mean in decency and well-adjustedness. They just did all their decent and well-adjusted things while smoking. All of them had started smoking at puberty (at the latest), and one could say of almost any of them what Louis MacNeice once said of W. H. Auden: "Everything he touches turns to cigarettes." My mother would make our favorite breakfasts coughing *hyerkh-kakh-hokh*, and my father would throw me endless batting practice with a stogie in his teeth, and my grandmother would rock my little sisters to sleep singing "You Ah My Sunshine . . ." with a Viceroy dangling out of the corner of her mouth, dribbling ash. Smoking has never struck me as evidence of either particular goodness or particular badness. It was just something adults *did*, like go to work in the morning, use deodorant, or talk about money.

But it would be an error of logic to assume that the crusade *against* smoking is therefore morally neutral as well. It is hard not to feel political qualms about this movement. The intolerance of its adherents is a strike against it. Antismoking language—from euphemisms like "smoke-free environment" to bureaucratese like "please extinguish all smoking materials"—is a constant temptation to dishonesty. And dishonesty, plain-and-simple, is a cornerstone of much of the antismoking political platform—like the claim that early deaths from smoking are a burden on government treasuries rather than a windfall.

The worst thing about today's antismoking mania is that it emboldens even selfish and depraved people to declare themselves morally superior to people like my "Sunshine"-singing grandmother, on the flimsy grounds of: *Well, at least I don't smoke!* It replaces real morality with something that looks like morality but is actually valetudinarianism. So when the leading antismoking crusader in Montgomery County, Maryland, pled guilty to diddling a boy in the lavatory of the National Cathedral two years ago, I was not among those you could have knocked over with a feather.

Last month, I brought to a close, for the usual reasons, my own fairly illustrious smoking career with what I assume is the usual mix of pride and regret. I had no real craving for tobacco, but I did spin into an immobilizing, feel-like-sobbing depression the likes of which I hadn't felt since I was young, broke, and alone. Conquering that made me proud. And my children—who have been taught in school that "mortality" is merely a synonym for "emphysema"—are now under the impression I'll live longer. On the other hand, I somehow feel I've broken solidarity with the sophisticated 17-year-old smoking beauties I first started smoking to impress. Or maybe it's that I've broken solidarity with the person I was when my head was full of poetry and my heart of song.

But I certainly fear that I've broken solidarity with people like my grandmother, and transferred it to the power-mad moralizers who have devised our present antismoking regime—that I have, in fact, done something socially irresponsible, by giving aid and comfort to those who wish my fellow citizens ill and, as someone put it in another context, "hate our freedoms." It occurs to me that if not for this menace I would have quit smoking long ago. Once these heirs to the hypocritical church ladies of yore succeed in expunging cigarettes from the landscape, they'll simply move on to something else.

So, while I'm moderately happy to be free of smoking myself, it would be irresponsible not to remind people—particularly impressionable teenagers—that good people continue to do it, and that smoking marks one as a person of independence. Now more than ever, in fact. And what—all propaganda to the contrary notwithstanding—could be cooler than that?

# DOG'S WORST FRIEND

## By Andrew Ferguson

*(Originally published, as "Buster Blues," in the March 1, 2004, issue)*

I LIKE DOGS IN THE ABSTRACT, as a class. I like dog-lovers, too, and think them superior to other men, because I admire their capacity for fellow feeling and their willingness to claim the mantle of stewardship to which all of us are called, so the Bible says. I like movies about dogs. I can watch the broadcast of the Westminster dog show for an hour at a time. That charitable organization that brings doggies to retirement homes so the elderly residents can be revivified by canine companionship—I would give money to that organization if I could remember what it is called.

In other words, as a general proposition, I am down with dogs. It's just Buster himself—Buster *en se*, as the metaphysicians would put it—that I have a problem with.

The picture you see to the right is a picture of Buster. It is an idealized rendering, in my opinion. It was drawn by my daughter, Emily, who is 10 years old and rather more fond of Buster than I. We all approach works of art differently, of course, and *vive la différence*, but when I look at this picture I see a warm, playful, cuddly pup, a bundle of joy, a creature to lift a lowering heart and make gentle the face of the world. Then when I put the picture down and look across the room I see the real Buster. He doesn't appear this way to me at all. Far from a joy-bundle, he is scruffy, panting, half-asleep; or, in the alternative, he is scruffy, panting, and hyperactive, heckling me with gurgles and whimpers and even outright barks until I drop whatever it is I'm doing—gathering the remains of a tissue box he's torn apart, for example—and devote all my energies to keeping him occupied. And this is not easy. I've met teenage boys with longer attention spans than Buster's.

There are many clear and rational reasons why Emily and I see Buster so differently, beyond the obvious one that she is a sweetly disposed 10-year-old girl and I am a tired, cranky, middle-aged man. Buster is not really a man's dog. Men have hounds and setters, Dobermans and mastiffs—fearless beasts with giant heads that burrow deep into the underbrush to roust a covey of quail, graceful beasts that swing great ropes of spittle onto a passerby as they leap for Frisbees or footballs in parks and open fields.

Buster, by contrast, neither burrows nor leaps, neither does he roust. He is a bichon frise—a breed, like the Pekingese, that does not seem to have any males within its genotype. Bichons were bred (by the French) to be show dogs, so it's no surprise they sashay like fan dancers at the Folies Bergère. Every bichon appears to be a girl dog; and every bichon looks as if it should be owned by a girl. When I first told a friend we might buy a bichon and sent him a picture so he could see what we were getting, his reply, by e-mail, was swift: "Wo. I'm not sure I'd want to be seen in public with one of those, you know?"

Oh, I do know, I do. At this season our neighborhood is filled with construction crews repairing streets and renovating houses. They crack apart concrete slabs and bust up walls. When it comes time for Buster's mid-morning walk, I descend from my home office, where my soft, pale hands have been tap-tap-tapping out my little articles, and with only the greatest reluctance do I escort fluffy Buster past these Doberman-owning fellows, these mastiff-loving men. They stop what they're doing and stare as we pass. I feel like one of the Gabor sisters.

The resentments pile up. Buster wasn't an impulse buy, exactly, but in retrospect it's plain I didn't think things through. When we brought Buster home, late last fall, I didn't expect, for example, that I would never again be able to sleep past six in the morning, when Buster's internal alarm goes off, nor did I expect him to eat half a dozen books and the lining of my overcoat. I never thought I would become a student of bowel movements, of their intensity, size, and frequency, and it didn't occur to me that I will have to write many, many more of my little articles than I would have done otherwise, just to pay bills from clinics, sitters, kennels, and Petsmart.

I have tried to keep these resentments from Emily, for the most part, and of course Buster himself doesn't care whether I'm showering him in praise or giving him the high hat. Emily meanwhile has embarked on a writing project of her own: a series of adventure stories starring Buster. *Buster in Autumn* has been followed by *Buster in Winter.* She says she's waiting for the seasons to be upon us before she undertakes *Buster in Spring* and *Buster in Summer.*

"*Buster in Spring* will be totally different from *Buster in Winter*," she tells me. "Really?" I say. "Do you think so?"

# BROTHERLY LOSERS

## By Jonathan V. Last

————

*(Originally published in the July 26, 2004, issue)*

MY NAME IS JONATHAN and I'm a Philly fan. In 1995, the Philadelphia Eagles made what was then the biggest free-agent signing in sports, acquiring star running back Ricky Watters for $6.9 million per year. In his first game as an Eagle, Watters found his new team losing in the fourth quarter to the Tampa Bay Buccaneers. Quarterback Randall Cunningham (I once went trick-or-treating at his house) threw a pass to him over the middle, and as the ball neared, Watters saw defensive backs fast approaching. He put his hands up and, instead of catching the ball, batted it down.

After the game reporters asked Watters why he'd refused to catch the pass. He explained that the Eagles were probably going to lose and he saw no reason to take a hit in the middle of the field. "For who?" Watters asked defiantly. "For what?"

It was an enduring and iconic moment. Today, Watters is long gone, but "For who? For what?" remains part of the city's lexicon.

No town has a more heartbreaking sports history than Philadelphia. For 21 years—nearly my entire life—the city has gone without a champion of any kind—in baseball, basketball, football, or hockey. Growing up just a few minutes from the City of Brotherly Love, I am one whose heart has been broken so often that it has hardened to stone. Or maybe disappeared altogether.

When Smarty Jones faltered in the final stretch of the Belmont Stakes a few weeks ago, failing to win the Triple Crown, no one from Philadelphia was surprised; we're used to losing.

We've taken it hard. Philadelphians are renowned for acts of ill humor, and in my time I've seen a lot. Top prospect J. D. Drew was drafted by the Phillies. He chose to sit out of baseball for a year rather than play for Philadelphia. Eventually he signed with the St. Louis Cardinals, and fans threw batteries at him the first time he came to town.

Others got worse. On a trip to Philadelphia in the early 1980s, the unofficial mascot from the Washington Redskins was accosted and wound up in the hospital.

In December 1989 the Dallas Cowboys visited, and Eagles fans— encouraged by city district attorney (and future mayor and governor) Ed Rendell—pelted their coach with snowballs.

In 2001, the '76ers made it to the NBA Finals, where they faced the Los Angeles Lakers. During Game 4, fans booed the halftime entertainment because one of the singers was wearing a Lakers jersey. The Lakers won the series behind the sharp-shooting of Kobe Bryant. Bryant grew up just outside of Philadelphia. That didn't help. The following season, the NBA held its all-star game in Philly and Bryant was awarded MVP honors. The fans booed him mercilessly.

The best-known booing in sports history took place in Philadelphia on December 15, 1968, as the Eagles battled the Minnesota Vikings. Santa Claus was circling the field, waving to the crowd. Fed up with another losing season (the Eagles were 2–12), the fans booed Santa.

So Matthew Scott should have known what he was getting into. Matthew had undergone the first successful hand transplant performed in the United States and was chosen by the Phillies to throw out the first pitch of the 1999 season. His toss fell short of home plate. The good people of Philadelphia booed him, too.

All of which made the recent news almost comforting in its constancy. The Phillies opened a splendid new ballpark this season, and the architects tried hard to make it "fan friendly." For instance, they built the home team's bullpen 10 feet away from a balcony, so that people could gather to watch their favorite pitchers warm up. The visitors' bullpen was tucked away, out of sight.

This arrangement lasted but a short while. Instead of admiring the home-team pitchers, fans congregated around the bullpen and heckled them. Just two games into the *preseason*, Phillies pitchers had had enough of their fans. The hecklers were so *vicious to their own team* that the Phillies swapped bullpens with the visitors. Only in Philadelphia.

There are questions of cause and effect: Do Philadelphians boo because they lose, or do they lose because they boo? The Catholic in me wants to believe that losing is cosmic justice; that we deserve it because we're so prickly.

But the romantic in me believes that all the booing and heckling and battery-throwing is caused by the losing. The fans boo because they still care. Philadelphians may be the last people in America who have refused to accept sport as mere entertainment. They believe that the games mean something, that athletics is an important part of the human condition, and that it matters whether you win or lose.

# About the Authors

**Fred Barnes** is executive editor of *The Weekly Standard*.

**Jeffrey Bell** is a principal at Capital City Partners in Washington, D.C.

**David Brooks** is a columnist for the *New York Times* and was formerly a senior editor at *The Weekly Standard*.

**Joseph Bottum** is editor of *First Things* and was formerly Books and Arts editor of *The Weekly Standard*.

**Christopher Caldwell** is a senior editor at *The Weekly Standard*.

**Paul A. Cantor** is the Clifton Waller Barrett professor of English at the University of Virginia.

**Tucker Carlson** is a *Weekly Standard* contributing editor and author of *Politicians, Partisans, and Parasites: My Adventures in Cable News.*

**Eric Cohen** is editor of the *New Atlantis* and a resident scholar at the Ethics and Public Policy Center.

**Terry Eastland** is publisher of *The Weekly Standard*.

**Noemie Emery** is a *Weekly Standard* contributing editor.

**Joseph Epstein** is a *Weekly Standard* contributing editor. His most recent book is the short story collection *Fabulous Small Jews*.

**Andrew Ferguson** is a senior editor at *The Weekly Standard*.

**David Frum** is a *Weekly Standard* contributing editor and the *Reader's Digest* resident fellow at the American Enterprise Institute.

**David Gelernter** is a *Weekly Standard* contributing editor and professor of computer science at Yale University.

**Reuel Marc Gerecht** is a *Weekly Standard* contributing editor and a resident fellow at the American Enterprise Institute.

**Ethan Gutmann** is author of *Losing the New China: A Story of American Commerce, Desire and Betrayal*.

**Stephen F. Hayes** is a senior writer at *The Weekly Standard*.

**Gertrude Himmelfarb** is professor emeritus of history at the City University of New York.

**Christopher Hitchens** is a contributing editor to *Vanity Fair* and writes a regular column for the *Atlantic Monthly*.

**Alan Jacobs** is professor of English at Wheaton College.

**Donald Kagan** is Sterling Professor of Classics and History at Yale University.

**Robert Kagan** is a *Weekly Standard* contributing editor, senior associate at the Carnegie Endowment for International Peace, and columnist for the *Washington Post*.

**Charles Krauthammer** is a *Weekly Standard* contributing editor and syndicated columnist.

**Irving Kristol,** former editor of the *Public Interest,* is author of *Neoconservatism: The Autobiography of an Idea.*

**William Kristol** is editor of *The Weekly Standard.*

**Matt Labash** is a senior writer at *The Weekly Standard.*

**Jonathan V. Last** is online editor of *The Weekly Standard.*

**Tod Lindberg** is a *Weekly Standard* contributing editor, research fellow at the Hoover Institution, and the editor of *Policy Review.*

**Harvey Mansfield** is William R. Kenan Jr., Professor of Government at Harvard University.

**Victorino Matus** is an assistant managing editor at *The Weekly Standard.*

**Wilfred M. McClay** is a professor of history and holds the SunTrust Bank Chair of Excellence in Humanities at the University of Tennessee in Chattanooga.

**Larry Miller** is a Los Angeles–based writer and actor who regularly contributes to *The Weekly Standard*'s online edition.

**Jay Nordlinger** is managing editor of *National Review* and was formerly associate editor of *The Weekly Standard.*

**P. J. O'Rourke** is a *Weekly Standard* contributing editor. His most recent book is *Peace Kills: America's Fun New Imperialism.*

**John Podhoretz** is a *Weekly Standard* contributing editor and was formerly the magazine's deputy editor. He writes a regular column for the *New York Post.*

**Matthew Rees** is chief speechwriter and senior adviser to the chairman of the Securities and Exchange Commission. He was formerly a staff writer at *The Weekly Standard*.

**Robert Royal** is president of the Faith and Reason Institute in Washington, D.C.

**Stephen Schwartz**, a frequent *Weekly Standard* contributor, is author of *The Two Faces of Islam: The House of Sa'ud from Tradition to Terror*.

**David Skinner** is an assistant managing editor at *The Weekly Standard*.

**Richard Starr** is a managing editor at *The Weekly Standard*.

**Irwin M. Stelzer** is a *Weekly Standard* contributing editor and a senior fellow at the Hudson Institute.

**David Tell** is opinion editor of *The Weekly Standard*.

**Philip Terzian** is Books and Arts editor of *The Weekly Standard*.

**James Q. Wilson** is the Ronald Reagan Professor of Public Policy at Pepperdine University and past president of the American Political Science Association.

**Claudia Winkler** is a managing editor at *The Weekly Standard*.